Chinese Research Perspectives on the Environment, Special Volume

Chinese Research Perspectives: Environment

International Series Advisors

Steven A. Leibo, *Russell Sage College*
Li Yang, *Natural Resources Defense Council*

VOLUME 2

SSAP

The titles published in this series are listed at brill.com/cren

Chinese Research Perspectives on the Environment, Special Volume

Annual Review of Low-Carbon Development in China (2011–12)

Edited by

QI Ye

BRILL

LEIDEN • BOSTON
2013

This book is the result of a co-publication agreement between Social Sciences Academic Press and Koninklijke Brill NV. These articles were selected and translated into English from the original《中国低碳发展蓝皮书 (2011–12)》(*Zhongguo ditan fazhan lanpishu 2011–2012*) with the financial support of the Chinese Fund for the Humanities and Social Sciences.

This publication has been typeset in the multilingual "Brill" typeface. With over 5,100 characters covering Latin, IPA, Greek, and Cyrillic, this typeface is especially suitable for use in the humanities. For more information, please see www.brill.com/brill-typeface.

ISSN 2212-7496
ISBN 978-90-04-25116-8 (hardback)
ISBN 978-90-04-25182-3 (e-book)

This book is printed on acid-free paper.

CONTENTS

FOREWORD

Steven A. Leibo and Li Yang

At first glance, the *Annual Review of Low Carbon Development in China (2011–12)*, a product of the Climate Policy Initiative at Tsinghua University, looks to be a book of arcane information of interest to the few whose professional lives require knowledge of the evolution of the People's Republic of China's energy use. Nothing could be farther from the mark. Indeed the success or failure of China's effort to "green" its economy has to rank as one of the most important issues in human civilization for a very simple and obvious reason. Human civilization and more recently modern civilization are products of two core factors. First, the stabilization of the global climate that developed some twelve thousand years ago and is commonly known as the "long summer," which facilitated humanity's evolution from simple hunting and gathering societies to the larger settled civilizations we are familiar with in ancient Mesopotamia and Egypt. Second, the global civilization that has emerged since the eighteenth century is a direct product of mastering fossil fuels as an extraordinarily significant new energy source. Disastrously, these two different phenomena have become intertwined as humanity's prolific burning of heat-trapping fossil fuels has unexpectedly disrupted the climatic stability of the last millennium. In short, carbon based fuels, the core of modern civilization, disrupted the climatic stability that allowed civilization itself to develop.

But it is not really the entire human community that has been especially involved both in the problem and the potential solution. Ironically, the planet's oldest and largest continuous civilization, China, and its newest major civilization, the United States of America, have become the central players in creating the climate problem. They are, hopefully through an enhanced level of cooperation, the best positioned to lead humanity in its fight to confront a dramatically and a dangerously changing climate.

America remains the single largest cause of the destabilization of the planet's heat balance, and the consequent and the dramatically altering global climate environment. And that is not surprising given the size of the United States and its long history of industrialization.

Thus when one couples the size of the America with almost two centuries of fossil fuel burning, and the reality that CO_2 in the atmosphere can last over a hundred years, the importance of America's role becomes self-evident. England might have been industrialized longer, but its population is far smaller. India might be far larger in numbers, but its people have not been vigorously burning greenhouse gas producing fossil fuels nearly as long. In short, given the longevity of greenhouse gases and especially CO_2, what matters is the cumulative effect.

But that hardly takes China off the hook. China may not have been burning massive amounts of carbon based fuels for very long either. Even today its per capita carbon emissions are a mere fraction of the average American. Still, in terms of annual emissions China passed the United States in 2007. And while it remains true that per capita the Chinese still do not especially emit that large a percentage of the world's greenhouse gases, the sheer size and extraordinary growth the nation has experienced over the last few decades has made it a major contributor to the problem. And it is not just the enormous amount of coal that is used to provide the energy to fuel China's explosive growth, but the construction industry, which is seeing not just new buildings but entire new cities emerge. Since 2007, China has engaged in over 2 billion square meters of building construction each year, comprising half of all new construction in the world. That is more than the total existing building floor space in Canada, which means China has been building a new Canada every year for the past six years, and is projected to continue doing so until 2020. That requires massive amounts of cement, the production of which is especially problematic in terms of greenhouse gases production. Since the country's carbon footprint skyrocketed in 2011, China's per capita carbon emissions are now on par with those of Europe. However, that doesn't mean the majority of Chinese people are enjoying quasi-European living conditions. Such fast growth in energy consumption primarily comes from two sources. One is the large amount of embedded energy in the "Made-in-China" products that China manufactures for the world; the other is the handful of highly developed cities like Beijing, Shanghai, and Guangzhou that lie along China's eastern and southern coast. On the other hand, given China's huge geographic disparity in economic development, the enormous urbanization process has just started in mid-western areas. People from these areas shouldn't be denied the right to economic prosperity. As a result, China's contributions to the greenhouse effect are projected to grow enormously over the next few decades.

Thus, across the planet, and at a faster rate than previously expected, the industrially augmented build-up of greenhouse gases is beginning to transform planetary conditions in an enormous variety of ways. Some examples are melting mountain glaciers to increasing desertification, and the spread of invasive species that are having a significant impact on both animals and vegetation. More dramatically and often more immediately dangerous is that rising heat and thus the atmosphere's ability to hold more water in recent decades is allowing storms, from hurricanes to tornadoes, to carry more destructive power. And that enhanced destructive power comes in the form not only of more powerful winds but also more regular flooding of the sort that is more and more common and increasingly referred to as "extreme" weather though it probably represents less a series of weather anomalies than a new emerging climate realty.

While experienced differently in different parts of the world these foreboding developments challenge the United States and China as countries that happen to cover extraordinarily diverse regions.

In China's case, the challenges have been quite different depending on the region. Southern China, always vulnerable to damaging typhoons, experienced a growing threat of floods of unprecedented dimensions, but even that threat is not confined to the south. Indeed as we wrote this foreword, Beijing was experiencing some of the worst flooding in generations. Meanwhile yet another problem has emerged dramatically. North Central China, long a relatively dry region is seeing an explosion of desertification, which is already creating large numbers of climate refugees. Yet the most fatal challenge, very often neglected, is the water crisis caused by accelerated glacier melting on the Qinghai-Tibet Plateau, which is the headstream of China's two largest rivers.

Meanwhile, the United States has also found itself especially vulnerable to rising ocean levels. In fact, as a nation it has an especially large percentage of its most valuable infrastructure built directly along the coast, from Boston and New York to the famous Silicone Valley of the San Francisco Bay Area. Meanwhile, the American Southwest, somewhat like North Central China, is particularly vulnerable to the threat of desertification—a threat that seemed especially real during the summer of 2012 as the United States experienced a particularly devastating drought across much of the country's heartland. As in China and elsewhere, that drying out process dramatically weakened crop production yields.

Clearly between the two national communities, the United States and China can be seen a significant part of the reason humanity has found itself so climatically challenged in the early twenty-first century. But it is equally true that the healthier a cooperation that can be forged between Beijing and Washington on this absolutely critical challenge, the greater chance humanity will be able to stave off the worst environmental disruptions. But that cooperation requires as much transparency as possible thus the importance of the current *Annual Review of Low Carbon Development in China* (*2011–12*) report. In short, the more Americans understand the extent of China's commitment, the better America itself will be able to devise its own aggressive plan to take up the challenge of a dramatically changing climate.

Unfortunately, while many regions of the world have been especially conscious about the necessity of taking up the challenge of anthropogenic climate change neither the United States nor China, the two most important nations in this struggle, have been at the forefront of the battle. Indeed, each nation for reasons associated with their different histories has been especially reticent until recently to even directly confront the challenge.

In the United States, the single largest impediment to confronting the climate crisis has absolutely nothing to do with the chemistry or physics of greenhouse gases, but the evolution of the nation's economic philosophy. Residue of the Cold War creates an inner conflict between the nation's capitalist model and its generational struggle with communism. And while that battle was often dramatic and at points threatened nuclear annihilation at its core was the most basic of questions: "What should the relationship be between government and the economy?" That core speculation stands at the core of the struggle between democratic capitalism and the command economies of communism, fueled the Cold War and certainly since the administration of Franklin Delano Roosevelt, most of the differences between the Republican and Democratic parties in the United States. In short, Republicans tend to be very suspicious of a government role in the economy while Democrats, hardly Social Democrats in a European sense, see a more active creative role, especially in those areas that profit making economic activities short change such as free public libraries In lies the rub. Although the efforts of individuals matters a great deal in confronting the climate crisis and the transition to greener, safer energy sources only governments have the broader regulatory power to make much more meaningful changes. Thus, an acceptance of the challenge of anthropogenic climate change requires an ability to accept more government regulation of the energy economy, a bridge simply too far for many Americans. Thus, it has tended to be easier to deny the science than change one's ideology, a problem that has often been at the root of American problem with carrying out a vigorous response to a changing climate brought about by humanity's prolific burning of fossil fuels.

For China, the difficulty in confronting the climate crisis has had very different origins. Taking on national regulations to confront the climate crisis or almost anything else has hardly been a problem even in today's China, which has moved so far from its earlier roots in the command economies of communism. No, for the Chinese, the challenge has been quite different and quite simply because China has had other priorities, most immediately improving the living standards of its enormous population while building the national infrastructure necessary to operate a modern twenty-first century developed society.

When one adds the historic reality that the destabilization of the planet's heat balance has largely been the "accidental accomplishment" of the developed world, a world that until the last generation used enormously greater amounts of fossil fuels than China, it is not hard to see why China has been reticent to distract itself from the great work of modernization, largely on the Western model by activities that were until just a few years ago simply seen as activities likely to likely to slow down that accomplishment.

The combination of the two nations' reticence to take on the climate crisis was, despite their different origins, quite complementary. Each nation, the United States and China has long used the other as an excuse to do little or nothing about the challenge, or at least did so until recently when over the last several years the situation began to evolve again.

For the United States, it was the arrival to power of an administration more willing to invest in green energy solutions and public transportation projects albeit reluctant at times to speak openly of the challenge of climate change directly. And while the American Obama administration has not spoken with the commanding voice many would have liked, its spending priorities especially associated with the stimulus programs designed to counteract the 2008 global recession have certainly shown a commitment to both short and long term green energy conversion efforts.

While China, stirred perhaps more immediately by a deepening national anxiety about the environmental price it has paid for its rapid industrialization effort. At times significant social discontent deeply disrupted China and forced the nation's leaders to rethink some of their priorities. Perhaps more importantly within China's leadership, a greater sense than perhaps yet exists in the United States, has emerged that converting to clean energy may not be the drag on the nation's development but a new and successful way forward to the jobs of the twenty-first century. Nevertheless, while the national leadership has increasingly shown a deep commitment to sustainable development and to confronting the climate crisis, it is also true that much of the mid-level of the Chinese establishment, both private and government, has been significantly less committed.

Perhaps especially important is the deep sense of China's emerging identity crisis. The question of just what China represents in the twenty-first century: the extraordinarily accomplished imperial civilization of earlier centuries, the recently humiliated nation of the era of Western colonialism, the explosively industrializing heritage of the Deng Years or perhaps something entirely new? Most interesting is the question of whether China will return to the time honored tradition of sifting carefully through the western heritage for what is useful and rejecting what is not. If the issue in the last years of the Qing dynasty was to maintain the Chinese Confucian essence even as western technology was borrowed or more recently, during the early Deng era a focus on retaining the socialist essence even as western market skills were adopted, is the core question of today whether or not China wants to slavishly adopt western consumerism as the only model of modernity or to carve out for itself and perhaps the rest of the world a different vision of post-industrial society. The latter approach would surely place China in the future, as it has stood so often in the past at the forefront of the human civilizing experience.

Certainly for both countries one can easily put together an impressive list of green energy accomplishments that have emerged over the last few years. Whether they are enough to significantly respond with the urgency needed is quite another issue. And most importantly, individual measures by each nation however important are not substitutes for the two countries offering global leadership in the challenge.

One of the biggest challenges in accomplishing a level of Sino-American cooperation commensurate with the problem is the exchange of accurate information itself. Sadly information about China's energy goals and accomplishments like almost everything else about China thought to be known in America is often enormously misinformed. That of course is not surprising. China, one of the largest countries in the world has been changing at a pace unprecedented in human history. That Americans have trouble sifting the incorrect from the merely obsolete is quite understandable. Indeed, the Chinese often have the same problem understanding developments in their own country for the same reasons. And yet a healthy and vigorous Sino-American cooperation on the climate crisis absolutely requires as much transparency as possible. The two largest greenhouse gas emitters in the world need better communication about what each country is actually committed to and accomplishing. This includes better understanding of each other's measurements, mutual trust, and acknowledgement of each other's greenhouse gas accounting and of evaluations of the results of their own actions. They are also obliged to provide better public information on the real levels of their current green energy spending and investment. That is where the current *Annual Review of Low Carbon Development in China (2011–12)* by Tsinghua University comes in. It opens the door for Americans to get in touch with the heated ongoing discussions on energy and climate policies among Chinese academics and policy makers.

This brings us to the question of authority carried by the document itself. Although less well known in the West, Tsinghua University is usually ranked as perhaps the first or second highest ranking university in China, a role perhaps akin to that held by the Massachusetts Institute of Technology in the United States. Tsinghua University is known to have nurtured some of the most outstanding scholars and scientists, eminent entrepreneurs and political leaders in modern China. Tsinghua alumni include the internationally well-known mathematician Shiing-Shen Chern, the current Chinese president Hu Jintao, and China's new president Xi Jinping. Thus the report itself comes with influential credentials.

The Annual Review of Low-Carbon Development in China (2011–12) is a pioneering volume on China's low-carbon development efforts, challenges, plans, trends, and policy recommendations, all based on research conducted by the Climate Policy Initiative at Tsinghua, an independent, experienced, and professional research group based at the university. In this work, key results in China's 11th Five-Year Plan are explored by reviewing China's performance against targets, while the implementation of key policies and institutions are described, with a focus on the effectiveness of low-carbon development policies in China during the period of 2005–2010. Additionally, focus is placed on key indicators of low-carbon development such as energy consumption, CO_2 emission, and low-carbon technologies. The compilation offers not only insights on facts, but also introduces discussion on some of the more controversial issues China faces as it works to meet climate and energy challenges up through 2020.

While some articles delve in great technical detail, the articles selected for this volume particularly convey the larger picture and complementary detailed facts, statistics, and policy examples. Issues range from the low-carbon transformation of China's economy to innovative low-carbon technologies, from low-carbon financing and incentive policies to changes in the business sector and consumer behaviors. While reviewing the achievements and lessons learned during the 11th Five Year Plan (FYP) period, this volume also evaluates the outlook of the 12th FYP and beyond.

The articles contain analyses and interesting details that researchers and members of the Western public need to understand and offers answers to many questions and points of uncertainty. The report's authors have also introduced details on how the national carbon emission reduction targets break down into provincial and local targets, while discussing challenges in meeting the goals set forth in the 12th FYP. They give examples of different models of low carbon city initiatives in China and identify the opportunities for undertaking policy reform or make new policies that are most cost effective to promote low carbon development. They also discuss the current status of China's Clean Development Mechanism (CDM) projects and the challenges and uncertainties China is facing in the international carbon market.

As a distinguished group, the authors of the report include members of the national expert committee on climate change and the research group at Tsinghua University. Given the type of individuals involved, policy makers are very likely to adopt their views in the future. Moreover, some of the experts themselves are policy makers. For example, some authors are advisors to the National Development and Reform Commission (NDRC), a macroeconomic management agency under the Chinese State Council, which has broad administrative and planning control over the Chinese economy.

In short, the better Americans and other Westerners understand the extent of China's commitment the better the United States itself will be able to devise its own aggressive plan to take up the challenge of a dramatically changing climate. But the United States cannot take up that challenge without understanding China's energy evolution and that can hardly be accomplished without the expert information available in the *Annual Review of Low Carbon Development in China (2011–12)*.

CHAPTER ONE

CHINA'S TRANSITION TOWARD A LOW-CARBON ECONOMY DURING THE 11TH FYP PERIOD

Abstract: During the 11th Five-Year Plan (FYP) period, China successfully curbed its rapidly growing trend in energy intensity and achieved a 19.06% reduction in energy intensity, which translates into total energy savings of 630 megatons of carbon equivalent (Mtce), and a CO_2 emissions reduction of 1,550 megatons (Mt). Increased energy efficiency was a major force driving the reduction of carbon emissions, contributing to 87% of the total. Among China's total energy savings, technological factors contributed 64%, structural changes yielded 28%, and about 8% was a result of reduced consumption.

China's low-carbon development is characterized by decreasing emissions intensity, while total emissions keep growing rapidly. During its 11th FYP period, China surpassed the U.S. to become the largest greenhouse gas emitter in the world while its carbon emissions intensity decreased by 20.8%. As a result, China faces mounting pressure both domestically and internationally to pursue further emissions reductions.

During the 11th FYP period, China announced extensive policies intended to save energy, develop renewable energy sources, and cope with climate change. China's low-carbon development policies featured a variety of tools and a tremendous amount of government funding. Despite the commitment of various stakeholders, China's low-carbon development still lacks a solid foundation. While exploring the transition to a low-carbon economy, local governments face challenges posed by expansion-based economic growth; Although energy-intensive key enterprises achieved notable success in energy saving and carbon mitigation, medium and small enterprises (SMEs) find it difficult to meet their energy saving and carbon mitigation targets; residential and transportation-related energy consumption witnessed rapid increases, posing a major challenge to China as it pursues low-carbon development strategies.

Energy-intensive industries displayed a trend to migrate from eastern China towards the west during the 11th FYP period, resulting in an improved industrial structure in the east and an expansion of heavy industries in central and western China. This is another challenge that China still needs to address in order to improve its economic structure. With the global financial crisis impacting China in the mid-11th FYP period, the growth of energy-intensive sectors slowed down. However, the economic stimulus package issued in 2008 again drove up the development of energy intensive sectors, which imposed pressure on China as it struggles to achieve its energy savings target in the 11th FYP, and the projected target for the 12th FYP period.

1.1 The 11th FYP Period: China's Transition toward Low-Carbon Development

1.1.1 *China Successfully Met Its Energy Savings Target, Curbing the Rapid Growth of Its Energy Intensity*

In 1980, energy shortage became the bottleneck for China's economic development; as a result, energy saving became a national strategy. In 1980, the then National Planning Commission[1] and National Economy Commission[2] started to develop five-year energy saving plans. From 1980 to 2000, China issued specific targets for the reduction of energy consumption per unit of GDP in its four consecutive National Economy and Social Development Five Year Plans (hereafter referred to as the Five Year Plans, or FYP). The 10th FYP (2001–2005) was the only one without any explicit energy savings target, though the National Economy and Trade Commission in 2001 published a "10th Five Year Plan for Energy Saving And Resource Comprehensive Utilization," proposing a 4.5% annual reduction in energy consumption per unit of GDP in between 2001 to 2005. With the implementation of energy-saving policies from 1980 to 2002, China's energy consumption per 10,000 RMB GDP dropped from 3.401 tons of carbon equivalent (tce) to 1.162 tce (2005 RMB), a decrease by 65.82% (Figure 1.1). In the period from 2002–2004, China's trend in energy intensity reversed: energy intensity increased from 1.162 tce/10,000RMB in 2002 to 1.285 tce/10,000RMB in 2004. As a result, the 10th FYP period became the only period with increasing energy intensity among all Five-Year periods since 1980.

In order to reduce energy consumption, China announced the "National Economy and Social Development 11th Five-Year Plan" (hereafter referred to as the "11th FYP") in March 2006, raising a mandatory target of "reducing energy consumption per unit of GDP by about 20% in 2010 relative to the 2005 level." This was the first time that China proposed and legalized a binding target for energy intensity reduction, reflecting China's strong political commitment to achieve energy savings. With the implementation of energy saving policies during the 11th FYP period, China's

[1] The National Planning Commission was established in 1952, and renamed the National Development and Planning Commission in 1998. In 2003 it was reformed into the National Development and Reform Commission (NRDC).

[2] The National Economy Commission was established in 1956, dissolved in 1970, reinstituted in 1978, dissembled again in 1988, reassembled again in 1993 and renamed the National Economic and Trade Commission. It was dissolved again in 2003.

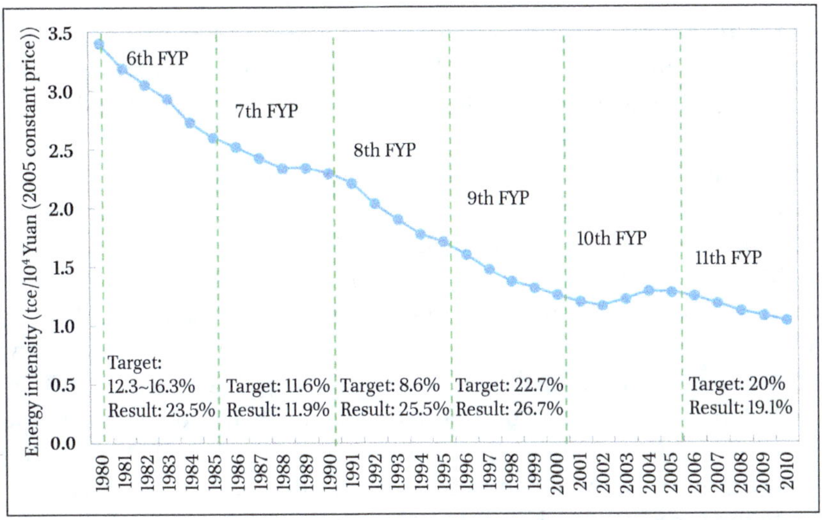

Source: "National Economy and Social Development 6th Five-Year Plan (1981–1985)," "National Economy and Social Development 7th Five-Year Plan (1986–1990)," "National Economy and Social Development 8th Five-Year Plan (1991–1995)";[3] "National Economy and Social Development 9th Five-Year Plan (1996–2000)," and "National Economy and Social Development 11th Five-Year Plan"; Energy consumption per unit of GDP data adopts the 2005 RMB value and was calculated from data found in "China Yearbook 2012"; 2010 data was derived from the 19.06% in reduction during the 11th FYP.

Figure 1.1. Change in China's energy intensity from 1980–2010.

energy consumption dropped from 1.276 tce/10,000RMB in 2005 to 1.033 tce/10,000RMB in 2010, a decrease by 19.06%. China has successfully met its 11th FYP energy savings target.

Most economic sectors also met their energy savings targets (Table 1.1). In the 11th FYP period, thirteen key industries, including steel, electrolytic aluminum, copper, cement, and flat glass, achieved a significant reduction in energy consumption per unit of production, with an average of 18%. Some enterprises in those industries achieved energy efficiency on a par with or even above advanced international standards. The "Top-1,000 Enterprises Energy Efficiency Program" yielded energy savings of 150 Mtce, the "Ten Key Industry Energy Savings Program" 340 Mtce, the "Phasing-out Obsolete Capacities Program" over 110 Mtce. These three programs successfully met and surpassed energy savings targets. During the 11th FYP

[3] In March 1993, China published the "China Communist Party's Suggestions for Adjusting Several Targets in the 8th FYP" which changed the original energy-saving target from 1.8% to 3.7%. As a result, the energy-saving target for the whole 8th FYP period was about 17.2%.

period, the total area of energy efficient buildings reached 4.08 billion m^3, nearly twice the original target. The railway and the civil aviation sector reduced their energy consumption per unit of service by 23.8% and 11.3%, respectively, both surpassing their 11th FYP targets.

Road and water transportation failed to meet their energy savings targets, as these two sectors have a large number of enterprises, transportation equipment, and personnel, making energy saving management rather difficult. Moreover, these two sectors committed rather high targets in the 11th FYP. In the "12th FYP for Energy Saving and Emissions Reduction in Road and Water Transportation Sector," issued in 2011, the Ministry of Transportation assigned lower targets for these two sectors: 10% in reduction per unit of road transportation, and 15% in reduction per unit of water transportation in 2015 relative to 2005. The 12th FYP targets for energy intensity in road and water transportation sectors again demonstrate the excessively high targets set in the 11th FYP.

The 11th FYP raised the target of "increasing the share of the tertiary sector in GDP by 3% by 2010 relative to 2005," which was not met according to empirical data. The share of the tertiary sector in GDP reached 43.4% in 2009, but dropped to 43.1% in 2010 due to the implementation of the economic stimulus package. Despite the failure to meet its target, China's tertiary sector witnessed faster development in the 11th FYP period compared to the 10th FYP period. China still has a long way to go in order to transform its economic growth pattern by adjusting the sectoral structure.

Table 1.1. 11th FYP's energy savings targets by sector.

Sector	Energy savings targets	Results	
Overall target	Energy consumption per unit of GDP reduced by about 20% in 2010 relative to 2005	Actual reduction: 19.06%	Completed
Industry	In 2010, key industries achieved energy efficiency (in terms of unit production energy consumption) comparable to international best practices	Unit production energy consumption experienced significant decrease in all major energy-intensive industries	Completed

Table 1.1 (*cont.*)

Sector	Energy savings targets	Results	
	"Top-1000 Enterprises Energy Efficiency Program" energy saving target: 100 Mtce	Actual saving of 150 Mtce during 11th FYP period	Completed
	"Ten Key Industry Energy Saving Program" energy saving target: 240 Mtce	Actual saving of 340 Mtce during 11th FYP period	Completed
	Phasing out obsolete capacities, promoting industrial energy saving	Actual saving of 110 Mtce in four years	Completed
Buildings	Building sector energy saving target: 101 Mtce, total area of energy efficient buildings: 2.15 billion m^3	Total area of energy efficient building by 2009: 4.08 billion m^3	Completed
Transportation			
Road	Energy consumption per 100 ton*km decreased by 20% in 2010 relative to 2005	No data	Failed
Water	Energy consumption per 100 ton*km decreased by 20% in 2010 relative to 2005	No data	Failed
Railway	Energy consumption per unit income decreased by 20% in 2010 relative to 2005	Energy consumption per unit income decreased from 6.48 in 2005 to 4.94 tce/Mt equivalent *km, a decrease of 23.8%	Completed
Civil aviation	Oil consumption per ton*km decreased by 10% in 2010 relative to 2005	Oil consumption per ton*km decreased from 0.336 kg in 2005 to 0.298 kg in 2010, a decrease by 11.3%	Completed

Table 1.1 (*cont.*)

Sector	Energy savings targets	Results	
Agriculture	Over 50 Mtce in savings by reducing energy consumption in agriculture and households, and developing alternative and renewable energy such as biomass, solar, wind, and micro-hydro	No data	
Tertiary sector	The share of tertiary sector value-added increase from 40.3% to 43.3%	The share of tertiary sector in 2010 was 43.1%	Failed

Source: General target and tertiary sector target data are from "National Economy and Social Development 11th Five-Year Plan," data of target completion are from "China's National Economy and Social Development Report (2010);" industry energy savings target is from "Energy Development 11th FYP," completion data are from National Development and Reform Commission (2011a); "Top-1000 Enterprises Energy Efficiency Program" target is from "Top-1000 Enterprises Energy Efficiency Program Action Plan," completion data are from National Development and Reform Commission (2011b); Ten Key Industry Energy Saving Program target is from "11th FYP Ten Key Industry Energy Saving Program Implementation Schedule," completion data are from National Development and Reform Commission (2011c); phasing out obsolete capacity target is from "Comprehensive Work Schedule for Energy Saving and Emissions reduction," completion data are from National Development and Reform Commission (2010); building energy savings target is from "Construction Sector 11th FYP," completion data are from (China Government Website, 2010); Road and water transportation energy savings targets are from "Guidelines on Developing Efficient Transportation;" Railway energy savings target is from "Railway Sector 11th FYP for Energy Saving and Resource Comprehensive Utilization," completion data are from Ministry of Railway (2011); Civil aviation energy savings target is from "China Civil Aviation Development 11th FYP," completion data are from China Civil Aviation Administration, 2011); agriculture energy savings target is from "Guidelines for Energy Saving and Emissions reduction in Agriculture and Rural Area by Ministry of Agriculture."

During the 11th FYP period, energy intensity followed the decreasing trend it had displayed since 1980. However, in a historical context, achieving the energy-saving targets during this period was rather remarkable. During the 11th FYP period, three factors posed major challenges to energy saving: the rapid development of heavy industries, accelerated urbanization, and the rather high share of exports in the overall economy. Since 1999 (Figure 1.2), the growth of heavy industry notably accelerated, reaching 15.6% in the 10th FYP and 15.7% in the 11th FYP period, a rate much higher than that of the light industry (growing 13.0% and 13.1%, respectively). Before 1999, however, average growth rates of the light industry in China were higher than that of the heavy industry. The 10th FYP proposed

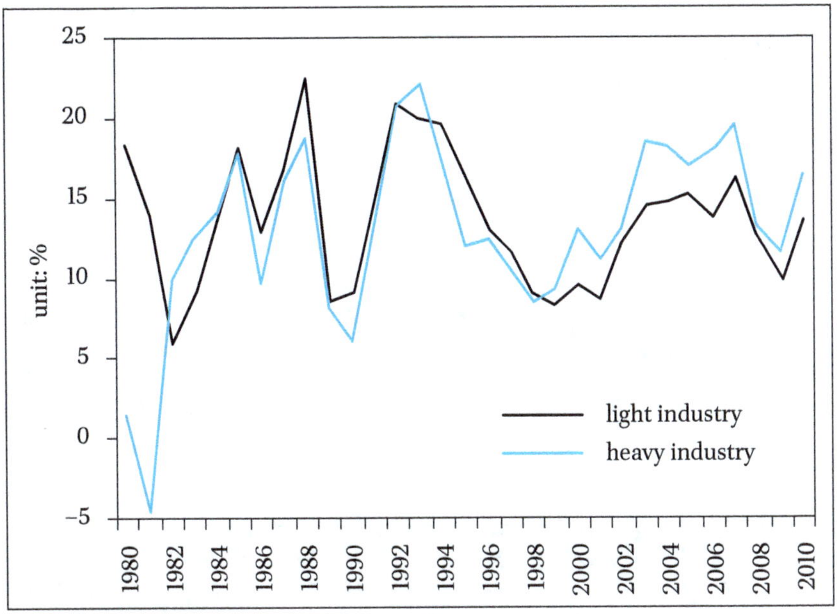

Source: Calculated from data in the "National Economy and Social Development Yearbook" (1980–2010).

Figure 1.2. Growth rates of heavy industry and light industry in 1980–2010.

to pursue an adjustment and upgrade of China's economic structure, and the 11th FYP proposed to "increase the quality of economic growth and pursue further economic structural optimization and upgrades"; however, the idiosyncrasies of China's industrialization make economic structural adjustments unusually difficult. China's capital-intensive, energy-intensive, and resource-demanding economic growth pattern will continue to persist for a long period of time and even intensify. The growth of heavy industries is a major contributing factor to the rapid growth of total energy consumption in the 11th FYP period.

Starting from 1996, China's urbanization rate increased annually by 1% on average, reaching 49.95% in 2010 as compared to 36.22% in 1995 (Figure 1.3). Rapid urbanization drove infrastructure development such as roads, buildings, water supply and treatment facilities, and electric infrastructures. As a result, energy-intensive industries such as steel and cement experienced rapid development, stimulating energy consumption. During the 11th FYP period, about 10 million people on average migrated into urban areas annually, necessitating the construction of large numbers of residential buildings. The increasing pervasiveness of urban lifestyles

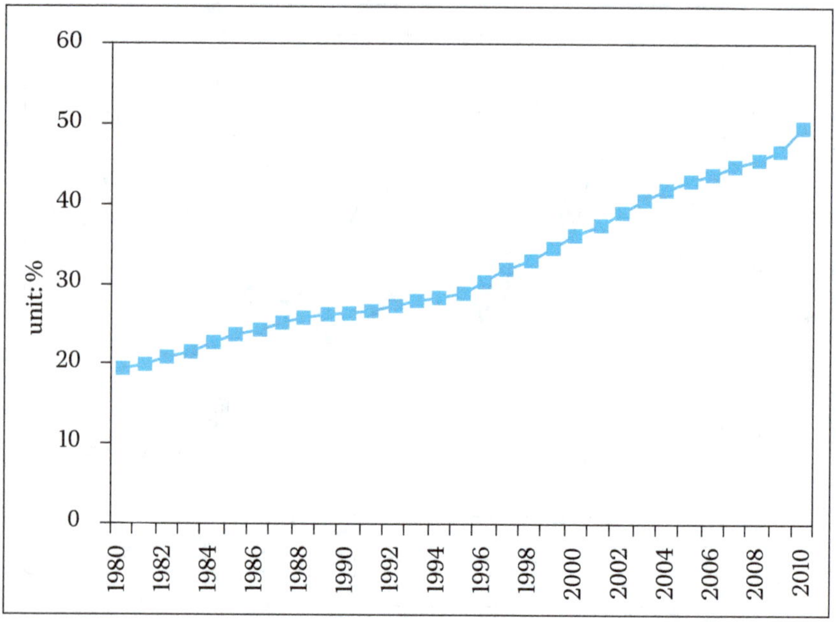

Source: "China Yearbook 2010," 2010 data see: "6th Nationwide Population Census 2010 (No. 1)".

Figure 1.3. Urbanization rate in 1980–2010.

also resulted in increased demand for more energy services. In conclusion, urbanization was one of the primary driving forces behind China's rapid growth of energy consumption.

Export was another major driver for China's rapid growth of energy consumption and carbon emissions. Export is one of the three factors that supports China's economic development and plays a crucial role in its economy. Export has been increasing since 2000, from 2.06 trillion RMB in 2000 to 10.04 trillion RMB in 2008. Affected by the global financial crisis, export decreased in 2009, yet the total volume was still as high as 8.20 trillion RMB. In 2010, it increased to 10.70 trillion RMB. The growth rate of export in the 11th FYP period was unprecedented in China's history. Its tremendous trade volume also stimulated China's rapid growth in energy consumption and carbon emissions. Assuming export commodities have the same carbon emissions intensity as China's carbon emissions per unit of GDP, the embedded carbon emissions in export during the 10th FYP and 11th FYP period were 5,540 megatons of carbon dioxide ($MtCO_2$) and 9,424 $MtCO_2$, respectively, accounting for 26.8% and 30.8% of China's total carbon emissions during the same period (Figure 1.4). Research shows that the average annual growth rate of China's total carbon emissions between

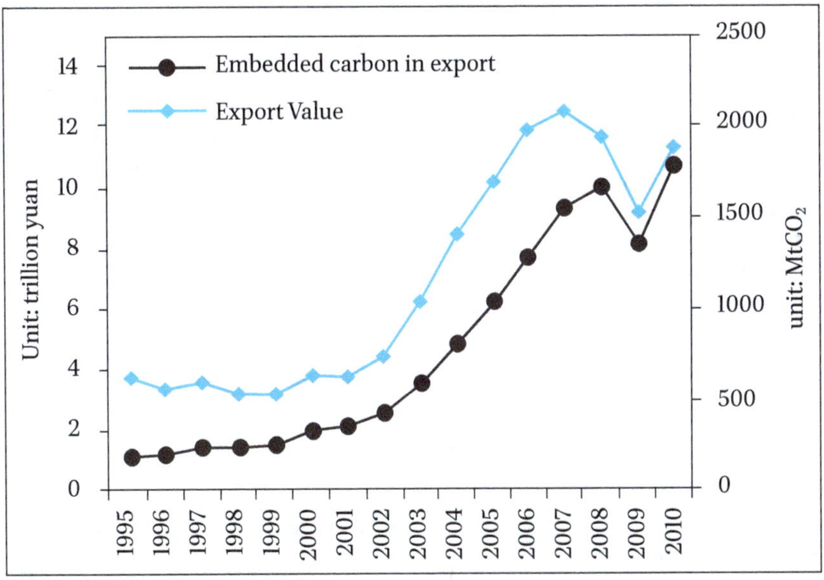

Source: 1995–2009 export data from "China Yearbook 2010"; 2010 export data from "China National Economy and Social Development Report 2010."

Figure 1.4. Embedded carbon in export during 1995–2010.

1997 and 2007 would be 4.39% if embedded carbon emissions in export products were to be excluded, much lower than the figure of 7.82% that factors in export-related carbon emissions.[4]

In the context of unprecedented industrialization, urbanization, and increasing exports in the 11th FYP period, it was rather remarkable for China to curb its growing trend in energy intensity in the 10th FYP period and achieve a 19.06% decrease in energy intensity. During the 11th FYP period, China developed a series of policy tools and institutions aimed at energy saving and emissions reduction. In terms of energy management institutions, China developed supporting mechanisms such as the "energy savings target accountability, measurement, and verification system," the "energy saving assessment and verification system for fixed assets investment," the "energy reporting system for key industries and major consumers," and others. In terms of policy, a wide range of tools such as tax incentives, pricing, financing, and government procurement policies were adopted. China also imposed energy consumption quotas, enforced energy labeling, and implemented mandatory energy audits for

[4] Li Huimin 李惠民 (2009): The Issue of Embedded Carbon in Trade in China's Climate Change Response (Doctoral Dissertation). Beijing, Beijing Normal University.

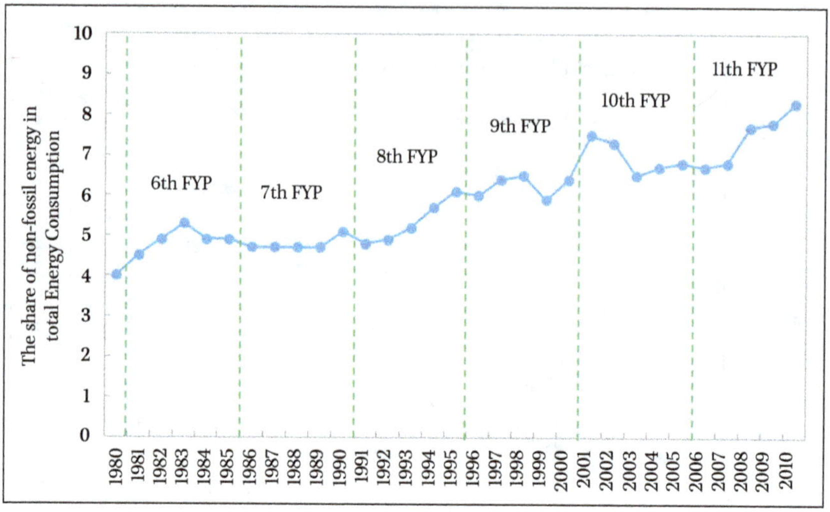

Source: 1980–2009 data is from "China Energy Yearbook 2010," and 2010 data is from "China Yearbook 2011."

Figure 1.5. The share of non-fossil energy in total Energy Consumption in China (1980–2010).

key energy consumers. These policies and institutions were reflected in the newly revised "Energy Conservation Law of the PRC (2007)." Although there is still room for improvement regarding such policies and institutions, they undoubtedly established a solid foundation for China's low-carbon development in the long term.

1.1.2 *Unprecedentedly Rapid Development of Non-Fossil Energy*

From 1981 to 1990, the overall share of non-fossil energy in China's total energy consumption increased from 4% to 5.1%, a very modest annual growth rate of 0.1% on average over 10 years (Figure 1.5). Not until the 8th FYP period (1991–1995) did non-fossil energy development accelerate. During this time, installed hydro capacities increased from 36.046 gigawatt (GW) in 1990 to 52.184 GW in 1995, i.e. an annual growth rate of 7.7% on average (Figure 1.6). The Qinshan and Dayawan nuclear plants went into operation, and installed nuclear capacity reached 2.10 GW in 1995 (Figure 1.7). The share of non-fossil energy use increased from 4.8% in 1991 to 6.1% in 1995, an increase of 1.3 percent over five years. Although installed capacity of hydro and nuclear power expanded rapidly, this was counteracted by the even higher growth of total energy consumption. The share of non-fossil energy in total energy consumption remained stable,

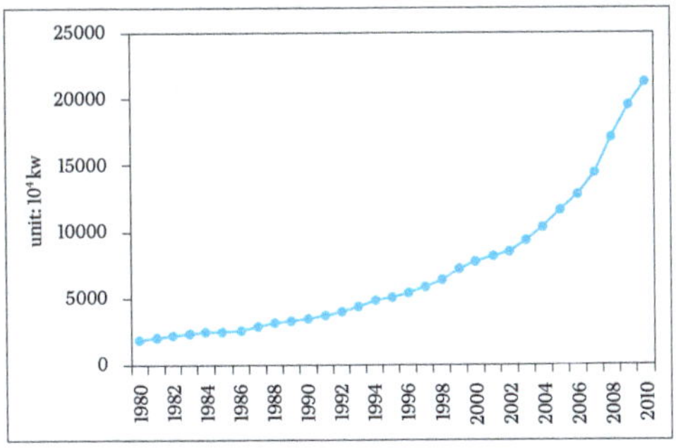

Source: State Power Information Network, "Linian dianli zhuangji he fadianliang de goucheng bi (1952–2001) 历年电力装机和发电量的构成比 (1952–2001) [Structural Comparison of Installed Capacity and Generation Capacity of Electric Power for the Years 1952 to 2001]" (2008-01-15), URL: http://www.sp.com.cn/dlsc/dltj/qgdtsj/dzhzb/200805/t20080516_104965.htm [accessed: 2011-07-01].

Figure 1.6. Installed hydro capacity (1980–2010).

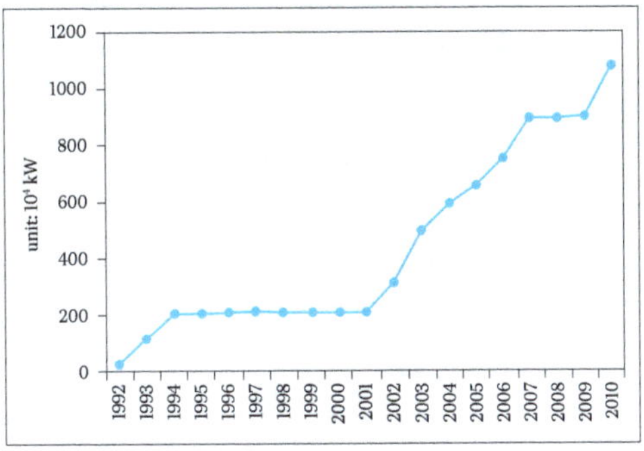

Source: United Nations Statistics Division, "Energy Statistics Database; Electricity Industry Yearbook (2009–2010)."

Figure 1.7. Installed nuclear capacity (1992–2010).

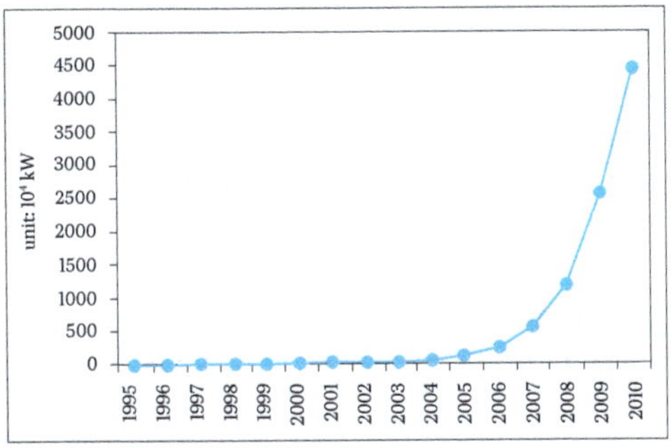

Source: 1995–1999 data from "China Alternative Energy Website" (2004); 2000–2009 data from "China Medium and Long-Term Energy Development Strategies" (2011); 2010 data: Li Junfeng 李俊锋 et al., 2011 Zhongguo fengdian fazhan baogao 2011 中国风电发展报告 [2011 Report on Wind Energy Development in China] (Beijing: Zhongguo huanjing kexue chubanshe, 2011).

Figure 1.8. Installed wind capacity (1995–2010).

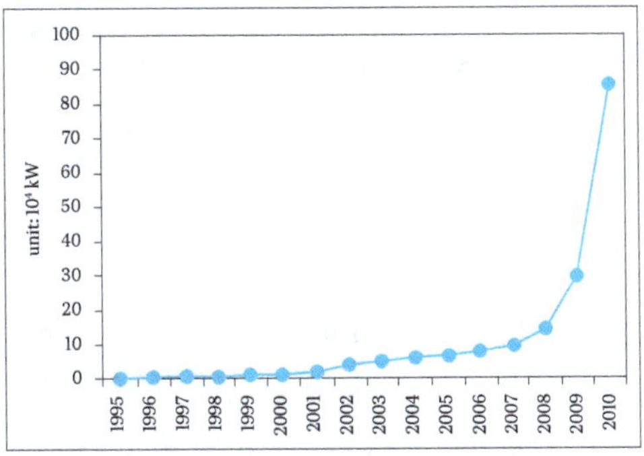

Source: 1995–2003 data from China Renewable Energy Development Program. (2004); 2004–2009 data from "China Medium and Long-Term Energy Development Strategies" (2011); 2010 date: Li Junfeng 李俊锋 et al., *2011 Zhongguo fengdian fazhan baogao* 2011 中国风电发展报告 [2011 Report on Wind Energy Development in China] (Beijing: Zhongguo huanjing kexue chubanshe, 2011).

Figure 1.9. Installed solar capacity (1995–2010).

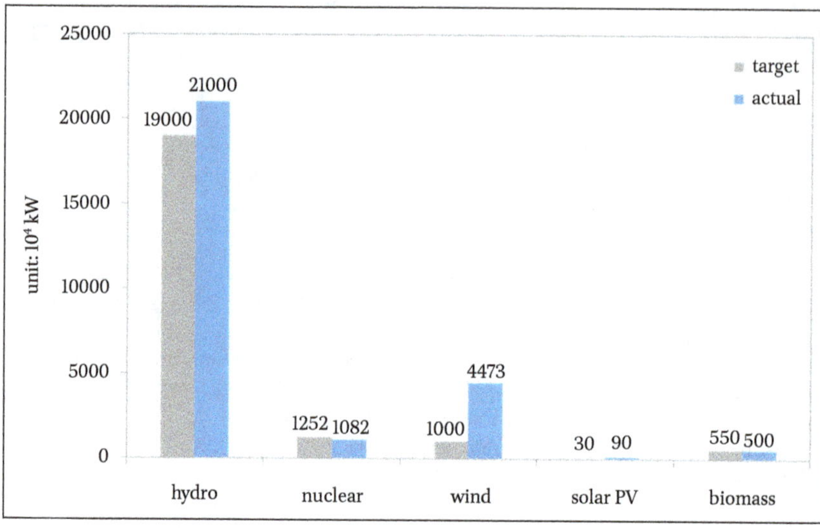

Source: Nuclear target from "Medium and Long-Term Development Plan for Nuclear Energy (2005–2020)"; other renewable targets from "Renewable Energy Development 11th FYP." Hydro and nuclear installed capacity in 2010 from "China Electricity Council" (2011). Wind installed capacity in 2010, see: Li Junfeng et al. (2011); According to "Electricity Sector Yearbook (2010)," grid-connected wind installed capacity in 2010 was 31.070 GW. Solar installed capacity in 2010, see: Li Junfeng et al. (2011), however, according to China Electricity Council (2011) it is 240 MW. It was difficult to find data on biomass generation installed capacity in 2010. According to Yu (2011), "the State Energy Administration set the plan of biomass generation installed capacity to reach 1.3 GW in 2015, an increase by 160% relative to 2010." (see: Yu Xiangming 于祥明, "Shengwuzhi fadian zhuangji rongliang wuniannei beizeng 生物质发电装机容量5年内倍增 [Installed Capacity of Biomass Power Generation Doubles Within Five Years]" (2011-07-21) http://finance.stockstar.com/SS2011072100000794.shtml [accessed: 2011-08-23]. As a result, we calculated 5 GW for 2010.

Figure 1.10. Targeted and actual development of alternative and renewable energy during 11th FYP period.

with a moderate increase from 6% in 1996 to 6.8% in 2005. During the 11th FYP period (2006–2010), installed capacity of hydro and nuclear power expanded faster, while wind and solar energy experienced exponential growth (Figure 1.8 and Figure 1.9). As a result, the share of non-fossil energy increased from 6.8% in 2005 to 8.6% in 2010, an increase of 1.8 percent over five years. In terms of both absolute growth and share in total energy consumption, the 11th FYP period undoubtedly witnessed the fastest growth of non-fossil energy since the 6th FYP period.

The National Development and Reform Commission (NDRC) released the "Renewable Energy Development 11th FYP" in 2008, raising targets for renewable energy (including hydro) development—"by 2010, total installed hydro capacity will reach 190 GW, wind 10 GW, biomass 5.5 GW, and solar 300 megawatt (MW)." In 2007, the NDRC also released a "Medium and Long-Term Development Plan for Nuclear Energy (2005–2020)," which specified the targets for nuclear development in the 11th FYP period: "nuclear power installed capacity will reach 12.5280 GW by 2010."

During the 11th FYP period, nuclear and biomass power failed to reach their assigned targets, meeting 86% and 91% of their targets, respectively (Figure 1.10). Installed hydro capacity increased from 117 GW in 2005 to 210 GW in 2010, exceeding its original target set at 190 GW. The decrease in hydro power plant capacity (from 3,642 utilization hours in 2005 to 3,429 utilization hours in 2010) is believed to be one of the factors that hindered the development of hydropower plants. Wind and solar energy expanded much faster than expected, meeting 447% and 300% of their original targets. Compared to actual achievements, the targets proposed in the "Renewable Energy Development 11th FYP" appear to have been rather conservative.

Each of the FYPs since 1980 took alternative energy development into account; however, those plans mostly focused on technological and industrial development. China released its first special plan for energy in 1995—the "Alternative and Renewable Energy Development Outlines (1995–2010)"—which set the strategic target of "consuming at least 390 Mtce of alternative and renewable energy by 2010 (including biomass utilization)." During this period, the main purpose of developing wind, solar, and other alternative/renewable energy sources was to provide electricity for remote areas and islands without access to electricity. The National Economic and Trade Commission released the "Alternative and Renewable Energy Industrial Development 10th FYP," setting strategic targets of "reaching 15 MW of solar photovoltaic (PV) capacity by 2005 while reaching 1.2 GW of grid-connected wind capacity." The 11th FYP defines the development of renewable energy as the development of a renewable industry.

During the 11th FYP period, China's alternative and renewable energy policies were significantly strengthened. The "Renewable Energy Law" took effect in 2006; The NRDC announced the "Medium and Long Term Development Plan for Renewable Energy" and the "Medium and Long Term Development Plan for Nuclear Power (2005–2020)" in 2007; in 2008, the NRDC issued the "Renewable Energy Development 11th FYP"; National Congress passed the revised "Renewable Energy Law" in 2009 which took effect in April 2010. The 11th FYP period was not only the time when alternative and renewable energy experienced rapid development,

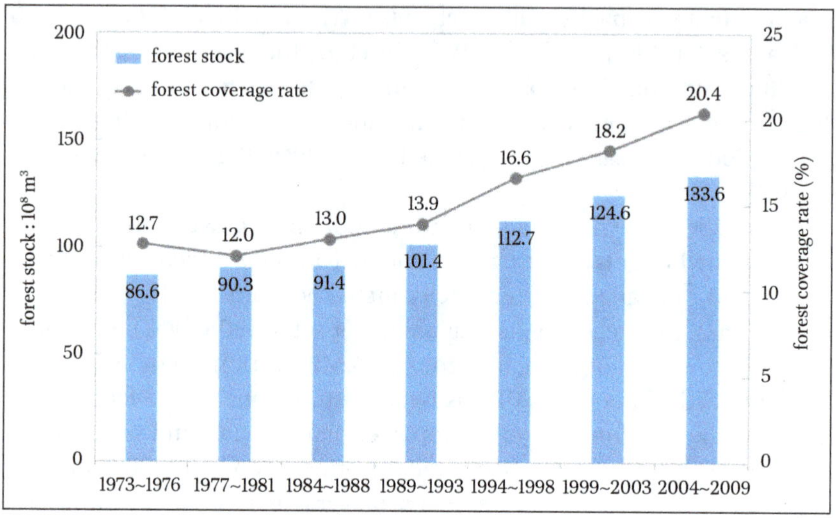

Source: State Forest Administration.

Figure 1.11. Forest coverage rate and stock in China.

but also a critical phase when China developed a legal, institutional, and a policy foundation for alternative and renewable energy development. In this sense, the 11th FYP can be regarded as a transitional period for China's alternative and renewable energy development.

1.1.3 *Forest Carbon Decrease Expanded, and Carbon Sink Strategy was Developed*

Forest coverage in China recovered slowly since 1978. A data comparison from the 2nd (1977–1981) to the 4th forest census (1989–1993) shows that forest storage increased by 1.11 billion m^3 (an annual increase by 93 million m^3) and forest coverage increased by 1.9% over 12 years, or 0.16% on average (Figure 1.11). After 1994, the expansion of forest cover accelerated. Comparing data from the 4th (1989–1993) to the 5th forest census (1994–1998), forest stock increased by 2.7% over five years, an annual increase by 0.54% on average. From 1999 to 2003, the expansion of forest cover slowed down, yet the coverage rate still increased by 0.32% annually. Since 2004, China has started to implement six major forestation programs, and forest has increased by 900 million m^3 in stock (or 2.2%) over 5 years, an annual increase by 180 million m^3 or 0.44% in coverage rate. In sum, the forest area expansion rate during 11th FYP period was the second highest after the 9th FYP period.

In 1998, China started to implement six major forest programs including "Converting Farmland Back to Forest" and "Conservation of Natural Forests," with the purpose of improving its ecological environment. The State Forest Administration set up a Carbon Decrease Management Office in 2003, indicating that the question of carbon emissions was beginning to shape forest management practices. In 2007, the State Forest Administration established the "Climate Change, Energy Saving, and Emissions Reduction Task Force" responsible for managing forest sectors with the specific goal of addressing climate change. In 2008, the State Council published "Comments on Promoting Stable Development of Agriculture and Increasing Farmers' Income in 2009" which for the first time proposed to "develop modern forest industries featuring mountainous forest products, eco-tourism and carbon decrease." The notion of carbon decrease specified in this policy reflected a new strategic perspective concerning forestry. The central government held a working meeting in June 2009 in which policy-makers made the decision to "re-think the critical role of forest in the context of climate change." In conclusion, the 11th FYP period was not only a time forests experienced rapid development in China, but also a transitional period in which carbon sink became a strategy for China's forest development.

1.2 Analysis of China's Low-Carbon Development during the 11th FYP Period

1.2.1 *Increase in Energy Efficiency Played a Critical Role in China's Low-Carbon Development*

In order to evaluate the roles energy efficiency and energy mix play in China's carbon emissions patterns, we use the following formula while neglecting the factor of forest carbon decrease:

$$\textit{Total } CO_2 \text{ emissions} = GDP \times \text{energy consumption per unit of GDP} \times CO_2 \text{ emissions per unit of energy consumed}$$

Compared to the baseline scenario, CO_2 emissions reduction per year is

$$CO_2 \textit{ emissions mitigation} = GDP_{actual} \times \textit{energy consumption per unit of GDP}_{baseline} \times CO_2 \text{ emissions per unit of energy consumed}_{baseline} - GDP_{actual} \times \textit{energy consumption per unit of } GDP_{actual} \times CO_2 \text{ emissions per unit of energy consumed}_{actual}$$

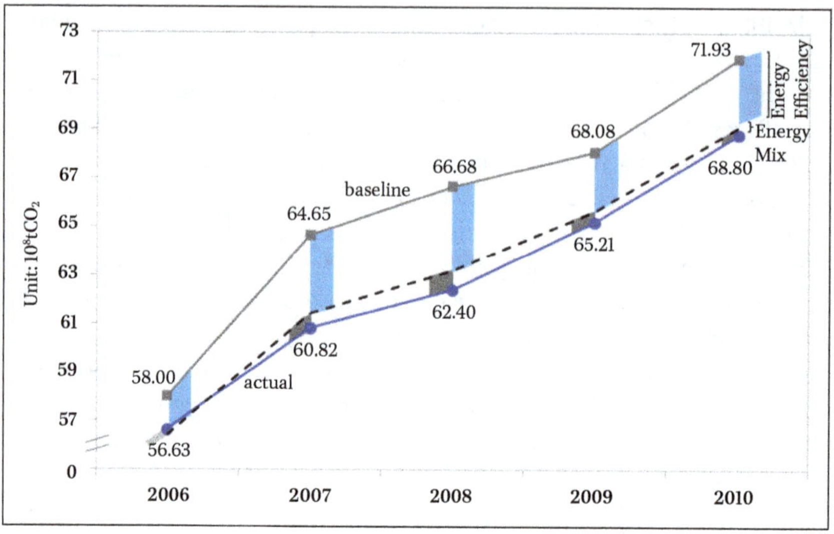

Figure 1.12. Contribution of energy intensity and energy mix in CO_2 emissions reduction.

Using LMDI (Logarithmic Mean Disivia Index, LMDI) methodology,[5] we can calculate the impact of energy intensity and energy mix on carbon emissions (Figure 1.12).

Using the previous year as a baseline for each year, data shows that during the 11th FYP period, China achieved a total CO_2 emissions reduction of 1.55 billion tons. The largest reduction was achieved in 2008: about 427 $MtCO_2$, accounting for 27.6% of total reductions; the 2007 reduction was the second largest one, accounting for 24.8% of the total; while the 2006 reduction was the smallest one, accounting for 8.9% of the total. Improvement in energy efficiency was the major factor for CO_2 emissions reduction, contributing to 87.2% of total reduction on average. Improvement in the

[5] B.W. Ang, "The LMDI Approach to Decomposition Analysis: A Practical Guide," *Energy Policy* 33, no. 7 (2005): 867–871. There are various methodologies to evaluate the impact of each factor in a variable. Two examples are Laspeyres Index and Paasche Index. However, these two cannot decompose multiple variables at the same time, or usually yield large residues. In comparison, LMDI methodology has several advantages. (1) LMDI is more convincing as it decomposes factors in a reasonable way and does not yield in residues; (2) The sum decomposing and multiply decomposing in LMDI methodology are related and convertible; (3) In LMDI methodology, the sum of sub-sector correspond to the total. Due to these advantages, LMDI methodology is becoming more and more popular in energy and carbon emission researches.

energy mix did not yield significant contributions to CO_2 emissions reduction, with even a negative contribution in 2006.

1.2.2 *The Factor of Technology Innovation and Structural Change*

China's energy consumption is analyzed in two categories: production and household consumption. As all production value is reflected in total GDP while consumption does not generate added value, lower growth of consumption than that of production will result in lower energy intensity. Energy saving from the change of share of consumption is:

$$\Delta En_{consumption} = GDP \cdot \left(\frac{En_{baseline}}{GDP_{baseline}} - \frac{En_{actual}}{GDP_{actual}} \right) - GDP \cdot \left(\frac{E^{production}_{baseline}}{GDP_{baseline}} - \frac{E^{production}_{actual}}{GDP_{actual}} \right)$$

Energy saving calculated in this formula does not mean the saving from equipment efficiency improvement and lifestyle changes.

Energy consumption from production is:

$$En_{production} = \Sigma(V_i \cdot I_i)$$

Where V_i is the value-added in sector i,

I_i is the energy consumption per unit value-added in sector i.

Energy saving on the production side has two variables: energy savings resulting from changes, i.e. technological energy saving in the broad sense, and energy saving from proportional changes in total added value, i.e. more specifically energy saving from structural change. As a matter of fact, there are other factors affecting energy saving besides these two, such as the increase of unit product value-added, product portfolio changes, etc. In order to eliminate non-technological factors, we use the following formula to calculate technological energy saving:

$$\Delta En_{technology} = \Sigma(P_i \cdot \Delta\rho_i)$$

Where P_i is output of product i,

$$\Delta\rho_i$$

is the change of energy consumption per unit product for product i.

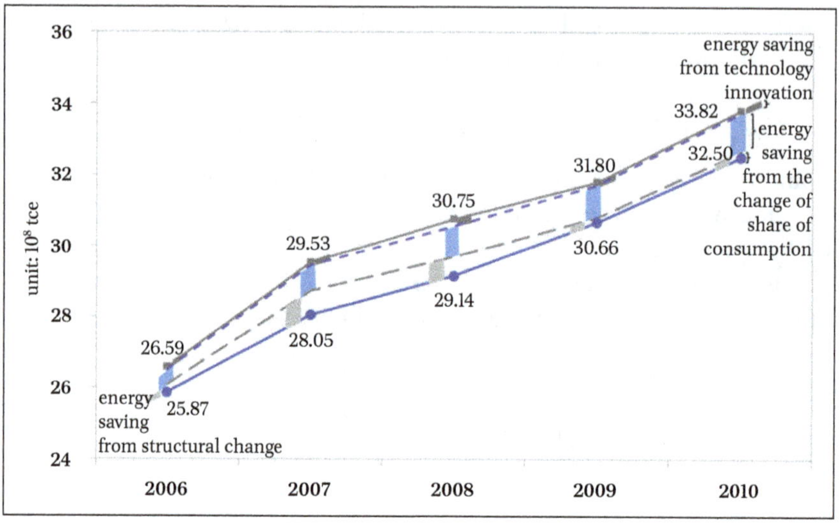

Note: Structural change shown in this figure include not only changes in industrial structure and sector structure, but also those in product range and increase in unit product value-added. energy saving from the change of share of consumption is theoretical saving, which did not mean savings from equipment efficiency improvement and lifestyle changes.

Figure 1.13. Contribution of technology and structural change in energy saving.

Structural changes in energy saving are calculated by deducting technological energy saving from total energy saving on the production side. Energy savings resulting from structural changes in this context include changes in industrial structure, sectoral structure, product portfolio, and even the increase in added value of unit/product. See Figure 1.13 for the structural change energy saving during the 11th FYP period.

According to our calculations, China achieved an energy savings total of around 630 Mtce during the 11th FYP period. Technological factors contributed about 64% in total energy saving, structural change contributed about 28%, while the remaining 8% came from the consumption side. From 2006–2010, technological energy savings increased steadily, from 43 Mtce to 108 Mtce; while the results from structural changes varied, achieving 0.68 Mtce in 2007 (nearly 45.8% of the total), but only 13 Mtce and 17 Mtce respectively in 2009 and 2010 (only 11.6%, and 12.8% of the total). Chapter 3 presents further analysis on energy savings resulting from structural changes.

1.2.3 *Provinces Completed their Energy Savings Targets, Especially those in Middle and Eastern China*

The State Council released "Comments on the 11th FYP Targets of Energy Consumption Reduction per unit Production for Provinces" in September 2006, setting energy savings targets for each province in the 11th FYP period. All provinces were assigned a target of 20% in reduction, except seven (Qinghai, Tibet, Yunnan, Guangdong, Guangxi, and Fujian) which received targets of under 20%, and four (Shandong, Shanxi, Inner Mongolia, and Jilin) with targets of higher than 20%. Targets for Shanxi, Inner Mongolia, and Jilin were set at 25%, 25%, and 30% respectively. Later, these three provinces found it too challenging to meet their targets and re-negotiated with the central government which agreed to set new targets at 22% each.

By the end of 11th FYP period, all provinces completed their target with the exception of Xinjiang (Figure 1.14). 28 provinces exceeded their targets,

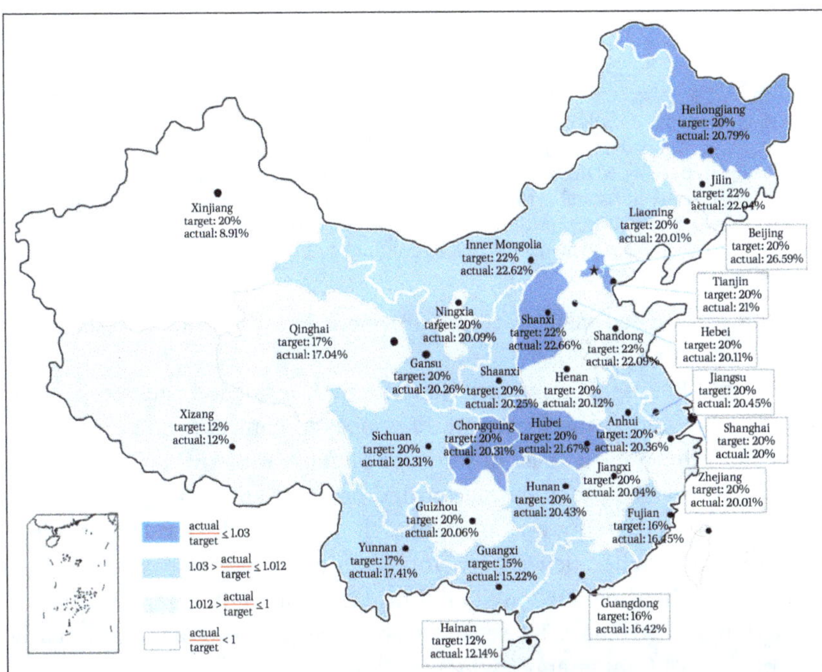

Source: Data of all provinces except Xinjiang is from "11th FYP Energy savings target Completion Report by NDRC and national Bureau of Statistics"; Xinjiang data is from "12th FYP Energy Saving and Emissions reduction Comprehensive Work Schedule."

Figure 1.14. Completion of 11th FYP energy savings targets by provinces.

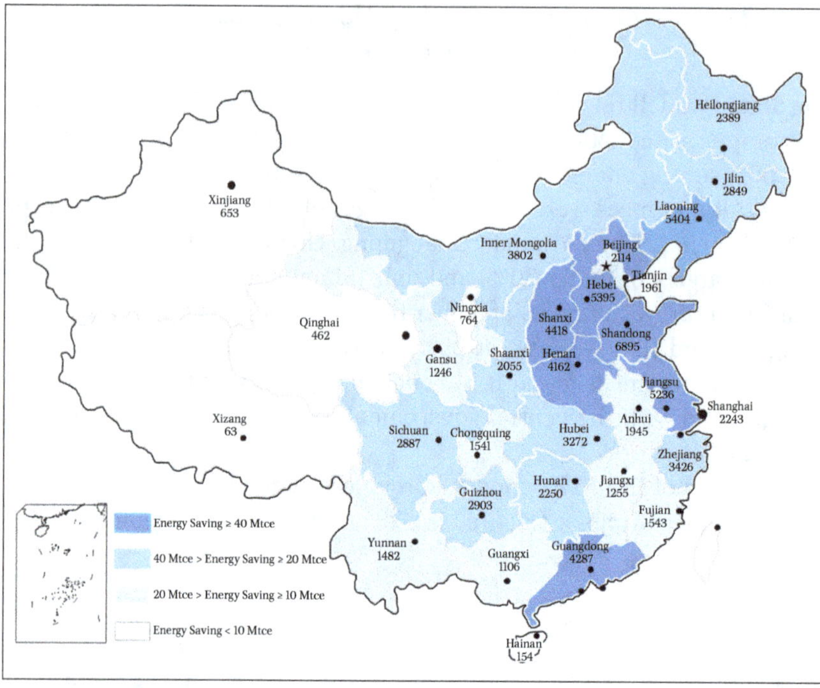

Note: Energy saving data is calculated via energy consumption per unit of GDP and total energy consumption data in each province according to "11th FYP Energy Saving Target Completion Report by NDRC and national Bureau of Statistics" and "China Energy yearbook 2010." Due to statistical errors, the sum of energy consumption does not add up to "total national energy consumption" in "China Energy Yearbook 2010." As a result, the sum of energy saving in each province in this figure does not add up to total national energy saving either.

Figure 1.15. Actual energy saving in provinces during the 11th FYP period.

for example Beijing, Hubei, and Tianjin which exceeded their respective targets by 6.59, 1.67, and 1 percent. Provinces in central China performed better than those in the east in terms of target completion. However, the central and eastern provinces performed better than the western ones in terms of absolute energy saving because their economic development levels were much higher. Several provinces achieved an energy saving figure of more than 40 Mtce, such as Shandong, Liaoning, Hebei, Jiangsu, Shanxi, Guangdong, and Henan. These seven provinces together achieved nearly 45% of China's total energy savings.

Table 1.2. Low-carbon indicators in electricity sector (2005–2010).

		2005	2006	2007	2008	2009	2010
Gross heat rate	gce/kWh	343	342	332	322	320	312
Station consumption ratio	%	5.87	5.93	5.83	5.90	5.76	5.43
Line loss	%	7.21	7.04	6.97	6.79	6.72	6.53
Net heat rate	gce/kWh	370	367	356	345	340	333

Source: 2005–2008 data is from "China Energy Yearbook 2010," 2009–2010 data is from China Electricity Council (2010, 2011b). China thermal power plants data refers to units above 60 MW.

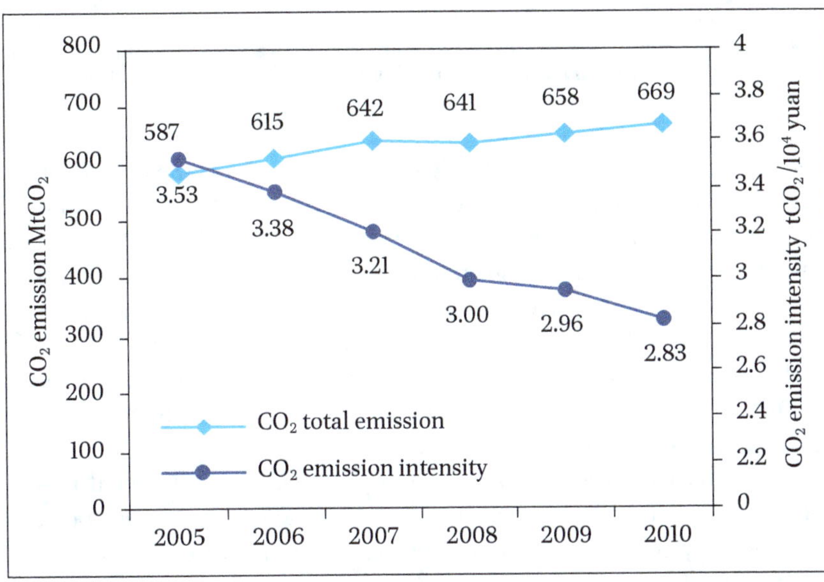

Figure 1.16. CO_2 emissions and intensity in energy industry (2005–2010).

1.2.4 *Sectoral Transition to a Low-Carbon Development Trajectory*

1.2.4.1 *Decarbonization of Energy Industry*

China's energy industry consists of five sectors: coal mining and dressing, petroleum and natural gas extraction and processing, coke processing and nuclear fuel, electricity, heat production and supply, and natural gas production and supply. The energy industry was responsible for 229 Mtce, or 10.14%, of total energy consumption, and 587 Mt, or 11.4%, of national

Table 1.3. Generation mix in China (2005–2010), unit in TWh.

	2005	2006	2007	2008	2009	2010
Thermal	2047.3	2369.6	2722.9	2790.0	3011.7	3416.6
Hydro	397.0	435.8	485.3	585.2	571.7	686.7
Nuclear	53.1	54.8	62.1	68.4	70.1	74.7
Wind etc.	1.3	2.7	5.6	13.1	27.8	49.7
Total	2500.3	2865.7	3281.6	3466.9	3681.2	4227.8

Source: 2005–2009 wind and solar installed capacity data are from "China Medium and Long term Energy Development Strategies"; 2010 data: Li Junfeng et al. (2011). 2005–2008 installed capacity data are from "National Electricity Industry Statistics," generation data are from "China Energy Yearbook 2009." For 2009 installed capacity data and generation data: China Electricity Council, "Dianli hangye 2010 nian fazhan qingkuang zongshu 电力行业 2010 年发展情况综述 [Summary of the Development of China's Power Industry in 2010]" (2011b-6-27) URL: http://tj.cec.org.cn/niandufazhanbaogao/2011-06-27/58873.html [accessed: 2011-8-16]; and: China Electricity Council, "Zhongdianlian fabu 2009 nian quanguo dianli gongye niandu tongji shuju 中电联发布 2009 年全国电力工业年度统计数据 [China Electricity Council Publishes 2009 Statistical Survey on China's National Power Industry]" (2010-11-17), URL: http://tj.cec.org.cn/tongji/niandushuju/2010-11-17/160.html.

CO_2 emissions in 2005; in 2009 however, it was responsible for 268 Mtce (9.2%) of national energy consumption and 649 Mt 9.2% of national CO_2 emissions, a significant decrease from the 2005 level in relative terms. During the 11th FYP period, the energy industry's CO_2 emissions witnessed a significant decrease—dropping from 3.53 tCO_2/10,000 RMB in 2005 to 2.83 tCO_2/10,000 RMB in 2010, a decrease by 21.14% (Figure 1.16).

The electricity sector is of major significance for the energy sector overall. During the 11th FYP period, the electricity sector was remarkably successful in achieving decarbonization.

Large improvements in thermal power plant efficiency: Thermal power generation accounts for over 80% of China's total generation. As a result, the efficiency increase in thermal power plants is critical to achieve a decarbonization of the sector on a national level. During the 11th FYP period, China closed down 76.830 GW of small thermal plants. Meanwhile, a lot of 60–1000 MW super-critical and ultra-super-critical plants went into operation. Among China's thermal fleet, the share of units with capacity of and above 300 MW increased from 48.25% in 2005 to 72.68% in 2010.

The overall heat rate of China's thermal power plants with 6 MW capacity or above decreased from 343 gce/kWh in 2005 to 312 gce/kWh in 2010, a decrease by 31 gce/kWh (Table 1.2). From 2005 to 2010, station consumption ratio decreased from 5.87% to 5.43%, and line loss decreased from 7.21% to 6.53%. With the increasing generation efficiency and decreasing station consumption and line loss from 2005 to 2010, the net heat rate of thermal power plants decreased from 370 gce/kWh to 333 gce/kWh, a decrease by 37 gce/kWh. Meanwhile, CO_2 emissions intensity of thermal generation decreased from 936.4 gCO_2/kWh in 2005 to 866.3 gCO_2/kWh in 2009, a decrease by 7.48%.

Decarbonization of the generation fleet: During the 11th FYP period, non-fossil energy witnessed rapid development, improving China's generation fleet. Non-thermal generation reached 813.2 terawatt hours (TWh) in 2010—an increase by 80% compared to the 2005 level. Such growth rates exceeded that of thermal generation (67%) (Table 1.3). In 2010, China's non-fossil energy generation accounted for 19.2% of total generation, an increase by 1.1% relative to the 2005 level.

Hydro power expanded with the largest margin among all non-fossil energies. With an increase of 20 MW new capacity per year, total installed capacity of hydro power reached 210 GW, 2.5 times more than during the 10th FYP period. The growth rate of installed wind capacity was the fastest. From 2005 to 2010, with an annual average growth rate of 105%, overall installed capacity reached 44.73 GW, accounting for 22.4% of global total installed capacity. Solar power also experienced exponential growth, with installed capacity reaching 0.86 GW in 2010, 12.29 times as much as in 2005. The development of nuclear power was also significant: By the end of 2010, there were a total of 13 nuclear units in commercial operation, and another 28 units under construction.

The increase of generation efficiency and overhaul of the generational fleet lead to a rapid decrease in CO_2 emissions intensity—from 766.7 gCO_2/kWh in 2005 to 695.6 gCO_2/kWh in 2009, a decrease by 9.27%.

1.2.4.2 *Manufacture Industry Achieved Significant Carbon Mitigation*

In this report, the term "manufacturing industry" refers to all industries in the second sector except for energy industries. In 2010, energy consumption by the manufacturing industry was about 1.37 gigatons of carbon equivalent (Gtce), accounting for 66.5% of end energy consumption, or 42.1% of total national energy consumption; CO_2 emissions amounted to ca. 4.30 Mt, accounting for 62.0% of national total energy-related emissions. From 2006 to 2010, total energy consumption and CO_2 emissions by the manufacturing industry had been increasing; however, its energy intensity and CO_2 emissions intensity experienced a rapid decrease. From 2005 to 2010, the manufacturing industry's energy consumption per 10,000 RMB of added value decreased by 23.2%, an annual decrease by 5.14% on average—a higher rate of decrease than the national average of 4.15%. This translated into an energy savings total of 329 Mtce,

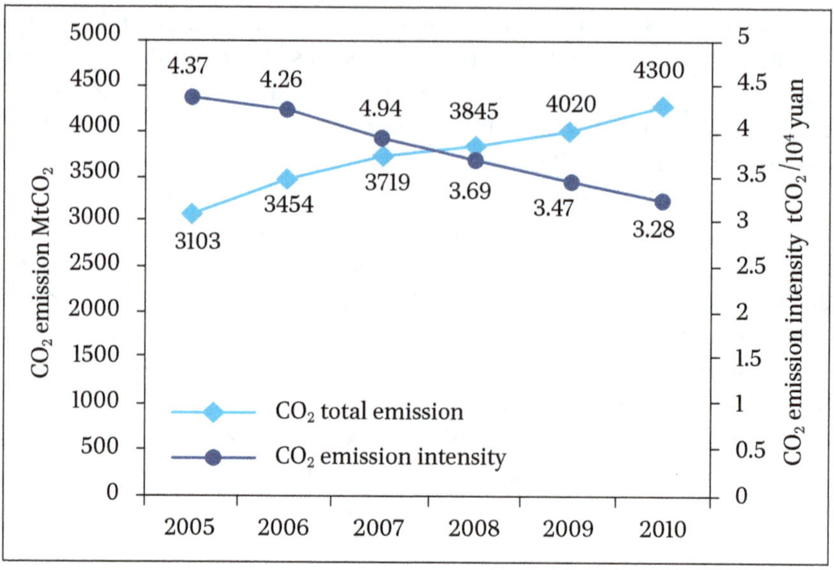

Figure 1.17. CO_2 emissions and intensity in manufacture industry (2005–2010).

accounting for 52.5% of the national total. CO_2 emissions per 10,000 RMB of added value decreased from 4.37 tons in 2005 to 3.28 tons in 2010, a decrease by 25.1% or an annual decrease by 5.6% on average (Figure 1.17). Cumulative CO_2 emissions reduction was 1.16 Gt, accounting for 74.8% of national total reduction.

Unit product energy consumption dropped significantly: During the 11th FYP period, unit product energy consumption experienced a rapid decrease. Energy consumption per ton in large and medium enterprises dropped from 732 kWh/t in 2005 to 701 kWh/t in 2010, a decrease by 4.2%; AC electricity consumption of electrolytic aluminum decreased from 14680 kWh/t in 2005 to 13979 kWh/t in 2010, a decrease by 4.8%; energy consumption for copper smelting decreased from 780 kgce/t in 2005 to 500 kgce/t in 2010, a decrease by 35.9%; for cement from 167 kgce/t in 2005 to 126 kgce/t in 2010, a decrease by 24.6%; for glass plates, alkali, pure alkali, calcium carbide, compound ammonia, paper, and paper board, a decrease of 25.9%, 22.4%, 19.9%, 3.2%, 13.9%, 31.1%, respectively. During the 11th FYP period, the manufacturing industry achieved an energy savings total of 311 Mtce through energy efficiency improvements per unit/product, accounting for 94.6% of total savings achieved by the manufacturing industry, or 49.6% of total national energy savings.

Table 1.4. China phasing out obsolete capacity during 11th FYP period.

Sector	Measure	Unit	Target	Completion
Electricity	Phasing out small units under "Replacing Small Units with Large Ones" policy	MW	50	76.825
Iron	Phasing out blast furnace below 300 m3	Mt	100	111.72
Steel	Phasing out small furnace with annual capacity of 200 Kt and below	Mt	55	66.83
Electrolyte aluminum	Phasing out Small pre-baked cells	Mt	0.65	0.80
Iron	Phasing out submerged arc furnace below 6300 kVA	Mt	4	6.63
Calcium carbide	Phasing out furnace below 6300 kVA	Mt	2	3.60
Coke	Phasing out small scale metallurgy with calcinated room lower than 4.3m	Mt	80	105.38
Cement	Replacing vertical kiln clinker	Mt	250	330
Glass	Phasing out backward float glass	Million heavy boxes	3000	3800
Paper	Phasing out straw pulp with annual capacity below 34,000 tons and chemical production line with annual capacity below 17,000 tons, and closing down recycled paper processing factories that fail to meet emissions requirements	Mt	6.50	10.30

Source: Phasing out obsolete capacity targets, see "Circular on Publishing Energy Saving and Emissions reduction Work Scheme" (State Council [2007]15); Electricity data are from National Development and Reform Commission (2011d); Electrolytic aluminum and calcium carbide data, see: Ministry of Industry and Information Technology of the PRC, "2010 nian Zhongguo jingji yunxing baogao 2010 年中国工业经济运行报告 [2010 Report on the Dynamics of China's Industrial Economy]" (2011-02-28), URL: http://www.gov.cn/gzdt/2011-02/28/content_1812626.htm [accessed: 2011-08-09]. Remaining data: National Bureau of Statistics of China, "Woguo jingji jiegou tiaozheng qude zhongyao jinzhan 我国经济结构调整取得重要进展 [Adjustment of China's economic structure makes important headway]," URL: http://www.stats.gov.cn/tjfx/ztfx/sywcj/t20110311_402709772.htm [accessed: 2011-08-09].

Many low-carbon technologies were adopted during the 11th FYP period. The rate with which coke dry quenching technology was adopted increased from 35% to 83% in the steel industry, capacity produced by new dry process clinker in cement industry increased from 40% to 80%, cement bulk rate increased from 39% in 2005 to 48%; the proportion of ion-exchange membrane alkali production increased from 34% to 76%; and the share of float glass sheets increased from 79% in 2005 to 86%. These and other cost-effective low-carbon technologies were adopted rapidly.

During the 11th FYP period, a large amount of obsolete capacity was phased out (Table 1.4), leading to technological adjustments in the manufacturing industry and a rapidly increasing consolidation of the sector. In the steel sector, for example, the prevalence of large blast furnaces with capacities of 1000 m^3 or above increased from 21% in 2005 to 34% in 2010, while crude steel production by the top-ten steel enterprises increased from 34.8% in 2006 to 48.4% in 2010. In the glass sector, the share of float glass production with a daily capacity of 600 tons and above increased from 32% in 2005 to 60% in 2010, while the share of production by the top-ten enterprises increased from 45.8% in 2005 to 57% in 2010. In the cement sector, the share of large daily capacities of 4,000 tons and above increased from 38% in 2005 to 56.9% in 2010, while the share of production claimed by the top-ten enterprises increased from 16.7% in 2005 to 30% in 2010.

Upgrades of the sectoral structure and product portfolio: During the 11th FYP period, the manufacturing industry witnessed structural changes and upgrades in product portfolios that helped to save energy. From 2005 to 2010, the value-added growth rate of the six major energy-intensive industries decreased by 2.3%, and their total share decreased by 1.3%. Meanwhile, the high value-added machinery manufacturing industry continued to experience rapid growth. During 11th FYP period, the share of five major machinery manufacturing industries in total value-added increased by 2.9%.

Product portfolios also displayed a trend towards increasing the share of high value-added products. In the steel sector, the production of strip products that had long been in shortage increased faster than long product, from 41.78% in 2006 to 45.35% in 2010. In the building material industry, the glass deep-processing rate increased from 25% in 2006 to 36% in 2009, and the share of new wall materials increased from 40% in 2005 to 58% in 2010.

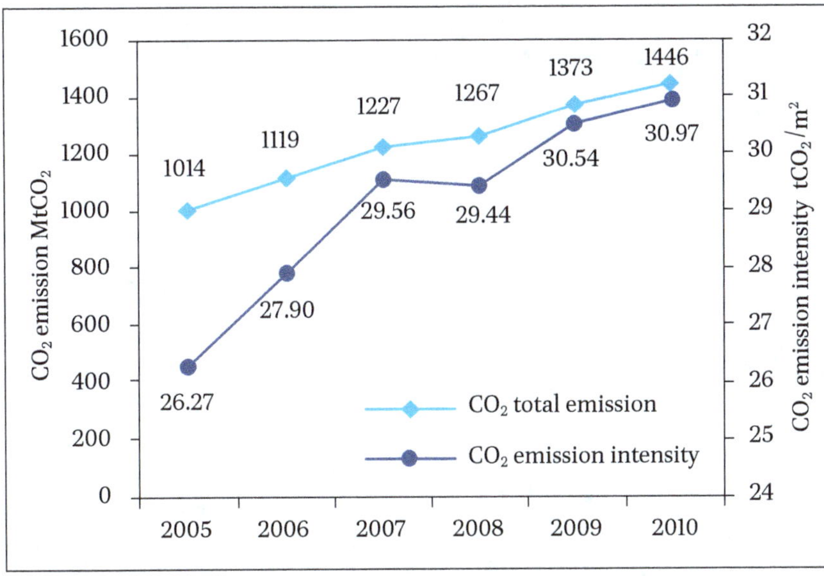

Figure 1.18. CO_2 emissions and intensity in building sector (2005–2010).

1.2.4.3 *Carbon Emissions Growth Rate in Building Sector Slowed Down, and Energy Efficiency Showed Certain Increase*

During the 11th FYP period, energy consumption and CO_2 emissions in the building sector continued to grow, but at a notably slower rate than during the 10th FYP period. From 2005 to 2010, unit area energy consumption increased by 19.7% over five years, or by 3.7% annually on average. At the same time, unit area CO_2 emissions increased by 17.9% overall or 3.3% annually on average (Figure 1.18). At present, China's unit building area CO_2 emissions are still by far lower than those of developed countries, less than one-third the figure of the U.S. level for example.

Among the four categories of energy consumption in the building sector, the "northern China urban central heating" category achieved the largest carbon mitigation with unit area energy consumption continuously dropping (Figure 1.19), from 17.78 kgce/m² in 2005 to 16.28 kgce/m² in 2010, a decrease by 8.41%. At the same time, associated unit area CO_2 emissions also decreased from 47.48 $kgCO_2/m^2$ in 2005 to 43.87 $kgCO_2/m^2$ in 2010, a decrease by 7.6%. This contributed to curb both unit area and China's total energy consumption and CO_2 emissions growth rate, as central heating in northern China accounted for nearly 25% of total energy consumption in the building sector nationwide.

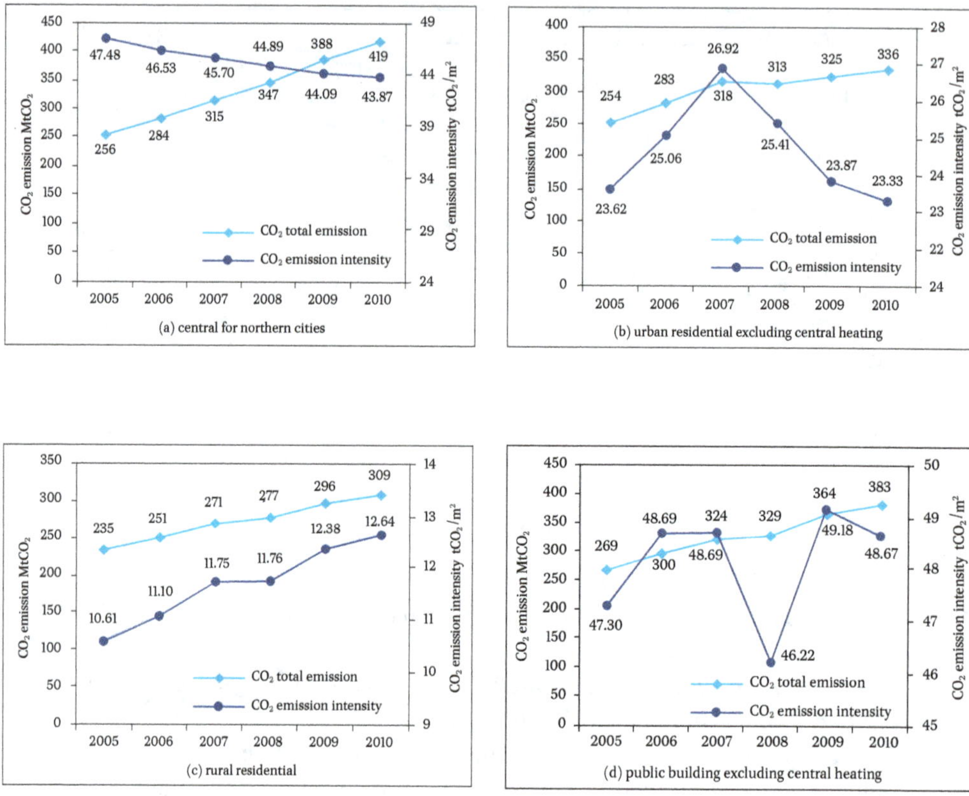

Figures 1.19a, b, c and d. Total CO_2 emissions and intensity in building sector by its four categories (2005–2010).

The building sector achieved notable success in energy saving. By the end of 2009, 99% of building designs and 90% of building constructions in China followed mandatory energy-saving requirements. Nationwide, the area of energy-efficient urban buildings reached 4.08 billion square meters, accounting for 21.7% of urban areas in total. During the 11th FYP period, the building sector achieved an energy savings total of 67.50 Mtce, equivalent to a 185 $MtCO_2$ emissions reduction, through installation of envelope weatherization systems, overhaul of central heating systems, upgrades in energy-saving lighting, and energy labeling for home appliances.

1.2.4.4 *Carbon Emissions in the Transportation Sector Increased Rapidly*

China's energy consumption in the transportation sector increased from 230 Mtce in 2005 to 300 Mtce in 2009, an increase of over 30%, higher than the average growth rate of energy consumption overall. However, compared to the 10th FYP period, the growth rate notably dropped. CO_2 emissions in the transportation sector increased from 510 Mt in 2005 to 710 Mt in 2010, an increase of about 39% (Figure 1.20). While transportation-related energy consumption increased rapidly, energy consumption per unit/transportation service decreased. Energy consumption in the railway sector per service unit decreased from 6.48 tce/Mt*km equivalent in 2005 to 4.94 tce/Mt*km equivalent in 2010, a decrease by 23.8%. Unit aviation service oil consumption decreased from 0.336 kg/t*km in 2005 to 0.298 kg/t*km in 2010, a decrease by 11.3%.

During the 11th FYP period, private vehicle ownership in China increased rapidly from less than eleven in every thousand people in 2005 to over thirty-six in every thousand people, an increase of nearly 250%. Travel via private vehicle increased to around 10% among all travel nation-wide. These developments posed challenges for the transportation sector to make advances in low-carbon development. After the financial crisis in 2008, China developed policy incentives to encourage the sale of small vehicles with mileage below 1.6l, which included tax incentives,

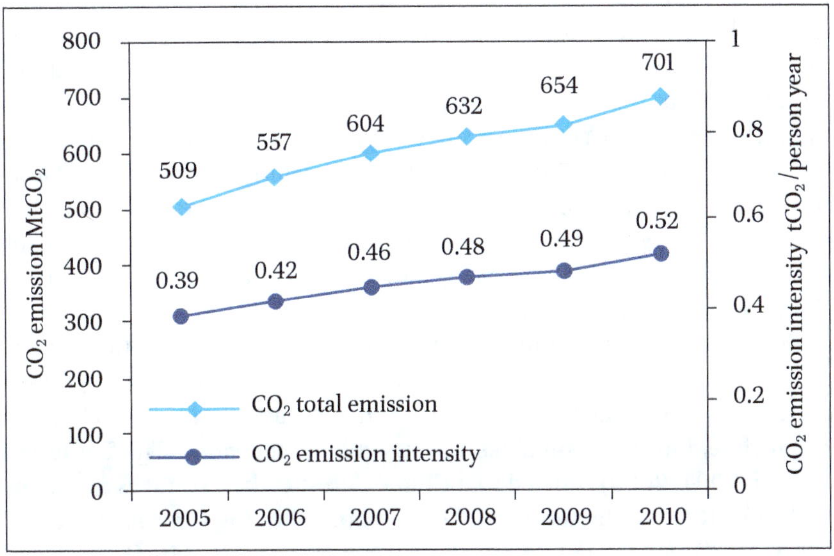

Figure 1.20. CO_2 emissions and intensity in transportation sector (2005–2010).

differentiated taxes for large vehicles, etc. With implementation of these policies, the share of small vehicles below 1.6l increased from 66.77% in 2005 to 68.77% in 2010, slowing down the transportation-related energy consumption growth rate to a certain extent.

1.2.4.5 *The Remarkable Low-Carbon Development in Agriculture and Forest*

Agriculture is both the foundation of the Chinese national economy and a strategically important sector for low-carbon development. Total fossil energy consumption by agriculture in 2010 was about. 37 Mtce, accounting for only 1.14% in national total energy consumption yet yielding 9.4% of national GDP. During the 11th FYP period, energy consumption by agriculture stabilized between 35–37 Mtce, while CO_2 emissions amounted to around 120 Mt. CO_2 emissions intensity in agriculture showed a notably decreasing trajectory, with a decrease of 16.4% in 2010 relative to the 2005 level (Figure 1.21).

It is important to point out that agricultural production material contains embedded emissions through the production of fertilizer and other input materials. In 2009, embedded emissions contained in agricultural production materials amounted to 325 $MtCO_2$, an increase by 13.1% relative to the 2005 level, and 2.6 times as much as the direct emissions caused by the agriculture sector. Fertilizer is the major source of embedded carbon emissions, accounting for about 86% of total embedded emissions. With the implementation of the "Tailor-made Fertilizer Application Plan for Different Soils Program," the growth rate of fertilizer application per unit/area during the 11th FYP period notably slowed down compared to that of the 10th FYP period, effectively contributing to an embedded carbon emissions reduction in this sector. Agricultural production is also responsible for the emissions of methane and nitrogen dioxide, which contributed to twice the amount of embedded carbon emissions (Figure 1.22). During the 11th FYP period, methane and nitrogen dioxide emissions were stable at around 600 Mtce.

China's forest development, especially large-scale forestations, contributed to an expansion of carbon sinks. National forest coverage rate reached 20.4% in 2009, meeting the 11th FYP target ahead of schedule. Meanwhile, forest carbon sinks witnessed large increases: The 7th forest Census (2004–2009) showed that China's forest carbon storages added up to 22.29 $GtCO_2$, an increase by 10.4% compared to the 6th census period (1998–2003), equivalent to an annual increase of 420 $MtCO_2$ on average. While the total global forest cover shrunk by 20,000 ha daily, China

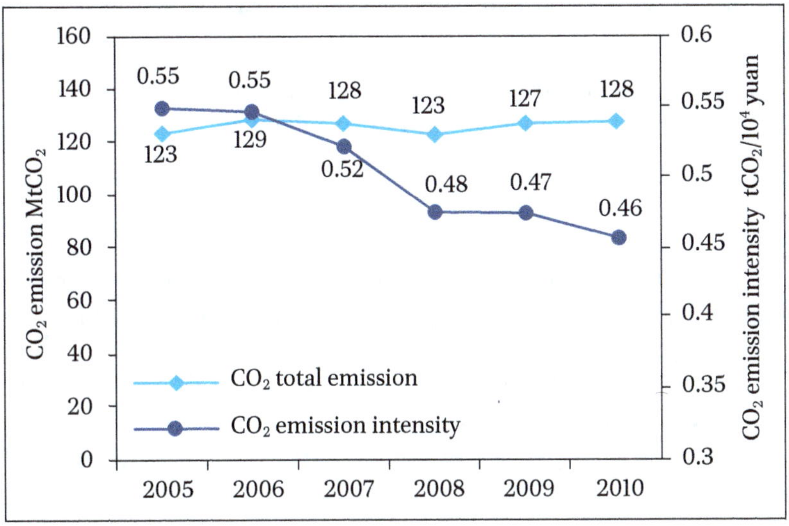

Figure 1.21. CO_2 emissions and intensity by agriculture (2005–2010).

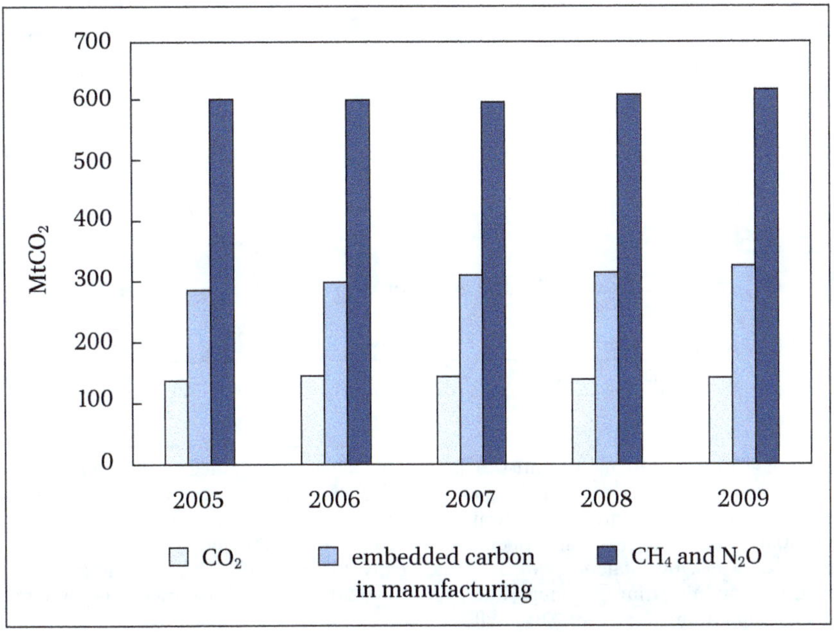

Figure 1.22. GHG emissions by agriculture (2005–2010).

undertook tremendous efforts in re-forestation and achieved a steady expansion of forest cover, contributing to a global CO_2 emissions reduction.

1.3 Characteristics of China's Low-Carbon Development during the 11th FYP Period

1.3.1 *Total Emissions Increased Rapidly while Emissions Intensity was Decreasing*

During the 11th FYP period, China's CO_2 emissions intensity decreased significantly, however, total CO_2 emissions still increased rapidly. The "X-shaped" graph formed by the intensity decrease curve and the total emissions increase curve was the primary feature of China's low-carbon development at present (Figure 1.23).

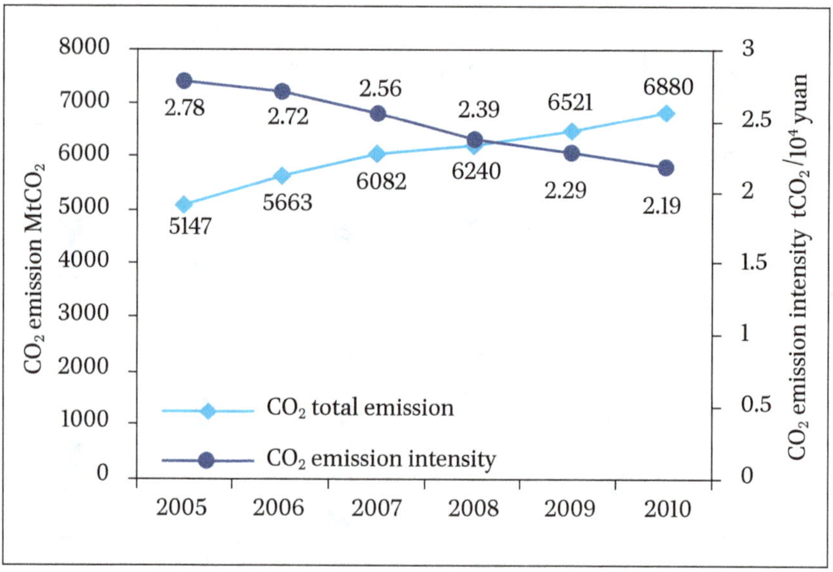

Source: National Bureau of Statistics of China, "Woguo jingji jiegou tiaozheng qude zhongyao jinzhan 我国经济结构调整取得重要进展 [Adjustment of China's economic structure makes important headway]," URL: http://www.stats.gov.cn/tjfx/ztfx/sywcj/t20110311_402709772.htm [accessed: 2011-08-09], "Annual Energy Review 2010," URL: http://www.eia.gov/totalenergy/data/annual/archive/038410.pdf; EIA (U.S. Energy Information Administration), Annual Energy Review 2011," URL: http://www.eia.gov/total energy/data/annual/archive/038411.pdf.

Figure 1.23. China's total CO_2 emissions and intensity (2005–2010).

China's energy intensity decreased by 19.1% during the 11th FYP period, the most significant decrease world-wide. In comparison, energy intensity in the U.S. decreased by only 6.1% between 2005 and 2010. During the 11th FYP period, China's nuclear power generation increased by 40%, the second-fastest rate in the world, only secondary to Romania.[6] In absolute terms, China's nuclear power generation increased by 21.6 TWh, the third-largest increase after the U.S. and Russia. At the same time, China's increase in hydro power generation amounted to 62.5% of the total global increase. The installed capacity of wind power in China experienced the largest increase in the world, newly installed capacity accounted for about 31% of global new capacity.[7] Compared to these figures, China's solar development was rather slow, with an increase rate lower than that of Italy, France, Spain, or Greece; however, it was still more than twice that of the global average rate.

Although China achieved remarkable successes in decreasing its energy intensity and developing non-fossil energy sources, its total energy consumption and CO_2 emissions still experience a large increase. During the 11th FYP period, China's total energy consumption increased by annually 178 Mtce on average, which translated into a total increase from 2360 Mtce in 2005 to 3250 Mtce in 2010, an increase by 37.7%, the fourth-largest increase after Vietnam, Singapore, and India, and contributing to 61.6% of global increase at the same period.[8] From 2005 to 2010, China's energy-related CO_2 emissions increased from 5.147 Gt in 2005 to 6.93 Gt in 2010, an increase of 34.64% in total or about 7% annually on average, a rate 3.38 times as high as that of the global average. Within the same period, China's CO_2 emissions increase accounted for about 72.0% of the global increase, making China the country with the largest increase during this period.

During the 11th FYP period, China moved from being the country with the fifth-largest GDP in the world to become the country with the second largest GDP, while it also became the country with the largest volume of CO_2 emissions. In 2007, China's emissions were on a par with U.S. levels, however with the impact of the financial crisis in 2010, U.S. emissions

6 BP, "BP Statistical Review of World Energy 2011," URL: http://www.bp.com/assets/bp_internet/globalbp/globalbp_uk_english/reports_and_publications/statistical_energy_review_2011/STAGING/local_assets/pdf/statistical_review_of_world_energy_full_report_2011.pdf. 2011.

7 Ibid.

8 Ibid.

dropped by 6.3% while emissions from China increased by nearly 14%. In 2010, China's CO_2 emissions already surpassed U.S. levels by 23%.

China's rapid increase in CO_2 emissions is due to the developmental phase it finds itself in. As China is undergoing processes of intense industrialization and urbanization, with its export-oriented GDP growing steadily, China's economic development is largely driven by investment and export, necessitating large amounts of energy consumption on the production side. While agricultural energy consumption is stabilizing (Figure 1.20), the production side—especially in the energy and manufacturing industries—witnessed rapid increases in energy consumption. These two industries account for over 70% of China's total CO_2 emissions. As a result, they were also the priority target for the implementation of China's energy saving policies during the 11th FYP period. These two industries witnessed both significantly decreasing energy and carbon intensity and notably increasing total energy consumption and emissions, just as was the case for China's overall story concerning energy and carbon trajectory (Figure 1.15, Figure 1.16).

During the 11th FYP period, China's per-capita GDP more than doubled, increasing from 14,185 RMB in 2005 to 29,762 RMB in 2010. Annual disposable income per capita for urban households increased from 10,493 RMB in 2005 to 19,109 RMB in 2010, an increase by 82.1%; rural per capita net income increased from 3,255 RMB in 2005 to 5,919 RMB in 2010, an increase by 81.8%.[9] Growth in per-capita GDP and per-capita income resulted in increasing energy consumption. Despite the use of energy-saving appliances and the increase in energy efficiency, total household energy consumption and carbon emissions still displayed a rapidly increasing trajectory.

1.3.2 *Diversified Policy Tools and Tremendous Government Investment*

Accountability for meeting energy saving targets was one of China's major energy policies during the 11th FYP period. Through this mechanism, energy savings targets were decomposed and assigned through different hierarchies. Combined with a system of reporting, measuring, and verifying energy savings, the targets assigned to each provincial government

---

[9] National Bureau of Statistics of China, "Woguo jingji jiegou tiaozheng qude zhongyao jinzhan 我国经济结构调整取得重要进展 [Adjustment of China's economic structure makes important headway]," URL: http://www.stats.gov.cn/tjfx/ztfx/sywcj/t20110311_402709772.htm [accessed: 2011-08-09].

and major energy-consuming enterprises were clear, concrete, and binding. The energy savings target accountability system is closely linked to China's administrative landscape; accountability for meeting energy savings targets can be enforced through administrative measures, e.g. through withholding of promotions for government officials or denial of development opportunities for enterprises.

Policy tools aimed at achieving energy conservation ranged from administrative regulations such as "assigning targets for major energy-consuming enterprises" and "phasing out obsolete capacities," to incentives such as "energy saving awards," "energy saving subsidies," and market instruments such as energy management contracts. These measures varied widely in terms of their efficiency and effectiveness. During the 11th FYP period, administrative tools contributed to a reduction of 472 $MtCO_2$, incentives to 784 $MtCO_2$, and markets to 14 $MtCO_2$ (see Chapter 4 for further analysis).

With the implementation of energy-saving policies during the 11th FYP period, China achieved a rapid decrease in energy intensity. Policies with a large amount of government funding proved to be most effective. During the same period, the central government allocated 85.1 billion RMB, while local governments spent 41 billion to implement energy-saving policies and programs, including "phasing out obsolete capacities," building energy efficiency, energy management contracts, and upgrades to more energy-efficient light bulbs, air-conditioners, and vehicles. In this period, the central and local governments invested 167 RMB for each ton of reduction in CO_2 emissions. For the "ten major energy saving programs" and "phasing out of obsolete capacities," the investment turned out to be highly effective, with tremendous carbon mitigation capacity built at lower-than average cost, which was respectively 125 RMB/tCO_2 and 104 RMB/tCO_2. The government invested 11.54 billion and 3.04 billion RMB respectively in upgrades for more energy-efficient air-conditioners and vehicles, while only forming a respective capacity of 690,000 tCO_2 and 696,000 tCO_2, equivalent to 16725 RMB/tCO_2 and 4368 RMB/tCO_2.

1.3.3 *Stakeholders Actively Pursue Effective Ways of Low-Carbon Development, but the Foundation of China's Low-Carbon Development Remains Fragile*

1.3.3.1 *Local Governments are Conflicted between Expansion-Based Development and Low-Carbon Development*

Local governments play an important role in implementing state policies. In the 11th FYP period, local governments at all levels effectively executed the national government's low-carbon development policies, such as actively enforcing the target accountability system, strengthening energy monitoring, inspection, and legal enforcement, setting up funds designated for energy saving, and enhancing advocacy and education. Meanwhile, some local governments actively pursued their own low-carbon development strategies. For instance, some local governments vigorously supported non-fossil energy businesses; others attempted to construct low-carbon cities. Local governments' active participation is crucial in meeting China's low-carbon development target.

Top-down policies, as e.g. the "target accountability system," to some extent reduce local governments' strive towards economic expansion, but fail to fundamentally alter local governments' preferences for certain development patterns. Local governments are conflicted between expansion-based economic development and low-carbon development. Historically, China's fiscal federalism and political incentives structure caused local governments to prioritize high GDP growth. Local governments, in order to attract regional investments, often join in a "race to the bottom," providing investors in the manufacturing industry with land at low cost and subsidized infrastructure, as well as low standards in terms of labor rights and environmental protection, all of which pose difficulties to low-carbon development. In the 11th FYP period, China entered an era of accelerated urbanization and industrialization. Intensive infrastructure construction by local governments stimulated the development of an energy-intensive consumption sector. Even though a certain degree of infrastructure construction by local governments is laudable, a large amount of redundant and ineffective construction work continues to be pursued. Quite a number of local governments blindly expand the area of their municipalities, wastefully investing in "prestige projects," and frequently demolish and construct buildings based on chaotic planning which leads to a much shorter lifetime of buildings compared to the design life and tremendous waste of energy and resources.

Due to sometimes excessive pursuit of economic development, the vast majority of local governments cannot easily find suitable low-carbon development patterns. To maintain rapid economic growth, many local governments find it difficult to abandon carbon-intensive industries, and do not strictly enforce the phasing-out of obsolete capacities; some local governments even enacted policies in support of energy-intensive industries. Except for a small number of developed regions, a large number of local governments have not yet achieved any structural adjustment of their economy due to the relatively small set of choices in terms of industries they are able to attract. Although local governments have strong incentives to support low-carbon development, they often find themselves in fierce regional competition, and thus grant priority to GDP growth and a much lower priority to low-carbon economic development. Limited by institutional factors and capacities, the vast majority of local governments are painfully conflicted between expansion-based development and low-carbon development. Hence, local governments' dedication to low-carbon development is still very fickle.

1.3.3.2 *Enterprises Meet their Energy Savings Targets under Pressure, but Enterprises' Foundation in Independent Energy Saving is Fragile*

Enterprises are the major factor in China's low-carbon development. In the 11th FYP period, different types of enterprises responded differently to the pressure exacted by local governments to save energy. Large-scale and highly competitive enterprises carried out extensive technological upgrades to save energy through internal financing schemes, green credits, and subsidized funding for energy-saving technical upgrades. These enterprises not only met their energy savings targets, but also lowered their costs and improved their market competitiveness. In the 11th FYP period, thousands of enterprises exceeded their energy savings targets, upgraded their manufacturing techniques, and improved profitability. Many medium-sized enterprises carried out technological upgrades under political pressure to save energy, but their progress was relatively slow due to constraints related to profitability and funding considerations. A large number of small enterprises were confronted with gradual closure due to their old and outdated equipment, unsatisfactory earning power, and lack of capital.

Through policies such as "phasing out obsolete capacities" and "replacing small units with large ones," a great number of small enterprises were merged or recombined, contributing to a rapid industrial concentration in sectors with energy-intensive products, displaying an evident trend in enterprise upsizing. Enterprise upsizing is beneficial for the dissemination and promotion of advanced technologies. In the 11th FYP period, even though a lot of obsolete capacity was phased out, overall capacity for energy-intensive sectors remains strong and shows an expanding trend. Capacity expansion to a large extend is caused by the expansion of production of large enterprises. The expansion of the capacities of energy-intensive industries poses a tremendous challenge for meeting energy saving targets in the 12th FYP period.

During the 11th FYP period, many enterprises implemented energy saving measures under governmental pressure, but the basis for independent energy saving efforts by companies is still fragile. On the one hand, energy pricing mechanisms are still not yet unitized: energy prices have not yet effectively demonstrated their leverage function when it comes to energy saving. On the other hand, financing channels for energy saving are impeded, and enterprises do not have easy access to technological upgrades. In addition, due to large-scale capacity expansion and under the influence of international markets, the overall profitability of energy-intensive enterprises is not promising, and enterprises might not be able to steadily sustain their energy saving investments. In the 11th FYP period, many low-cost energy-saving technologies were widely adopted, energy utilization rates of some enterprises met advanced international standards, and room for further improvement in energy efficiency became considerably narrower. As a result, meeting energy savings targets in the 12th FYP period will be a real challenge for businesses.

1.3.3.3 *Change in Lifestyle Habits Contributes to Continuously Rising Energy Consumption and Carbon Emissions, Policies have not yet Proved Effective*

The 11th FYP period has seen dramatic changes in Chinese lifestyle habits, leading to continuously rising energy consumption levels and carbon emissions. From 2005 to 2010, urban housing areas increased by 50%, private transportation leaned more and more towards individual automobiles, and energy consumption by urban residents became an even more notable driving force behind China's rising carbon emissions: It increased from 2,500 Mtce to 3,600 Mtce, which at 42% grew faster than the average national energy consumption (38%).

Growth in urban housing is primarily driven by basic living demands, and in the context of China's ongoing urbanization, demand for residential housing will continue to increase in the future. At present, per capita living space for urban residents in China is only 22 m^2, less than one-third of the U.S., lower than the 40m^2 per capita housing space in Europe, and lower than the 30 m^2 found in Korea. Therefore, there is still room for improvement when it comes to per capita dwelling space in China. Other than basic living needs, China's housing market displays a huge demand for investment, causing an increase in housing space. Chinese residents' modes of transportation are still dominated by bicycles and walking, but as private vehicles become more popular among residents, the share of private vehicles in transportation has continued to rise. In some large cities, the share of private vehicles in national transportation makeup has reached or even exceeded the counterpart in advanced countries. Given China's large population, insufficient per capita resources and energy resources, this lifestyle is leading China's energy consumption in a wrong direction.

Currently, the general public is virtually not subject to any restrictions concerning energy saving or carbon mitigation, while their demand for material goods is constantly rising. Based on the experience of advanced countries, consumption-driven carbon emissions will become the primary source of CO_2 emissions when individual income continues to rise. During the rapid economic development China is currently undergoing, the government needs to guide and restrict public consumption. Otherwise, excessively high growth in CO_2 emissions on the demand side will impose substantial pressure on China's low-carbon development.

1.3.4 *Regional Policies Drive Changes in Spatial Sectoral Patterns, which will have an Impact on National Performance and Trend in Low-Carbon Development*

Given the emphasis of new strategies such as "Westward Expansion," "Reviving Northeastern China's Old Industrial Base," or "The Rise of Central China," economic growth in western and northeastern China exceeded that of eastern China during the 11th FYP period. Since China's energy intensity ranks in the order of western—northeastern—central—eastern China, changes in regional economic patterns drive up the national energy intensity per unit of GDP by 0.5% t.

To facilitate economic development in western, northeastern and central China, the central government has enacted preferential policies for these regions in regards to e.g. taxation and investment, a strategy it hopes will attract more investment from within and outside of the country. In order to eschew the pressure caused by mandatory energy saving, cost increases, and labor and resource shortages, enterprises in the eastern parts of China have started to "go west." In this process, businesses not only upgrade their technical equipment, but also vigorously expanded their capacity in order to improve competitiveness. This shift pattern is facilitating the decrease in energy consumption per unit of product, but it also contributes to a further expansion of the energy-intensive sector in China. From a regional perspective, the westwards shift of the energy-intensive sector means an improvement of eastern China's sectoral structure, which helped the region meet its energy savings target. For the western and central parts of China, taking over the energy-intensive sector did not result in considerably higher pressure in terms of high energy intensity. On the contrary, it dramatically advanced economic development in these regions. Due to this mutually beneficial regional shift, overall sectoral structural change policies in China only had marginal effects.

1.4 Achievements and Challenges of China's Low-Carbon Development

During the 11th FYP period, China achieved tremendous progress in energy saving and CO_2 emissions mitigation. As the world's second biggest economy, China effectively reversed its energy consumption and CO_2 emissions trajectory while maintaining fast economic growth. Therefore, China's achievement in the field of low-carbon development is extraordinary and worthy of recognition. Starting in 2008, the global financial crisis impacted China's foreign trade and economic development, slowed the growth of its energy-intensive sectors, and slowed down the growth of energy consumption and CO_2 emissions. To cope with the financial crisis, China proposed a development strategy labeled "sustain economic growth, maintain employment, and boost internal demand," while implementing solid fiscal policies and relatively accommodating monetary policies. Within a short period of time, the central government invested 4,000 billion RMB, and all major banks issued 10,000 billion RMB in loans. Local governments also invested 1.8 billion in regional projects. These investments were widely applied to infrastructure construction projects which to some extent facilitated the development of the energy-intensive sector. As a result, energy consumption per unit of GDP in the first half year of 2010 increased rather than declined which made it more difficult to meet the energy savings target in the 11th FYP period. Research suggests that even though China's economic stimulus package will yield relatively large long-term benefits as far as energy savings and CO_2 emissions reductions are concerned, it will create significant pressure for China's in this regard before 2014.

Although China's low-carbon development achieved considerable progress during the 11th FYP period, the target of "16% decrease in energy intensity, and 17% decrease in CO_2 intensity" proposed in the 12th FYP period is still very ambitious. First, China's socioeconomic development will remain characterized by rapid industrialization and accelerated urbanization. The share of export in GPD has decreased, but the total export remains high. These factors will continue to drive the growth of China's energy consumption and CO_2 emissions. Second, after the low-carbon development in the 11th FYP period, low-cost and high-return energy-saving technologies have been utilized to a degree that makes further upgrades in energy efficiency more difficult to achieve. Meanwhile, due to the gigantic economic aggregate and the basic characteristics of industrialization, achieving energy savings through economic structural adjustment is not an easy task. Last but not least, the basis for low-carbon development established in the 11th FYP period is not yet solidified; local governments' impulse towards economic expansion is still evident. The predicted average GDP growth rates of all regions in the 12th FYP period range from 8% to 13%, which on average is much higher than the 7%, the central government is anticipating. Total energy consumption predicted by local governments is approximately 500 million tce higher than predicted by the central government. The discreapancy between predictions by the local governments and those by the central government indicates that limiting total energy consumption to 4.1 billion tce in the 12th FYP period will not be easy.

In the 12th FYP period, China will continue to experience fast economic development and rapidly rising energy demand and CO_2 emissions. Assuming a conservative GDP annual growth rate of 7% (more than 40% increase in 5 years) and a 17% decrease in carbon emissions per unit of GDP in 5 years, the total energy consumption in China in 2015 will be 3.83 billion tce, while energy-related CO_2 emissions will reach about 7.9 billion tce; based on projected GDP growth rates proposed by local governments and the decrease in carbon emissions per unit of GDP in different regions decomposed by the central government, total energy consumption and carbon emissions in 2015 will be 4.33 billion tce and 8.93 billion tce, respectively. Expected economic growth in the United States in the 12th FYP period is not promising: GDP in 2015 is expected to be around 16% higher than in 2010, and CO_2 emissions in 2015 will be 5.68 billion tce, approximately the same as the 2010 level.[10] Estimates based on aforementioned data suggest that energy-related CO_2 emissions in China in 2015 will be 39–63% higher than those of the United States. This implies that China will undoubtedly face tremendous pressure from the international community in future international climate negotiations. China is

[10] EIA (U.S. Energy Information Administration), "Annual Energy Review 2011," http://www.eia.gov/totalenergy/data/annual/archive/038411.pdf.

confronted with mounting challenges for future CO_2 emissions mitigation. The 12th FYP period is a key period for long term low-carbon development of China. China not only needs to meet its anticipated energy saving and CO_2 emissions reduction target, but also stabilize the basis of its low-carbon development, making preparations for greater emissions reduction targets in the future.

CHAPTER TWO

DEVELOPMENT OF LOW-CARBON TECHNOLOGIES DURING THE 11TH FYP

Abstract: During the 11th FYP, research on and utilization of low-carbon technologies in China mainly focused on six areas, including clean-coal combustion and power generation, energy efficiency upgrades within the industrial sector, energy efficiency upgrades in the building sector, wind and solar energy utilization, as well as carbon capture, utilization, and storage (CCUS). The strategic focus on these six technical areas demonstrates China's commitment to advance clean energy technology, as well as making technological preparations for CCUS and alternative fuel vehicles. Applications of low-carbon technologies during the 11th FYP were characterized by (1) large market size, (2) fast growth, and (3) significantly improved end-use energy utilization rates.

Given policy incentives and financial support, Chinese businesses witnessed a fast growth in energy efficiency upgrades due to the introduction of low-cost and high-return low-carbon technologies. Tremendous progress was achieved in renewable energy technologies, clean coal combustion, and power generation. Enhanced domestic manufacturing capacity of low-carbon technical equipment led to a higher import substitution rate. In particular, domestically manufactured supercritical and ultra-supercritical power generating units as well as circulating fluidized bed boilers (CFBBs) met advanced international standards. In short, as technological innovation in the low-carbon field in China advances, the gap between domestic and foreign technologies has effectively become narrower.

It is important to note that despite tremendous progress, China's overall low-carbon technology is still under-promoted. A number of core technologies remain to be mastered; innovation in the field of low-carbon technology requires further improvement in terms of scale, speed, depth and accuracy of the research.

During the 11th FYP Plan, China achieved tremendous success in the fields of energy saving and carbon emissions reduction. Energy savings achieved through technological progress accounted for 69% of total energy saved and steadily grew from 43 Mtce to 146 Mtce. Given policy incentives and financial support geared towards energy conservation, Chinese enterprises experienced a fast growth in energy efficiency due to the introduction of low-cost and high-return low-carbon technologies. In particular, application of energy-efficient technologies and equipment upgrades significantly contributed to improvements in energy efficiency. The 11th FYP period also witnessed a fast growth in technological innovations and their implementation in various fields such as renewable energy utilization.

Technological progress is an important basis for larger-scale low-carbon development in China. China carefully engages in technological innovations, strategic planning, and education in the field of low-carbon development; it aspires to be not only an active participant, but also a pioneer in the technological and industrial revolution that is low-carbon development.

It is important to note that despite tremendous progress, the utilization rate of low-carbon technology in China has not yet lived up to its full potential. Technological innovation in this field still requires further improvement in terms of both speed and scale.

2.1 Progress Made in Applications of Low-Carbon Technologies

During the 11th FYP, research on and utilization of low-carbon technology in China mainly focused on six areas, including clean-coal combustion and power generation, energy efficiency upgrades within the industrial sector, energy efficiency upgrades in the building sector, wind and solar energy utilization, as well as carbon capture, utilization, and storage (CCUS). The strategic focus on these six technical areas demonstrates China's commitment to advance clean energy technology, as well as establishing the technological basis for CCUS and alternative fuel vehicles. Applications of low-carbon technologies during the 11th FYP were characterized by:

(1) Large market size: In 2010, the total investment in the clean energy[1] market in China totaled 54.4 billion USD, 39% higher than in 2009 and accounting for 22.4% of global clean energy investments. This qualified China as the biggest clean energy market in the world in two consecutive years.[2]

(2) Fast growth: Growth rates of investment in clean energy averaged 88% during the 11th FYP; growth rates for installed capacity of renewable energy averaged 106%.[3] The gap between domestic wind turbine and photovoltaic manufacturing technologies and the global state of the art narrowed.

[1] Clean energy herein refers to renewable energy and energy efficiency technologies.

[2] PEW Environment Group, *Who's winning the clean energy race? (2010 edition). G-20 investment powering forward* (Washington, DC: The PEW Charitable Trusts, 2011).

[3] Ibid.

(3) Significantly improved end-use energy utilization rates, and considerably reduced per-unit energy consumption for energy-intensive products.

2.1.1 *Clean and High-Efficiency Fossil Fuel Utilization Technologies*

China's energy profile is dominated by coal. 91.7% of total energy consumption in 2010 in China came from fossil fuels, of which coal accounted for 70.9%. Coal combustion is a major source of emissions and pollution in China. Since restructuring the coal-dominant energy profile is taking significant amount of time and is unlikely to be completed in the short run, the key to China's low-carbon development lies in developing clean, high-efficiency coal (and other fossil fuels) utilization technologies. The process by which to achieve this goal involves coal separation and processing, clean coal combustion for power generation, and chemical processing of coal. All of these steps have been promoted on a large scale and made a remarkable contribution to China's low-carbon development. Research in certain technologies and their application is now on par with the global state of the art.

(1) Coal washing[4] is the first step in the clean utilization of coal. Washed raw coal amounted to 1.65 billion tons in 2010. The share of raw coal washed increased from 31.9% in 2005 to 50.9% in 2010.[5] Annual production of coal slurry[6] was 80 million tons in 2010, of which 30 million tons were used as fuel for power plants, industrial boilers and furnaces. 50 million tons were used as raw material for gasification, which can reduce coal consumption by 10% compared to direct coal combustion.

(2) In China, coal is predominantly utilized in thermal power generation. Relevant clean utilization technologies in coal-fired power generation involve both combustion and generation processes. A typical example of clean coal combustion technologies is Circulating Fluidized Bed Boiler (CFBB) technology, which has been widely adopted during the

[4] Coal washing removes 50%–80% of ash content in coal and 30%–40% of total sulfur (or 60%–80% of inorganic sulfur). Coal washing desulfurization costs only half as much as flue gas desulfurization.

[5] Wang Wen, "Fraction of raw coal washed in China at the end of the 11th FYP reached 50.9%," *Coal Processing and Comprehensive Utilization* 2 (2011a): 6.

[6] Coal water slurry (CWS) is a fuel which consists of finely ground coal particles suspended in water. CWS is made by blending 70% of fine dispersed coal particles with 30% of water together with trace amounts of dispersant and stabilizer, which avoids precipitation and deterioration. CWS is a substitute fuel for heavy fuel oil.

11th FYP. CFBB features higher combustion efficiency and lower pollution and emissions and makes better use of low-grade fuel. Total installed capacity of CFBBs in 2009 was 72.87 GW, higher than in any other country in the world. Supercritical and ultra-supercritical power generations are both advanced coal-fired power generation technologies. By 2009, 132 supercritical and ultra-supercritical power generating units had been built in China. In 2010, China led the world in the total number of ultra-supercritical generating units: thirty-three 1,000 MW ultra-supercritical generating units had been put into operation. In addition, Integrated Gasification Combined Cycle (IGCC), a high-efficiency clean coal power generation technology, has not yet been commercialized due to its high cost.[7] IGCC trial projects can be found in countries including the United States, the Netherlands, Germany, and Japan. Three IGCC power plants have been planned or are under construction in China, the first of which is expected to undergo its trial run by the end of 2011, with an anticipated technical adjustment period of 2 to 3 years. In particular, "dry pulverized coal pressurized gasification technology," a key factor within IGCC, has been invented with proprietary intellectual property rights during the R&D process of this power plant.

(3) Chemical processing of coal[8] is another way through which coal can be utilized cleanly and efficiently. During the 11th FYP, China achieved great successes in the chemical processing of coal by independently developing the world's first coal-to-alkene project with an annual production volume of 600,000 tons, and the first coal-to-ethylene glycol project with annual production volume of 200,000 tons. Both are demonstration projects as of now. In addition, China has mastered the technologies of Direct Coal Liquefaction (DCL) and Indirect Coal Liquefaction (ICL); it has independently developed the world's first large-scale DCL oil-producing equipment with a capacity of more than a million tons and three ICL oil-producing facilities with a 160,000 tons capacity. Last but not least, "multi-nozzle oppositely placed coal-slurry pressurized gasification technology" has been invented and applied in China with a daily processing capacity of 2,000 tons of coal.

[7] An integrated gasification combined cycle (IGCC) is an advanced power generation technology that turns coal into syngas and then removes impurities from coal gas before it is combusted in a high-efficiency combined cycle. It consists of two major processes, i.e., (1) coal gasification and purification, and (2) a gas-steam combined power generation cycle.

[8] In chemical processing of coal, clean synthesis gas (syngas) from gasification of coal is converted to liquid fuels and chemical products.

Table 2.1. Progress made in applications of clean coal technologies in 2010.

Technological indicators	2005	2010	Energy-saving performance
Percentage of raw coal washed (%)	31.9	50.9[a]	Every 100 million tons of raw coal washed reduces SO_2 emissions from coal consumption by 1–1.5 million tons, equivalent to saving the transport of 9.6 billion ton-kms of lower-grade coal[b]
Installed capacity of coal gangue and coal slime comprehensive utilization power plants (GW)[c]	9.57	25	Annual utilization of coal gangue and coal slime was 130 million tons, equivalent to 40 million tce
Annual production of coal water slurry (10,000 tons)	≤80 (2001[d])	8000	50 million tons was used as raw material for coal gasification, which reduces coal consumption by 10% compared to direct coal combustion
CFBBs put into operation (300 MW)	0	17	Compared to pulverized coal-fired boilers, CFBBs reduce SO_2 and NO_X emissions by 50%, coal consumption by 10%, and CO_2 emissions by about 10%
Supercritical and ultra-supercritical generating units	14	132[e] (2009)	Coal consumption in power generation for supercritical generating unit is approximately 308 g/kWh, 32 g/kWh lower than subcritical units; coal consumption for ultra-supercritical generating unit approximately 283 g/kWh, 57 g/kWh lower than subcritical units
IGCC demonstration plants planned or under construction	0	3	Net efficiency of IGCC power plants is as high as 43–45%, yet pollution emissions are only 1/10 of regular coal-fired power plants, water consumption 1/2. Desulfurization efficiency can reach 99%

Note: Except for specific citations, all data are provided by Qingyi Wang.

Source: China Electricity Council, 2010. Indicators for operational reliability of 100 MW thermal power generating units, 40 MW and larger hydropower generating units, and nuclear generating units in 2009; Wang Wen, "Fraction of raw coal washed in China at the end of the 11th FYP reached 50.9%," *Coal Processing and Comprehensive Utilization* 2 (2011a): 6; Wang Wen, "2010 status report of coal gangue for power generation and construction materials," *Coal Processing and Comprehensive Utilization* 2 (2011b): 6; Liu Jiongtian, "Discussion on low-carbon development of coal in China," *Journal of China University of Mining & Technology* 1 (2010): 5–11; Zhou Qiaoguang, "Reflections on development of coal water slurry," *Clean Coal Technology* 7, no. 2 (2001): 24–27.

China's coal-based national economy and increasing energy consumption pose considerable challenges to its transition towards a low-carbon economy as well as to its goals for emissions reduction, which makes research and pilot programs in the field of carbon capture, utilization and storage (CCUS) particularly important. At present, sixteen projects have been planned, constructed, or put into operation,[9] including four CCUS projects involving thermal power plants, two IGCC projects incorporating pre-combustion carbon capture, one project involving the chemical processing of coal, and others involving CO_2 Enhanced Oil Recovery (CO_2-EOR). Even though China has implemented all CCS technologies individually, it has not yet attempted to merge these technologies or study in depth the full spectrum of emissions sources. Nor has it ensured the long-term sustainability of underground carbon storages. In short, China still lacks practical experience regarding effective long-term carbon storage schemes on a large scale.[10]

2.1.2 *Nuclear Energy and Renewable Energy Technologies*

By 2010, installed capacity of nuclear and renewable power constituted 26.5% of total installed capacity in China. Installed capacity and power generation levels in 2010 were 2.1, and 2.2 times higher than their 2005 levels. During the 11th FYP, total emissions reductions through nuclear power, hydropower, wind power, and solar power amounted to 253 million tons of CO_2. Manufacturing technology for nuclear power equipment, wind turbines, and solar cells dramatically improved.

Newly installed capacity for nuclear power in 2010 stood at 1.71 GW, with 13 generating units in service with a total capacity of 10.84 GW. In addition, China leads the world in terms of total nuclear power currently under construction: 28 units with a total capacity of 30.57 GW have been approved and are currently being built.[11] With more than 20 years of history, China's nuclear power industry has moved from simply adopting foreign technologies towards inventing proprietary technology

[9] Department of Social Development and Technology in the Ministry of Science and Technology, *Carbon capture, utilization and storage in China* (Beijing: Administrative Centre for China's Agenda 21, 2010): 1–22.

[10] The Climate Group, *CCUS in China: 18 hot topics* (Beijing: The Climate Group, 2011).

[11] Han Wei, "How to support China's nuclear development—Interview with Zusheng Yu from the Committee of Experts on nuclear safety and nuclear environment, Ministry of Environmental Protection," *Energy Review* 31 (2011): 101–105.

of the second generation with superior security and economic efficiency. Innovation based on adopted technology and domestic inventions allowed China to obtain five key techniques of third generation AP1000 technology.[12] Construction of the first 200 kW nuclear power station, featuring a high-temperature gas-cooled nuclear reactor and security systems of the 4th generation commenced in Shandong Province in 2008; the facility is expected to be operational in 2013. In 2011, the first fast reactor independently designed by the China Institute of Atomic Energy (CIAE) was connected to the grid. China has become one of the few countries in the world that has mastered fast reactor technology.

China was also the world's most proactive promoter of wind energy in 2010: newly installed capacity for wind energy in 2010 amounted to 18.93 GW, making total installed capacity stand at 44.73 GW. Both numbers are the highest in the world.[13] Wind turbine manufacturing technology in China had fully been imported during the 10th FYP, but went through a fast development during the 11th FYP. In 2005, the first proprietary 1 MW wind turbine model was produced. Only five years later, the first domestically produced 5 MW model emerged on the Chinese market. During the 11th FYP, the localization rate of wind turbines rose up to 80%. Particularly noteworthy in this regard is China's innovative and fully proprietary "MW-level double-armature mixed excitation wind power generation system," featuring doubly-fed generators with fully domestically manufactured components and parts.

In 2010, photovoltaic production in China amounted to approximately 9 GW, more than half of global production.[14] Through continuous adoption of foreign technologies, China has mastered a number of key techniques in the field, including solar cell production and production of polycrystalline silicon. A higher share of the demand for photovoltaic equipment and raw materials is now met by domestic products instead of imports. The efficiency of commercial solar cells rose from 13.8% in 2004 to 16.8% in 2010 (Table 2.2). According to an industry survey, a one-percent increase in solar cell efficiency lowers power generating costs by 7%. Cutting techniques for silicon wafer continue to make breakthroughs: wafers become

---

[12] Qu Weiping, "Third-generation nuclear power technology and development," *China Electrical Equipment Industry* 6 (2010): 49–52.

[13] REN21 (Renewable Energy Policy Network for the 21st Century), *Renewables 2011 Global Status Report* (Paris: REN21 Secretariat, 2011).

[14] SEMI PV Group, China Photovoltaic Industry Alliance (CPIA), *2011 China Photovoltaic Industry Development Report* (2011).

Table 2.2. Technological evolution of leading Chinese crystalline silicon photovoltaic manufacturers.

	2010	2009	2008	2007	2006	2005	2004
Efficiency of polycrystalline silicon cells (%)	16.8	16.2	15.6	15.2	14.5	14	13.8
Thickness of silicon wafer (um)	180	180	180	200	210–240	240–280	280–325
Size of silicon wafer (mm*mm)	156*156	156*156	156*156	156*156 125*125	125*125 156*156	125*125	125*125
Silicon consumption (g/W)	5.8	6.3	6.8	7.5	8.5	11	14

Source: Industry surveys by Climate Policy Initiative at Tsinghua

thinner as size increases, and the amount of silicon consumed per watt in 2010 decreased by 58.6% compared to the 2004 level.

2.1.3 *End-Use Energy Utilization Technology*

During the 11th FYP, energy conservation technologies were vigorously promoted in the industrial, construction, and transportation sectors. As a result, end-use energy utilization efficiency increased remarkably. For instance, in the industrial sector, coal consumption in thermal power generation fell by 9% (Table 2.3), and energy consumption per unit of 13 key products dropped by 18.1% on average. Energy efficiency in leading thermal power generation, cement production, and electrolytic aluminum enterprises have met advanced international standards; average efficiency in all enterprises in these industries are close to advanced international standards.

2.1.3.1 *Energy-Saving Technologies in the Industrial Sector*

Major energy conservation technologies in the industrial sector include equipment upsizing, updates in manufacturing techniques, and waste-heat excess pressure recovery. The 11th FYP witnessed an evident trend in equipment upsizing in power, iron and steel, non-ferrous metal, and building material industries, and a markedly higher adoption rate of new high-efficiency manufacturing techniques in these industries as well as in chemical engineering. Furthermore, waste-heat excess pressure utilization technology has been widely adopted in industries such as iron and

Table 2.3. Decrease in energy consumption per unit of product in key industrial sectors during the 11th FYP.

	2005	2010	Decrease(%)	Advanced international standard
Coal consumption for thermal power generation (gce/kWh)	343	312	9.0%	312
Comparable energy consumption per ton of steel produced (kWh/t)	732	680	7.8%	610
AC power consumption in aluminum electrolysis (kWh/t)	14680	13979	4.8%	14100
Integrated energy consumption in copper smelting (kgce/t)	780	500	35.9%	500
Integrated energy consumption in cement production (kgce/t)	167	126	24.6%	118
Integrated energy consumption in flat glass production (kgce/weight case)	22	16.3	25.9%	15
Integrated energy consumption in crude oil processing (kgce/t)	114	100	12.3%	73
Integrated energy consumption in ethylene production (kgce/t)	1073	950	11.5%	629
Integrated energy consumption in sodium hydroxide production (kgce/t)	1297	1006	22.4%	—
Integrated energy consumption in sodium carbonate production (kgce/t)	396	317	19.9%	310
Integrated energy consumption in calcium carbide (kWh/t)	3450	3340	3.2%	3000
Integrated energy consumption in ammonia synthesis (kgce/t)	1700	1464	13.9%	—
Integrated energy consumption in paper and cardboard production (kgce/t)	528	364	31.1%	—
Integrated energy consumption in chemical fiber production (kWh/t)	1396	1034.5	25.9%	980

Note: 2010 energy consumption per unit of product for 13 industrial products is provided by: Wang Qingyi, *The China Sustainable Energy Program Reference Materials—2010 Energy Data* (Beijing: China Sustainable Energy Program of Energy Foundation (US), 2011).

Source: Ibid.

steel, non-ferrous metal, coal, building materials, chemical engineering, and textile.

The trend in equipment upsizing is illustrated in the power generation industry, iron and steel industry, and the electrolytic aluminum industry. For example, the share of installed capacity accounted for by thermal power generating units of 600 MW or more rose from 11.7% in 2005 to 36.4% in 2010. 1000m^3 and larger-scale blast furnaces accounted for 34% of total iron production capacity in 2010, compared to 21% in 2005. In the electrolytic aluminum industry, 90% of production came from 160 kA and larger-scale prebaked cells in 2010, compared to 80% in 2005 (Table 2.4). The upgrade of manufacturing techniques is exemplified by the fact that in 2010, 83% of coke production employed coke dry quenching technology (CDQ), 76% of sodium hydroxide production used ion-exchange membrane techniques, 80% of cement production employed dry processes, and 86% of flat glass production used float glass processes.

Waste-heat excess pressure utilization technology has been widely adopted in industries such as iron and steel, non-ferrous metals, coal, building materials, chemical engineering, and textile (Table 2.5). In the iron and steel industry, for example, waste energy recovery brought about tremendous economic and environmental benefits during the 11th FYP. Installed power generation capacity from recycled blast-furnace gas (BFG), coke oven gas (COG), and Linz Donawitz Gas (LDG) grew from 6.8 GW in 2005 to 16.8 GW in 2009; power generation grew from 36.4 billion kWh to 76.5 billion kWh, equivalent to saving 13.63 million tce, reducing CO_2 emissions by 34 million tons and SO_2 emissions by 300,000 tons, lowering the national electricity bill by 20 billion RMB.[15]

2.1.3.2 *Energy-Saving Technologies in the Building Sector*

Energy-saving technologies in the building sector apply to energy-consuming equipment and systems outside and within buildings. Typical technologies include those which improve the performance of building envelopes, technologies that improve energy efficiency in district heating, energy-saving electric appliances, and renewable energy applications in buildings (Table 2.6). Technologies that improve the performance of building envelopes and district heating efficiency are particularly attractive when

[15] Jiang Guocheng, *Ten key energy-saving projects created energy-saving capacity of 340 million tce* (2011/2/23), URL: http://www.gov.cn/jrzg/2011-02/23/content_1809161.htm [accessed: 2011-09-24].

Table 2.4. Equipment upsizing and manufacturing technique update in key industries.

	2005	2010	Energy-saving performance
	Electric Power Generation		
Fraction of installed capacity accounted for by 600 MW or larger thermal power generating units (%)[a]	11.7	36.4	Coal consumption for power generation by 600 MW generating units is 50 g/kWh less than by 100 MW units
	Iron and Steel		
Fraction of iron production capacity accounted for by 1000 m^3 and larger blast furnaces (%)	21	34	300 m^3 and smaller blast furnaces consume 20% more energy than 1000 m^3 and larger blast furnaces
Fraction of coke production accounted for by CDQ (%)	35	83	Processing 1 million tons of red-hot coke saves 100 thousand tce
	Non-Ferrous Metal		
Fraction of electrolytic aluminum production accounted for by 160 kA and larger-scale prebaked cells (%)	80	90	160 kA and larger-scale prebaked cells consume 9% less electricity than self-baked cells
	Chemical Engineering		
Fraction of sodium hydroxide production accounted for by ion-exchange membrane technique (%)	34	76	Ion-exchange membrane technique saves 123 kWh less electricity per ton of sodium hydroxide produced than diaphragm process
	Building Materials		
Fraction of flat glass production using the float glass process (%)	79	86	Integrated energy consumption by float glass process is 16% less than the vertical drawing process
Fraction of float glass production in melting furnaces with daily melting capacity higher than 600 tons (%)	32[b]	54.5[c] (2009)	600 tons melting furnaces consumes 18.4% d less energy (kJ/kg) on average than 300 tons melting furnaces
Fraction of cement production through dry process (%)	40	80	Heat consumption of large-scale dry process production lines is 40% less than mechanical shaft kilns

Note: All data provided by Wang Qingyi (2011) unless specifically noted.

Source: Wu Yin, *China's energy profile and energy development strategy for the 12th FYP*, URL: http://www.docin.com/p-213335728.html, [accessed: 2011-07-19]; China Building Materials News, *Chinese glass industry keeps healthy development in the 11th FYP. China* (2011-1-28), URL: http://www.glass.org.cn/hyzx/201101/28/252.html [accessed: 2011-08-02]; Xu Meijun, "Probing status quo of flat glass production capacity in China," *Glass and Enamel* 38, no. 3: 29–33; Wang Zongwei, "Upsizing of melting furnaces for float glass production and comprehensive utilization of energy," *Architectural Glass and Functional Glass* 8 (2008): 6–13.

Table 2.5. Penetration rates[16] of typical waste-heat excess pressure utilization technologies and their respective energy-saving performance.

Technology	Penetration rate (%) 2005	2010	Energy-saving performance
	Electric Power		
Waste heat recovery in flue gas desulfurization and operational optimization of fans	<1	3	2 × 1 000 MW generating units using this technology, for instance, consume 2.71 g/kWh less coal per generating unit, equivalent to saving 1.98 million kWh of electricity and 260 thousand tons of water per year[a]
In-depth waste heat recovery in thermal power plants flue gas comprehensive optimization system	<1	1	Coal consumption for electric power generation is reduced by 2–3 g/kWh
	Iron and Steel		
Sintering waste heat power generation	<1	10	Waste heat per ton of flue gas recovered generates 20 kWh on average[b]
Blast furnace top gas pressure recovery turbine for power generation	30	55	Maximum amount of electricity recycled per ton of iron is approximately 50 kWh
Regenerative combustion technology	10	20	Recovers heat loss from high-temperature flue gas, saving more than 30% energy
	Building Materials		
Pure low-temperature waste heat power generation in cement kilns	4	48	Power generating capacity of this technology is 30–40 kWh per ton of cement clinker
Steady-flow travelling cooler	0	2	Recovered heat is used to assist clinker calcination, which reduces electricity consumption by 20%
Waste heat power generation in glass smelting kilns	0	12	3 MW low-temperature waste heat power station saves 10,100 tce per year

[16] Penetration rate of a technology is defined as the fraction of production accounted for by the technology.

Table 2.5 (*cont.*)

Technology	Penetration rate (%) 2005	2010	Energy-saving performance
		Chemical/Petrochemical Engineering	
In-depth heat integration and low-temperature heat recovery in gas separation equipment	5	25	Integrated energy consumption for oil processing falls by 5%
Synthetic ammonia comprehensive energy efficiency upgrade	0	30	Energy consumption per ton of ammonia produced drops by 0.23 tce
Triple-effect reverse-flow falling film evaporation	0	13	This technology reduces steam consumption by > 0.2 ton than its dual-effect counterpart
Low-temperature heat recovery in medium and large-size sulfur-burning sulfuric acid devices	5	26	By recovering low-pressure steam and saving circulating cooling water, this technology saves 216,000 tce/year for a 400,000 t/year sulfur-burning sulfuric acid device
Waste heat recovery and tail gas power generation (heat supply) in black carbon production	<15	35	Tail gas per ton of black carbon produced can generate 1120–1440 kWh electricity[d]
		Industry	
Closed total-heat-recovery of waste steam and condensed water	10	40	A wholly remodeled system saves 5–30% water and 10–40% energy[e]

Note: Penetration rate data, see: Energy Research Institute of the NDRC, *Analysis and evaluation of key energy-saving and emissions-reduction industrial technologies* (Beijing: Energy Foundation, 2011). Energy-saving performance of technologies also comes from the same source except for specific indications.

Source: Ministry of Industry and Information Technology of P.R. China, *Promotion of sintering waste heat power generation in iron and steel enterprises—working plan*, URL: http://www.miit.gov.cn/n11293472/n11293832/n12843926/n13917027/n14033242.files/n12980623.pdf [accessed: 2011-9-22]; Liu Chengxiong, Li Deliang, He Weibo, "Low-temperature waste heat power generation technology for glass melting furnaces," *Glass* 32, no. 4: 48–50; Black carbon branch of China Rubber Industry Association, *Report on energy-saving and emissions reduction and environmental protection by means of electric power generation and heat supply by tail gas of black carbon production—designated funding and preferential policies for comprehensive utilization of tail gas of black carbon production* (2011-7-22), URL: http://www.cncb.org.cn/A/?C-1-272.html [accessed: 2011-9-22].

it comes to lowering carbon emissions. A carbon reduction capacity[17] of 46.53 $MtCO_2$ was realized through an improvement in the performance of building envelopes in 2010. (Please refer to calculation methodologies in the appendix.) Carbon reduction capacity due to superior district heating technologies (such as improved efficiency in heat source conversion) reduced heat losses in pipe networks, and effective and waste-reducing end-user metering totaled 31 $MtCO_2$ in 2010. It is important to note that despite the fact that these technologies are now widely installed in new buildings (cumulative dissemination rate[18] during the 11th FYP was 82%, meaning that 82 m^2 out of every 100 m^2 in new building space during that period had been equipped with the aforementioned technologies), only 3% of overall existing buildings did so. The low dissemination rate of energy-saving technologies in existing buildings illustrates the enormous potential for future carbon reductions within the construction sector in China.

Further technological advances in energy-consuming appliances in buildings have also yielded significant results in carbon reductions. In the lighting sector, the penetration rate of energy-saving lighting fixtures reached 75% in 2010, leading to a carbon reduction capacity of 27.5 $MtCO_2$. Cumulative area covered with solar-powered water heaters

---

[17] Carbon reduction capacity in this book is calculated using 2005 as the base year.

Carbon reduction capacity of building efficiency technologies: if we take energy-saving buildings as an example, then carbon reduction capacity of energy-saving buildings = (carbon emissions from ordinary buildings – carbon emissions from energy-saving buildings) × (retained area of energy-saving buildings in 2010 – retained area of energy-saving buildings in 2005). It is important to note that carbon reduction capacity refers to MAXIMUM carbon reductions that can be achieved in a particular year by new energy-saving buildings built during the 11th FYP. Carbon reduction capacity measures the ability of energy-saving buildings to reduce carbon, not actual carbon reductions. In the building sector, actual carbon reductions to a large depend on how residents use energy-consuming equipment and systems in buildings.

Carbon reduction capacity of energy-saving technologies in the industrial sector: refers to maximum carbon reductions that can be achieved in a particular year after adopting the technology. It measures the ability of energy-saving technologies to reduce carbon, not actual carbon reductions. Actual carbon reductions depend on how a particular technology is used.

Carbon reduction capacity of transportation technologies: refers to maximum carbon reductions that can be achieved in a particular year after implementing fuel economy standards. It measures the ability of energy-saving technologies to reduce carbon; actual carbon reductions depend on actual usage of vehicles. Carbon reduction capacity of ethanol-fuel and reduction capacity of alternative fuel vehicles are cumulative carbon reductions from 2006 to 2010 (Base year = 2005).

Carbon reduction capacity of nuclear and renewable energy technology: refers to cumulative carbon reductions from 2006 to 2010 (Base year = 2005).

[18] Penetration rate of a technology is defined as the fraction of buildings that have applied the technology.

amounted to 168 million m^2 in 2010; the total annual biogas consumption from household biogas pits and medium-to-large biogas projects was 14 billion m^3.[19] Renewable energy utilization in buildings via solar- and biogas-powered water heaters accounted for a total carbon reduction capacity of 25.5 $MtCO_2$.

Table 2.6. Key building efficiency technologies and their respective performance.

Technology	Penetration rate (2010)	Advanced technologies involved (including energy-saving performance)
Technologies which improve the performance of building envelope—new buildings	82%	Wall insulation technology: in cold regions, according to building standards in different provinces and cities, this technology saves energy by 50%–65%[a]
Technologies which improve the performance of building envelope—existing buildings	1%	High-efficiency energy-saving doors and windows: doors, window frames and glasses with superior thermal properties use 40%–60%[b] less energy than conventional doors and windows Sunshade technology for buildings: technologies such as "cool roofs" and interior and exterior sunshade lower the load of air conditioners by 10%–50%[c,d]
Technologies that improve energy efficiency of district heating—new buildings	82%	Combined Heat and Power (CHP): generates heat and electricity simultaneously, and is more than 20% more efficient than generating electricity and heat separately[e]
Technologies that improve energy efficiency of district heating—existing buildings	3%	Boiler combustion technology: improves combustion efficiency by 10% or more[f] Reducing heat losses in district heating pipe network: heat losses from directly buried heating pipe is less than 10%[g] Household-based heat metering: upgrades heat metering, realizes 10%–15% energy savings in the system[h]
Energy-saving lighting fixtures	75%	Regular energy-saving lighting fixtures: save energy by 70%, last more than 2 times longer than incandescent lamps
		LED lighting fixtures: have not been widely adopted in the building sector. LED lighting fixtures have higher energy efficiency than regular energy-saving lighting fixtures and last longer[i]

[19] Zhao Yongqiang, Shi Jingli, Gao Hu, "Status quo, prospects and policy measure recommendations for renewable energy development in China," *China Energy* 33 (2011): 5–9.

Table 2.6 (*cont.*)

Technology	Penetration rate (2010)	Advanced technologies involved (including energy-saving performance)
Energy-saving air conditioners	14%	Technologies such as compressor frequency conversion, throttle valves (e.g., economizer), high-efficiency heat exchanger contributes to higher energy efficiency standards of domestically produced air conditioners. Coefficient of performance (COP) of air conditioners is improved by 15%
Solar water heaters	21%	Regular vacuum tube solar water heaters can meet 40% of household bathing hot water demand
Biogas	18%	Biogas pits for households of all sizes, or medium to large-scale biogas pits for centralized use
Energy-saving technologies for air conditioning systems in large-scale public buildings	5%	Utilization technologies of natural hot and cold sources: technologies such as ground source heat pump (GSHP), water source heat pump (WSHP), and seawater source heat pump take advantage of natural hot and cold sources. The energy-saving performance depends on individual project. High-efficiency water chilling units: COP of water chilling units significantly improved, saving energy by 20%–30% Motor frequency conversion: frequency conversion control devices in fans and water pumps can save energy by 40%–50%[1]

Source: Penetration rates of different technologies are calculated by Climate Policy Initiative at Tsinghua; Luo Wanjun, Xu Chengli, Ma Hongting, "Current status and application prospects of a few major building efficiency technologies" *Anhui Agricultural Science Bulletin* 13, no. 11 (2007): 168–171; Yuan Liying, "Multi-dimensional analysis of Building efficiency technologies," *Shanxi Architecture* 34, no.14 (2008): 248; Zhang Wenxiao, "Discussion of building efficiency technologies," *Information of China Construction* 6 (2007): 36–38; Liu Linghua, Chao Qianggang, Zhang Yu, "Contribution of building sunshade to green architecture," *West China Development* 2 (2011): 39; Tsinghua University Building Efficiency Research Center (THUBERC), *Annual research report of China Building Efficiency Technology* (Beijing: China Architecture And Building Press, 2011); Ji Rongwei, "High-efficiency coal combustion technology for boilers for heat supply," *Resources Economization and Environment* 3 (2011): 72–75; Deng Xiaoli, "Analysis of energy-saving technologies in boilers for heat supply," *China E-Commerce* 5 (2010): 167; Luo Xueying, Dong Fulin, Wang Suilin, "Analysis of application status survey of China's heating metering technology development," *Gas and Heat* 31, no. 2 (2011): 34–36; Liu Ai, Liao Xiaohui, "LED lighting fixtures—Future of lighting fixtures," *Jiangxi Energy* 1 (2009): 64–67; Climate Policy Initiative at Tsinghua (2011): 1; Yang Xiwei, Hao Bin, Zheng Ruicheng, "Application and development of key building efficiency technologies in China," *Wall Materials Innovation and Energy Saving in Buildings* 9 (2007): 43–47.

2.1.3.3 *Energy-Saving Technologies in the Transportation Sector*

The transportation sector is comprised of road, rail, air, and water transport, which accounts for the large spectrum of energy-saving technologies applied in this sector. For major energy-saving transportation technologies and their performance, refer to Table 2.7.

A variety of technological innovations can improve fuel efficiency in automobiles, including the electronic fuel injection system (EFI), new radial-ply tires, and energy-saving tapered roller bearing. In 2010, the adoption of fuel consumption limits for M1 light passenger cars yielded 8.37 $MtCO_2$ in carbon reductions.[20] Additionally, as an example of alternative fuels, four aged grain fuel ethanol projects and cassava-based ethanol projects in Guangxi produced approximately 1.86 million tons of fuel ethanol in total in 2010.[21] At present, corn ethanol is the predominant form of fuel ethanol used in automobiles, which from a life-cycle perspective does not have noticeable carbon-reducing effects. Assuming a common heat value for corn ethanol, cumulative gasoline savings from 2006 to 2010 from fuel ethanol amounted to 1.7 Mt compared to 2005. In 2010, there were 7,000[22] alternative fuel vehicles in China with a cumulative emissions reductions capacity of approximately 3.14 $MtCO_2$.[23]

---

20 M1 light passenger car refers to cars with no more than 9 seats including the driver's seat and weighting less than 3500 kg.

Assume that M1 light passenger cars make up 80% of total passenger car sales and annual mileage of each M1 light passenger car is 20 thousand km. Methodology for calculating energy consumption: fuel consumption per 100km traveled*0.73*1.4714*annual mileage km*new passenger car sales *80%.

21 Zhao Yongqiang, Shi Jingli, Gao Hu, "Status quo, prospects and policy measure recommendations for renewable energy development in China," *China Energy* 33 (2011): 5–9.

22 According to the energy-saving and alternative fuel vehicle demonstration and promotion conference in July 2011 held by the Ministry of Industry and Information Technology (MITT), total alternative fuel vehicle ownership reached 10 thousand. Assuming 70% of alternative fuel vehicle ownership was achieved by the end of 2010, total number of alternative fuel vehicles in 2010 was 7000. Other sources have reported alternative fuel passenger car sales of 2417 from February 2010 to January 2011, within which 148 was electric sedans, 72 electric light buses, 2189 hybrid sedans, and 8 hydrogen-fuel lithium-battery sedans (http://zixun.wlstock.com/zixun/002001/2011/07271157475278.html).

23 Carbon reductions by alternative fuel vehicles mainly come from gas-fueled vehicles, i.e., CNG vehicles, LPG vehicles and LNG vehicles. Electric vehicles, hybrid vehicles and fuel cell cars have relatively low ownership, and their categorized statistical data are insufficient, therefore these three types of cars are excluded in this study.

Table 2.7. Some energy-consuming transportation technologies and their respective energy-saving performance.

Type	Technology	Energy-saving performance
Fuel economy	Electronic fuel injection (EFI)	Engine power improves by 5%–10%, fuel consumption decreases by 5%–15%, exhaust gas emissions drop by approximately 20%[a]
	Radial-ply tires	Abrasion resistance of radial-ply tires is 30%–70% higher than bias-ply tires, and rolling resistance 20%–30% lower. Radial-ply tires reduce fuel consumption by 5%–8%[b]
	Carburetor fuel saver during forced idle mode	The fuel saver reduces gasoline consumption by 5% in flat area, and 10% in mountainous area[c]
	Fuel stratified injection	Improves fuel economy by 40%[d]
	Energy-saving tapered roller bearing (TRB) technology	TRB reduces roller energy consumption by 30% and saves 2% more energy[e]
	Non-throttle valve control gear	Improves fuel economy by 10–20% compared to regular throttle valve driving gear[f]
	Exhaust gas turbocharging system	Improves fuel economy of car engine. When power increases by 30%–40%, fuel consumption drop by 5%[g]
Hybrid/ electric vehicles	Photovoltaic modules as auxiliary power systems for electric vehicles	If installed on hybrid vehicles, fuel economy can be raised by 17–29%[h]
	Energy recovery through regenerative braking	Reduces gasoline consumption by 5–12%[i]
Fuel cell vehicles	Hydrogen fuel cell engine	Real-life energy conversion efficiency of fuel cells is 60%–70%, 2 times as much as ordinary internal combustion engine. 100% hydrogen fuel (compressed hydrogen) plus a series hybrid drivetrain reduces fuel consumption by more than 25% and achieves zero emissions[j]

Source: Liu Chang, "Discussion of EFI-based automobile energy-saving technologies," *Journal of Higher Correspondence Education* 20, no. 2 (2007): 34–37; Cai Fengtian, Xie Yuanmang, *Factors affecting automobile fuel economy and automobile Energy and resource engineering research team, School of Engineering of Beijing University, 2011. Analysis of the trend in energy-saving technology application and promotion in energy-intensive industries: iron and steel and coal industries* (Beijing: Energy Foundation, 2006): 28–33; Zhou Zhonggan, "Study of carburetor fuel saver during forced idle mode," *Journal of Chongqing Jiaotong University* 19, no.4 (2000): 110–113; Yu Xiumin, Yang Shichun, Gao Ying, "Technology reducing emissions from automobile engines," *Automotive Engineering* 24, no. 5 (2002): 445–450; Mei Wen. "Energy-saving bearings—Improve automobile fuel economy," *Technology and Market* 35, no. 16 (2008): 45; Yang Yong, "New Automobile drivetrain technology," *Automobile and Parts* 26, no. 2 (2008): 26–29; Sun Zhanhong, Xu Zheng, Zhou Shu, Li Fengli, "Turbocharged intercooler technology for automobile engines," *Jilin Pictorial-New Vision* 12 (2010): 63–64; E.J. Simburger, *Solar Electrical Vehicles, Prius White Paper*, URL: http://www.solarelectricalvehicles.com/articles/prius-white-paper.shtml [accessed: 2011-09-24]; Zhang Xinghua, Xu Shuqin, "Study of influence of motor efficiency on fuel economy of hybrid vehicles," *Tractor and Farm Transporter* 35, no. 4 (2008): 74–76; Zeng Shuai, Zhang Na, Xu Zhaokun, "Fuel cell electric car technology," *Shanghai Automotive* 10 (2010): 12–15.

2.2 Characteristics of Low-Carbon Technological Development during the 11th FYP

2.2.1 *Low-Cost Technologies are Widely Adopted, yet Overall Technological Penetration Rate Remains Low*

Carbon reduction costs of various low-carbon technologies have always received considerable attention. Evaluating the cost-effectiveness of low-carbon technologies requires a comparison with regular technologies; therefore we will measure the costs of low-carbon technologies through incremental costs.[24] Carbon reduction costs of low-carbon technologies in this book are defined as the ratio of incremental costs of a technology as related to CO_2 reductions during the service life of the technology. The incremental cost of a particular technology refers to the additional cost of the technology in comparison to benchmark technologies. For instance, incremental costs of wind power generation refer to the additional costs of wind power in comparison to thermal power generation. Incremental cost equals the difference between the additional investment for applying a particular technology and the respective benefits. Investment refers to both the initial investment as well as operation and maintenance costs.

$$\text{Carbon reduction cost of a technology} = \frac{\text{Incremental cost of the technology during its service life}}{CO_2 \text{ reductions of the technology during its service life}}$$

Incremental cost of a technology during its service life
= (Investment in the technology – investment in benchmark technology)
– (Benefits of the technology during its service life)
– (Benefits of the benchmark technology during its sevice life)

If the carbon reduction costs of a particular technology are negative, it means that this technology is more cost-effective than the benchmark technology. If carbon reduction costs of a particular technology are positive, it indicates that this technology is less cost-effective than the benchmark technology. Carbon reduction costs for the usage of energy-saving

[24] Incremental costs: when comparing the costs of different proposals, we choose one proposal as the base proposal and calculate the incremental cost of a proposal as the additional cost of the proposal in comparison to the base proposal. That is to say, incremental cost is a form of differential costs measured by the cost differentials between two proposals.

lighting fixtures, for example, is negative when compared to the usage of incandescent light bulbs, implying superior cost-effectiveness for energy-saving lighting fixtures. Another example is the positive carbon reduction cost of wind power generation in comparison to thermal power generation, suggesting that wind power generation is less cost-effective than thermal power generation. Many technological upgrades in the industrial sector consist merely of retrofitting existing equipment with new components, in which case the basic technical proposal is to maintain the status quo and not implement technological upgrades (initial investment and benefits are both zero). Therefore, technological retrofits are cost-effective as long as benefits exceed investment.

This study evaluates the carbon reduction costs of 41 low-carbon technologies, 38 of which have carbon reduction capacities that exceeded one $MtCO_2$ in 2010. The remaining 3 technologies are solar photovoltaic, IGCC and energy-saving air conditioners. We pay close attention to photovoltaic and IGCC because of their promising carbon reduction potential; air conditioners were chosen because their cost-effectiveness has been controversial.

Figure 2.1 shows the "carbon reduction costs vs. carbon reduction capacity" curve for 41 technologies. The figure illustrates that end-use energy utilization technologies with negative carbon reduction costs have been widely applied within the industrial and buildings sectors. Low-carbon technologies in the industrial and buildings sectors usually belong to energy-saving technologies which were supported and sponsored by the policies and regulations enacted during the 11th FYP. Energy-saving technologies typically have attractive returns through lowering energy consumption, and effectively reduce CO_2 emissions. In the industrial sector, most end-use energy utilization technologies have negative carbon reduction costs, meaning their application would yield high returns. All energy-saving technologies in the buildings sector have negative carbon reduction costs (except for energy-saving air conditioners), as do technological retrofits in thermal power generation. Soil-testing based fertilizer prescriptions are also very cost-effective, since they reduce the demand for fertilizers as well as lower costs.

For the 31 technologies with negative carbon reduction cost whose carbon reduction capacity exceeded one $MtCO_2$ in 2010, the average carbon reduction costs amounted to -307 RMB/tCO_2 (except for the 3 technologies with economic benefits higher than -1000 RMB/tCO_2) Applications of these negative-cost technologies have proven effective in reducing carbon emissions. Total carbon reduction capacity of the 34 negative-cost technologies with carbon reduction capacities of more than one $MtCO_2$ reached 189 $MtCO_2$ in 2010.

Supercritical and ultra-supercritical technologies both have positive carbon reduction costs due to the current energy pricing mechanisms in China (Figure 2.1). Supercritical generating units have proven economical: their costs are already approaching those of subcritical units. Since coal prices have become market-driven (fuel costs of coal-fired power generation already exceed 60% of total costs in recent years), while electricity prices have historically remained state-controlled and relatively low, coal-fired power plants have suffered considerable economic losses in recent years. Carbon reduction costs of supercritical and ultra-supercritical power generation have remained high. Moreover, the Chinese government thinks highly of afforestation and reforestation projects as a tool to preserve water and soil and protect the environment. During the 11th FYP, carbon reduction capacity by afforestation and reforestation measures amounted to 101 $MtCO_2$. However, afforestation and reforestation are both capital-intensive with carbon reduction costs of 197 RMB/tCO_2. Environmental benefits such as water and soil preservation excluded, and equaling the costs of afforestation to carbon reduction costs, the carbon reduction cost of afforestation and reforestation becomes 197 RMB/tCO_2.

The cost-effectiveness of energy-saving air conditioners depends on their service life: these air conditioners are cost-effective only if they are in continuous operation for a long period of time. More specifically, calculations have shown that energy-saving air conditioners are cost-effective only if their annual service life exceeds 2,000 hours. However, in central and northern China, most air conditioners run less than 2,000 hours a year: In Beijing, for example, air conditioners run less than 500 hours a year,[25] making them not economical at all.

Although wind and solar power generation at present both have positive carbon reduction costs and higher costs than thermal power generation, these two technologies are highly important from a strategic perspective. Consequently, the Chinese government enacted a large number of policies during the 11th FYP promoting research and development, demonstration, and dissemination of wind and solar technologies. China set a benchmark feed-in tariff for wind power in 2009 and a feed-in tariff for

---

[25] Li Zhaojian, *Life-cycle energy consumption and resource consumption analysis of residential air-conditioners in urban China*. Doctoral dissertation (Beijing: Tsinghua University, 2009).

solar power in 2011, allowing wind power and solar power to be competitive on the energy market.

Furthermore, as an important energy strategy, China is well prepared for the most advanced low-carbon technologies. Even though IGCC, alternative fuel vehicles, and CCUS are known for their prohibitively high costs and therefore not ready for large-scale application, China has proactively built a number of demonstration projects and mastered the essential technological know-how. Low-carbon development is a long-term process which requires a stable and supportive policy environment, outstanding technological innovation capacity, and sufficient technological preparations. Although China has attempted to create a supportive policy environment and make early-stage technological preparations, achieving low-carbon development during its industrialization remains an enormous challenge.

Overall speaking, the penetration rate of low-carbon technologies is still generally low. In this study, we have looked at 113 low-carbon technologies because of their better data accessibility. Their average penetration rate is only 17.5%, implying the huge potential for further promotion of these technologies. Figure 2.2 presents the distribution in penetration rates of these 113 low-carbon technologies. Only 7 technologies, i.e., 6.2% of the sample, have penetration rates of 50% or higher. 14 technologies (12.4% of the sample) have penetration rates of 30–50%. 13 technologies, i.e. 11.5%, have penetration rates of 10–20%. 65 technologies, i.e. 57.5%, have penetration rates of 0–10%.

From a sectoral perspective, the average penetration rate of 82 low-carbon technologies in the industrial sector is 16.9%, and the average penetration rate of 19 low-carbon technologies in the thermal power generation sector is 10.5%. The wide spectrum of building efficiency technologies is classified into 10 categories with average penetration rate of 30.2%. The penetration rates of energy-efficiency technologies in new buildings are significantly higher than those in existing ones. The penetration rate of energy-saving lighting fixtures already reached 75%. In the agricultural sector, the penetration rate of soil-testing based fertilizer prescriptions[26] was 60.4%. Taking the difference between the anticipated forest coverage

[26] Penetration rate of soil testing-based fertilizer prescriptions = area where soil testing-based fertilizer prescriptions have been made/ area of total arable land in China. In 2010, area where soil-testing based fertilizer prescriptions have been applied to was 1.1 billion mu (1 mu = 0.0667 hectares), and total national arable land area was 1.82 billion mu (Agricultural Statistical Yearbook).

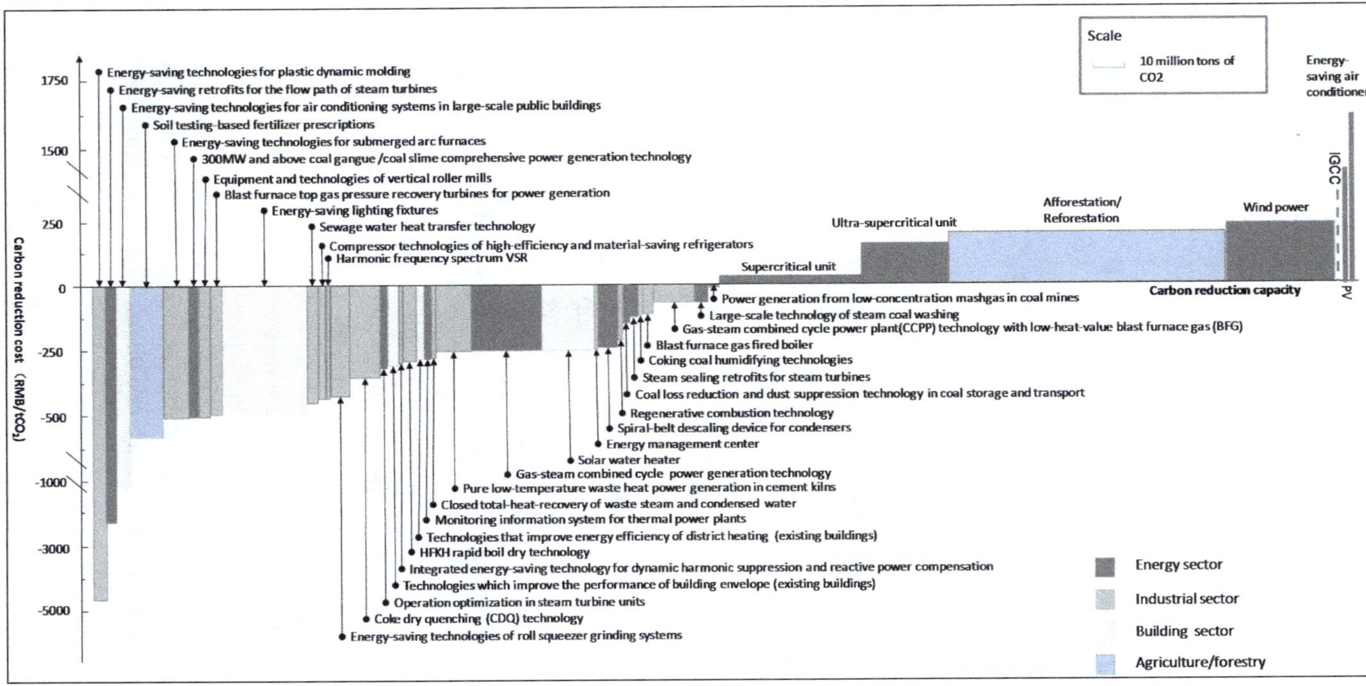

Note:

(1) 2010 carbon reduction capacities of photovoltaics, IGCC, and energy-saving air conditioners were 700,000, and 690,000 tCO_2, respectively.

(2) We use 2.71 tCO_2/tce as the CO_2 emissions index for coal equivalent. 2010 coal consumption for thermal power generation is 0.312 kg/kWh.

(3) Data source: data from industrial sources are calculated based on investment and energy-saving and carbon reduction benefits of typical case studies in "Analysis and evaluation of key energy-saving and emission-reducing industrial technologies." Other results are calculated by Climate Policy Initiative at Tsinghua University.

(4) Please refer to appendix for detailed methodologies for calculations.

Figure 2.1. Cost curve for technologies with carbon reduction capacities of greater than 1 million tCO_2 in 2010.

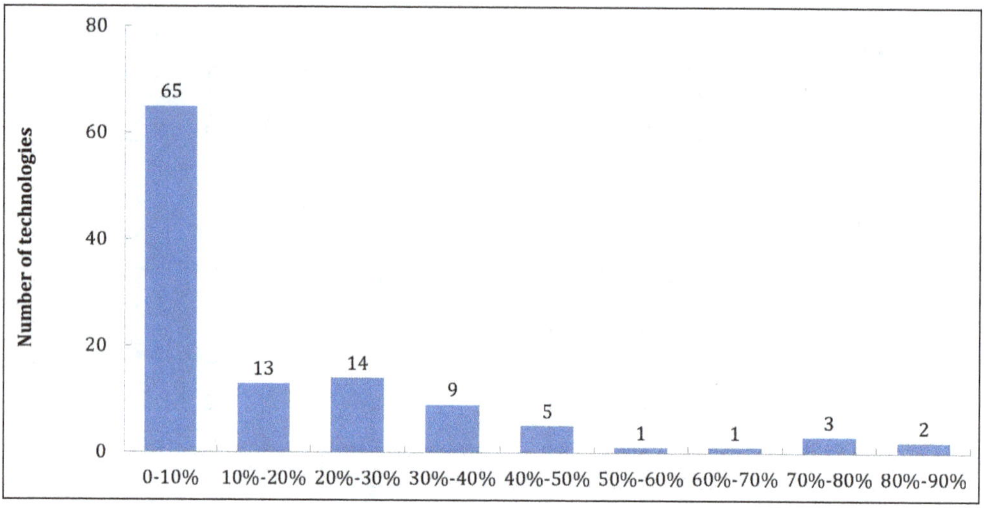

Note: All intervals on the horizontal axis in this figure include their upper limits. For instance, 0–10% includes 10%.

Source: Energy Research Institute of the NDRC, *Analysis and evaluation of key energy-saving and emissions-reduction industrial technologies* (Beijing: Energy Foundation, 2011). Penetration rates of building efficiency technologies and agricultural technologies are calculated results by Climate Policy Initiative at Tsinghua.

Figure 2.2. Distribution of penetration rates of 113 low-carbon technologies.

in 2050 (26%) and forest coverage in 2005 (18.21%) as the potential value, the penetration rate of afforestation/reforestation measures was 32.3%. The average penetration rate of 37 low-carbon technologies with carbon reduction capacities of more than one $MtCO_2$ in 2010 was 21.5%, considerably higher than average penetration rate of technologies across all sectors (17.5%).

Energy-saving technologies in the industrial, building, and thermal power generation sectors have relatively low penetration rates compared to conventional technologies, which suggests the huge potential for carbon reduction in these sectors. The industrial sector in China is characterized by low market concentration and the existence of a large number of small and medium-sized enterprises (SMEs). SMEs in energy-intensive industries rank far behind key enterprises in terms of scale, equipment capacity, and technological skills; SMEs' high energy consumption is a key factor inhibiting an overall energy efficiency upgrade of these industries.[27]

[27] Cai Fengtian, Xie Yuanmang, *Factors affecting automobile fuel economy and automobile Energy and resource engineering research team*, School of Engineering of Beijing

Since energy-saving policies during the 11th FYP are mainly oriented towards large enterprises, many SMEs find it hard to implement energy-saving technologies due to the shortage of financing opportunities. Therefore, a key task in the future is to encourage SMEs to adopt energy-saving technologies by creating proper incentives and financing opportunities by taking advantage of market instruments.

2.2.2 *Equipment Manufacturing Technique Improved*

During the 11th FYP, manufacturing capacity for low-carbon equipment improved dramatically. The localization[28] rate of low-carbon technical equipment is significantly higher (Table 2.8). Equipment manufacturing capacity exemplifies a country's technological capacity and core competitiveness. It is the prerequisite for the dissemination of low-carbon technologies and the fundamental way to lower costs of carbon reductions. Historically, the competitiveness of China's manufacturing industry has been built upon low labor costs and a huge domestic market. Domestically produced equipment will provide important competitive advantages for the development of low-carbon technologies. For example, domestic production of supercritical generating units lowers equipment costs by more than half. A growing localization rate of wind turbines had brought down the price of wind turbines from 9,000 RMB/kW in 2004 to 3,500 RMB/kW by the end of 2010.[29] Prices of domestically produced photovoltaic equipment are only one-third of prices of their imported counterparts.

The domestic manufacturing capacity for ultra-supercritical power generating units and circulating fluidized bed boilers (CFBBs) is on par with advanced international standards. From 1985 to 1990, the United States, the Soviet Union, Japan, Germany, and France started to design and manufacture ultra-supercritical generating units, many of which were put into operation after 1998. China imported the first 600 MW supercritical generating unit in 1998. The first domestically-produced 600 MW

University, 2011. *Analysis of the trend in energy-saving technology application and promotion in energy-intensive industries: iron and steel and coal industries* (Beijing: Energy Foundation, 2006).

28 Localization refers to the process through which domestically-held enterprises scrutinize and appropriate imported technologies, further develop it, and finally invent proprietary core technologies, thus domestically achieving the design, R&D, processing and manufacturing, installation, and maintenance of technological equipment.

29 Li Junfeng 李俊锋 et al., *2011 Zhongguo fengdian fazhan baogao* 2011 中国风电发展报告 [2011 Report on Wind Energy Development in China] (Beijing: Zhongguo huanjing kexue chubanshe, 2011).

Table 2.8. Increase in manufacturing localization rate of low-carbon technological equipment.

Technology	Localization rate during the 11th FYP (%)	Localization history during the 11th FYP
Ultra-supercritical generating units	Manufacturing localization rate reached 100%, materials localization rate reaches more than 90%[a]	863 project "Ultra-supercritical coal-fired power generation technology" was approved in 2002; the first ultra-supercritical power plant began construction in 2004
Circulating Fluidized Bed Boilers (CFBBs)	300 MW and smaller: 100%[b]	Smaller than 150 MW: 100% (2005)
Nuclear power	600 MW nuclear power technology: higher than 70%[c] CPR1000: 50–70%[d]	6000 MW nuclear power technology[e]: approximately 55% CPR1000: None
Wind turbine manufacturing	Whole unit: 85.7%[J]	750kW units: approximately 90%
Photovoltaics manufacturing (crystalline silicon)	Approximately 80%[f]	Production lines are virtually all imported
Ultra high voltage (UHV) power transmission and distribution	1000 kV power lines: 90%[g] (2009)	750 kV power lines: 90%[h] (2005)
Waste incineration power generation	90%[i]	Incinerators are imported

Source: Long Zhihui, "GW thermal power generating units in China all made-in-China," *China Energy News* (2010-2-1), URL: http://paper.people.com.cn/zgnyb/html/2010-02/01/content_440283.html [accessed: 2011-8-9]; Li Junyu Li, Zhou Chongming, *Supercritical circulating fluidized beds (CFB) will facilitate emissions reduction (frontier)* (2010-1-11), URL: http://paper.people.com.cn/zgnyb/html/2010-01/11/content_423906.html [accessed: 2011-8-9]; Zhang Guobao, "Self-led innovation in equipment manufacturing industry," *Qiushi* 8 (2008): 18–21; Sun Yuyao, *CPR1000 nuclear power generating units successfully connected to grid, improving localization rate* (2010-07-22), URL: http://info.machine.hc360.com/2010/07/221030204366-2.shtml; Xinhuanet, *Introduction to second phase project at the Qinshan Nuclear Power Plant* (2005-9-8), URL: http://www.China.com.cn/chinese/zhuanti/gjzdgc/964378.htm, [accessed: 2011-7-4]; Survey data by Climate Policy Initiative at Tsinghua; Zhou Fang, *China's Ultra high voltage (UHV) power transmission and distribution reached advanced international standard* (2009-4-17), URL: http://www.gzlps.gov.cn/art/2009/4/17/art_4183_110499.html [accessed: 2011-8-11]; State Grid Corporation of China (SGCC), *2005 Corporate Social Responsibility Report of SGCC* (Beijing: China Electric Power Press, 2006): 37; Wang Xuhui, *Domestic production solves the predicament of municipal solid waste power generation, localization rate of equipment reached 90%* (2010-7-29), URL: http://news.xinhuanet.com/environment/2010-07/29/c_12382710_4.html [accessed: 2011-8-10]; China Electricity Council, *Analysis of power generation equipment supply* (2011-2-21), URL: http://www.cec.org.cn/zhuanti/dianlixingyeshierwuguihua/kejichuangxinpian/2011-02-21/40577.html [accessed: 2011-9-10].

supercritical generating unit began its commercial operation in 2004; the first domestic one GW ultra-supercritical generating unit was commissioned in 2006. Both types of generating units have been put into volume production: 244 generating units (157.03 GW) had been produced by 2009, accounting for 45% of total production of thermal power generating equipment. Among these, 66 units (52.03 GW) are ultra-supercritical while the remaining ones (105 GW) are supercritical.[30] By 2010, China was producing 100% of ultra-supercritical units domestically, while more than 90% of manufacturing materials were provided locally (Table 2.4). It is important to note that shortage of materials is still the bottleneck in the production process of domestic supercritical and ultra-supercritical units in China. For a long period of time, R&D of high-end materials had been lagging behind the technological advancements made in power generation equipment. Materials have historically been the bottleneck inhibiting the manufacturing of ultra-supercritical units. P91 and P92, both materials needed for ultra-supercritical boilers, are mostly imported, which amounts to approximately half of the total costs of these boilers.

CFBB technology drew attention from all over the world during the 1980s and underwent rapid development. Since the first domestic 50 MW CFBB was commissioned in 1995, China has advanced from utilizing high-pressure to ultra-high-pressure circulating fluidized bed boilers (CFBBs). Since 2002, China has independently invented 100 MW, 200 MW, and 300 MW CFBBs. The localization rate of CFBBs with 300 MW and less is 100% at present. Domestic CFBBs have entirely saturated the market for domestic cogeneration of heat and power (CHP), and are also part of the power generation market. Seventeen 300 MW CFBBs went into operation in 2010, more than fifty are either under construction or in the planning stage. China is currently building the first 600 MW supercritical CFBB in the world. All design parameters meet advanced international standards. Overall speaking, even though China's manufacturing capacity has come a long way during the 11th FYP, China only qualifies as a "big manufacturing center" but not yet a "strong manufacturing center." Among the technologies listed in Table 2.8, manufacturers of ultra-supercritical units has not yet mastered the core technology in the development of high-end materials. Wind turbine manufacturing still depends on imported elements such as bearings, gearboxes, and frequency converters. China has not yet

---

[30] He Fa, "Status quo of development of large-scale clean and high-efficiency power generation equipment," *China Electrical Equipment Industry* 4 (2010): 20–23.

developed the full set of thin-film photovoltaic and key equipment such as vacuum pumps.[31] In addition, China depends on imports for some pieces of key equipment in nuclear power generation such as main coolant pumps. Substantial research and technological investments are required in order to manufacture nuclear power equipment that is "100% Chinese." Looking far into the future, China faces the task of dissecting and thus mastering imported core technologies, based on which China will then be able to build up its own innovation capacity, improve the technological standard in its equipment manufacturing, and achieve its goal of qualifying as a "strong manufacturing center."

2.2.3 *Gap between Domestic Technological Innovations and Foreign Counterpart Continues to Narrow*

The number of patent applications in a country is commonly seen as an indicator of the country's capacity for technological innovation. From 1991 to 2010, GDP growth in China averaged 10.1% per year; meanwhile, patent applications in China grew by 20.4% and 22.2% per year on average (Figure 2.3). In particular, growth rates of patent applications and patent applications for inventions in China during the 11th FYP reached 23.1% and 24.7%. The number of patent applications and number of patents for inventions in 2010 were 2.8 and 3 times of 2005 levels, and 7.7 and 11.1 times of 2000 levels. Since 2000, the percentage of domestic patent applications accounted for by patents for inventions has steadily grown from 18.1% in 2000 to 24.4% in 2005, and eventually 25.9% in 2010. From the perspective of total domestic patent applications and patent applications for inventions, the 11th FYP witnessed China's unprecedented progress in technological innovations.

Improving China's capacity for technological innovation is a long-term process that must be supported by steady market demand and sustainably growing investment. This process requires a free development of individual enterprises accompanied by government regulations. During the 11th FYP, the Chinese government funded more than 8,200 high-tech R&D projects, the "863" projects, and more than 4,800 technical support research projects. By 2010, there were 729 nationally authorized enterprise-based technical centers, twice as many as at the beginning of the 11th FYP. In 2010, R&D funding for nationally authorized enterprise-based

[31] UNDP, *China Human Development Report. 2009/10: China and a Sustainable Future: Towards a Low-carbon Economy and Society* (Beijing: China Translation & Publishing Corporation, China Publishing Group, 2010).

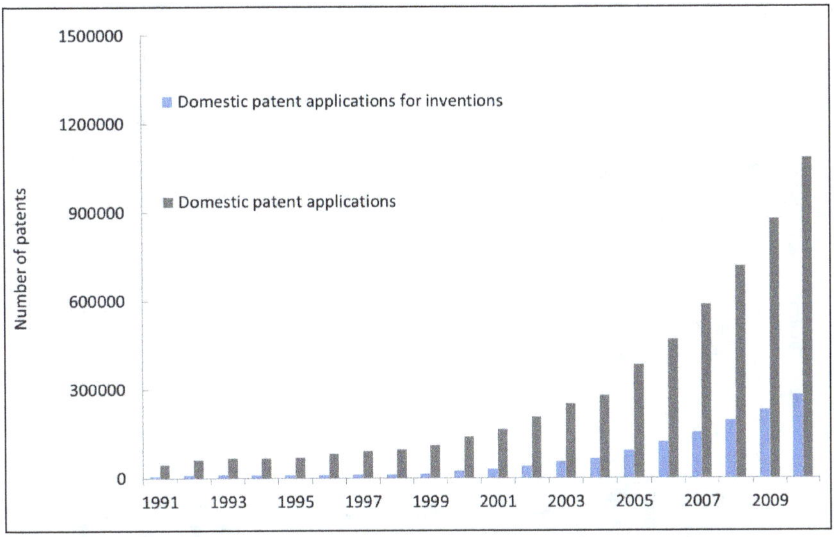

Source: National Bureau of Statistics of China, Ministry of Science and Technology of China, *China Statistical Yearbook on Science and Technology 1992–2010* (Beijing: China Statistics Press, 2011).

Figure 2.3. Growth in patent applications for inventions and domestic patent applications.

technical centers exceeded 180 billion RMB, 4.2 times as much as at the beginning of the 11th FYP.[32] The gap between domestic low-carbon technological innovation capacity and global state-of-the-art is continuously closing. In the wind turbine manufacturing industry, for example, patent applications experienced explosive growth since 2000. In 2009, patent applications for horizontal-axis grid-connected wind turbines reached 1,649, 6.1 times the 2005 level and 42.3 times the 2000 level. The global marketing strategies of wind power enterprises, however, also contributed to the global distribution of patents. A look at the top-ten wind turbine enterprises with the most patent applications reveals that more than half of them are foreign enterprises.[33] Meanwhile, the capacity for technological innovation of domestic enterprises dramatically improved. Figure 2.4 shows the number of patent authorizations for the top-ten

32 Department of Society and Science and Technology of National Bureau of Statistics of China, 2011.

33 Tian Zhanxiao, "Analysis of wind power patent status in China, Tianjin: 2011 Tianjin International Fair for Wind Power Industry Supporting (2011-3-8–2011-3-10).

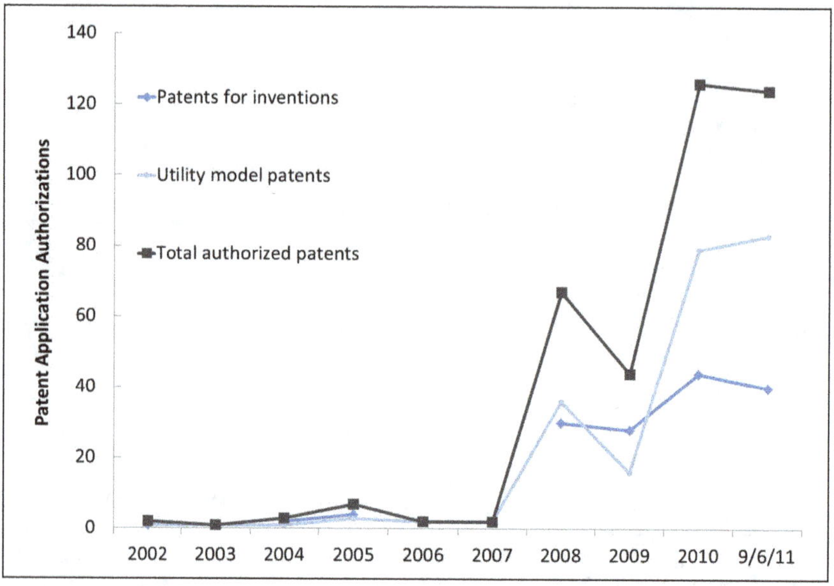

Source: Patent Searching Platform of State Intellectual Property Office.

Figure 2.4. Patent application authorizations of top ten domestic wind turbine manufacturers (up to September 6, 2011).

domestic wind power manufacturers. Before 2005, only one Chinese enterprise was issued three patents for inventions and utility model patents. Since 2008, authorizations for patent applications by domestic enterprises have significantly increased: patents for inventions grew steadily over the years; patents for utility models grew at the fastest rate. Total patent authorizations in 2010 were 18 times those of the 2005 figure, within which authorizations for patents for inventions and utility model patents were 11 times and 26.3 times that of 2005 levels, respectively.

Innovation capacity of domestic wind power enterprises significantly improved during the 11th FYP, narrowing the gap between domestic wind power manufacturing technology and its foreign counterpart. Figure 2.5 illustrates patent distributions of the top-ten domestic wind turbine manufacturers and of five prestigious European manufacturers, four of which were among the top-ten wind turbine manufacturers in the world in 2010. In terms of the total number of patents, the five European manufacturers held a total of 579 authorized patents up to December 31, 2009, while the top-ten Chinese manufacturers owned a total of 377 authorized patents by September 6, 2011. Therefore, there still exists a gap between Chinese wind power manufacturers' patent applications and those of prestigious

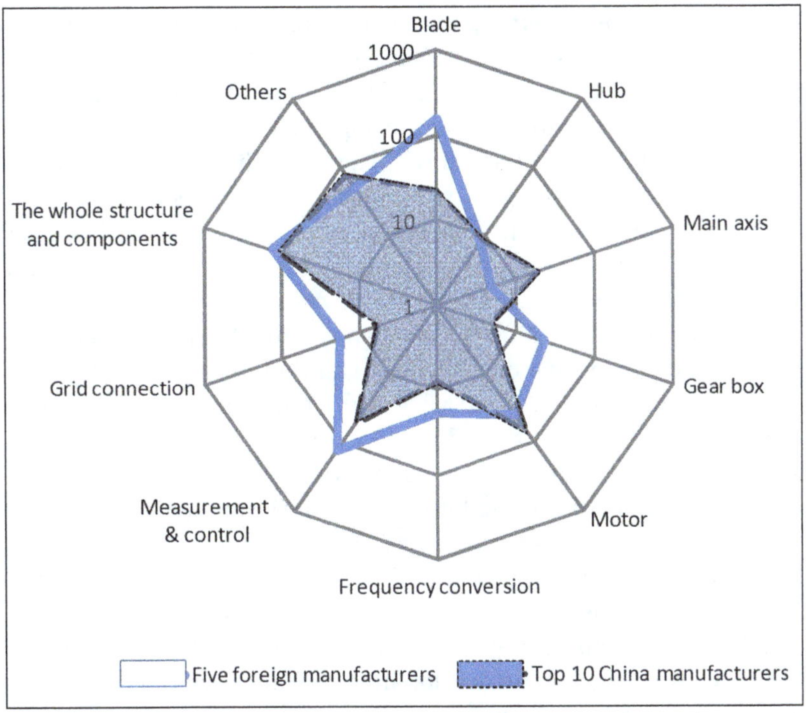

Note:

(1) Patent application data for top ten domestic wind turbine manufacturers are valid up to September 6, 2011. These ten manufacturers are: Sinovel Wind, XINJIANG GOLDWIND SCIENCE & TECHNOLOGY, DEC Wind Power, GUODIAN UNITED POWER TECHNOLOGY, MingYang WindPower, Hara XEMC wind power, Shanghai Electric, Zhejiang Windey Wind Power Engineering, SHENYANG CHINA CREATIVE WIND ENERGY, and CSIC (Chongqing) Haizhuang Windpower Equipment. Data source is Patent Searching Platform of State Intellectual Property Office.

(2) Patent application data for five prestigious foreign wind turbine manufacturers are valid up to December 31, 2009. These five manufacturers are Vestas (Denmark), Enercon (Germany), Gamesa (Spain), Repower (Germany), and Nordex (Germany), according to Thomson Innovation Database.

Source: Ma Jun, Cai Wengang, "World wind power industry through the lens of patent information analysis of European wind power manufacturers," *Machinery* 48, no. 555 (2010): 1–6.

Figure 2.5. Patent distribution of top ten Chinese wind turbine manufacturers vs. patent distribution of five prestigious European manufacturers.

European manufacturers. In terms of distribution of patent types, European manufacturers' patent applications are highly concentrated in three areas: blades, measurement and control systems, and overall structure and components. Patent applications in these three fields make up 71.8% of all applications, which illustrates the fact that European countries historically have been very strong in these technological fields. On the other hand, Chinese manufacturers' patent applications are relatively dispersed over a variety of fields. The top four fields with the most patent applications are: structure and components,[34] others, motors, and measurement and control systems.[35] Patent applications in these four fields make up 81.4% of total applications, which proves Chinese manufacturers' eligibility in whole units and parts production, as well as a deficiency in the R&D of new wind turbine models, large and new-material blades, and measurement and control systems.

In the context of the industry's rapid growth, the foremost objective for Chinese wind turbine manufacturers is to improve their capacity for technological innovation. European wind turbine manufacturers are known for their long history and their dedication to continuously improve their technology. Vestas from Denmark, for example, has been in the wind turbine R&D and manufacturing business for approximately 40 years. Vestas first attempted to manufacture wind turbines in 1971, and sold and installed the first wind turbine in 1979. The earliest Chinese wind turbine manufacturer, on the other hand, was created as late as 1998; the majority of domestic manufacturers started business even after the year 2000. Despite its short history, the development of China's wind turbine industry has been remarkable. In just more than a decade, China's wind turbine industry has undergone several stages of development: from adopting foreign technologies, to analyzing and mastering imported technologies, to eventually independently developing new technologies. Wind turbine manufacturing technology in China is catching up the most prestigious manufacturers in the world. Meanwhile, however, the explosive growth of the wind turbine industry has brought about a series of problems such as accidents involving fire and collapses in 2010, which constitute a big caveat to this rapidly growing industry. When firms blindly pursue an increase of per-unit capacity and engage in increasingly fierce price wars, the task to focus on product quality and R&D becomes increasingly difficult. As a result, China's wind turbine manufacturing industry, as well as policy makers, face the pressing challenge of having to relentlessly improve product quality and build up technological and innovational

[34] The whole structure and parts mainly include tower assist devices and installation methods of tower.

[35] Measurement and control system mainly refers to control system and operation methods of wind turbines.

capacity while the industry is catching up to the leading global wind turbine manufacturers.

Even though China has come a long way during the 11th FYP and is catching up with the most technologically advanced countries, one cannot overlook the weak innovation capacity of China as a whole. For example, China's scientific research is still at an initial stage; thus, funding of fundamental research remains at a low level. In 2009, the fraction of R&D funding allocated to basic research was only 4.7%, while in technologically advanced countries this percentage is usually higher than 10%.[36] Basic research is the foundation of scientific and technological development, and the key link in the entire "chain of innovation," thus deserving a very prominent position in China's national innovation strategy. Insufficient investment in basic research will unquestionably hinder technological development. According to the UNDP's "2010 China Human Development Report: China and a Sustainable Future. Towards a Low-Carbon Economy and Society," China needs at least more than 60 core technologies in order to achieve its goal of a low-carbon economy.[37] But China has not yet mastered the essential techniques for 42 of these technologies, including the development of new wind turbine models, application of large carbon fiber blades, high-end manufacture of photovoltaic equipment, as well as the design, materials, and manufacturing of the high-temperature parts used in advanced high-efficiency high-temperature gas turbines. Since technological innovation requires substantial funding and long-term preparations, the Chinese government is faced with the important task of increasing governmental funding for scientific research and stimulating a healthy market to empower enterprises to increase their research investment and cultivate scientific talent.

2.3 Summary

During the 11th FYP, China's research on and utilization of low-carbon technology mainly focused on six fields, including clean-coal combustion and power generation, upgrades in energy efficiency for the industrial

36 National Bureau of Statistics of China, Ministry of Science and Technology, *China Statistical Yearbook on Science and Technology, 2010* (Beijing: China Statistics Press, 2010).

37 UNDP, *China Human Development Report. 2009/10: China and a Sustainable Future: Towards a Low-carbon Economy and Society* (Beijing: China Translation & Publishing Corporation, China Publishing Group, 2010).

sector, energy efficiency upgrades in the building sector, wind energy utilization, solar energy utilization, as well as CCUS. The strategic choice of these six technical fields demonstrates China's commitment to develop clean energy technologies and end-use energy utilization technologies, as well as making technological preparations for CCUS and alternative fuel vehicles. Applications of low-carbon technologies during the 11th FYP were characterized by (1) large market size, (2) fast growth, and (3) a significantly improved end-use energy utilization rate.

With policy incentives and financial support in place, China witnessed a fast growth in energy efficiency upgrades in the business sector due to the development of low-cost and high-return low-carbon technologies. Tremendous progress was achieved in renewable energy technologies and clean coal combustion and power generation. Enhanced domestic manufacturing capacity of low-carbon technical equipment brought about a higher import substitution rate. In particular, domestically manufactured supercritical and ultra-supercritical power generating units and circulating fluidized bed boilers (CFBBs) have met advanced international standards. In short, as low-carbon technological innovation moves forward in China, the gap between domestic and foreign technologies has effectively narrowed.

It is important to note that despite the tremendous progress made, China's overall low-carbon technology is still underdeveloped. A number of core technologies remain to be mastered; low-carbon technological innovation requires further improvement in terms of scale, speed, as well as depth and precision of research.

Appendix 2.1 Methodology for Calculating Carbon Reduction Costs of Low-Carbon Technologies

1.

$$\text{Carbon reduction cost of a technology} = \frac{\text{Incremental cost of the technology during its service life}}{CO_2 \text{ reductions of the technology during its service life}}$$

2.

Incremental cost of a technology during its service life

= (Investment in the technology – investment in benchmark technology)
– (Benefits of the technology during its service life)
– (Benefits of the benchmark technology during its service life)

For technological retrofits, initial investment in benchmark technology and benefits of said benchmark technology during its service life are both zero.

3. Raw data used in calculating carbon reduction costs of electric power technologies:

	Per-unit Equipment cost (RMB/kW)	Share of total cost accounted for by equipment (%)	Per-unit Total cost (equipment + fuel + maintenance) (RMB/kW)	Service life (year)	Annual power generation per unit installed (kWh/kW)
Thermal power	3750	21%	17857	30	4832
Wind power	5900	75%	7867	15	1612
Photovoltaic	18000	80%	22500	20	1198
Supercritical	3833	21%	18252	30	4832
Ultra-supercritical	4500	21%	21429	30	4832
IGCC	11000	35%	31429	30	4832

4. Methodology for calculating carbon reduction costs of renewable energy technologies:

Carbon reduction cost of a particular renewable energy technology (RMB/tCO_2) = (Cost per kWh of the renewable power technology – Cost per kWh of thermal power technology) / (CO_2 emissions per kWh × 1000)
Cost per kWh of the renewable power technology = per-unit total installation cost / (Annual power generation per unit installed × service life)
CO_2 emissions per kWh = 0.318 kg/kWh (coal consumption for thermal power generation in 2010) × 2.71

5. Methodology for calculating carbon reduction costs of supercritical, ultra-supercritical and IGCC power generation:

We assume 4,832 annual service hours of supercritical, ultra-supercritical and IGCC power generating units, which are the average service hours of thermal power generating units in 2010. Coal consumption for subcritical power generation is 340 g/kWh, 308 g/kWh for supercritical power generation, 283 g/kWh for ultra-supercritical generation, and 255 g/kWh for IGCC.

Example: Supercritical generating units:

Carbon reduction costs for supercritical technology (RMB/tCO_2) = (per-unit total cost of supercritical technology – per-unit total cost of thermal power) / carbon reductions per installed unit within its service life.
Carbon reductions per installed unit within its service life = annual carbon reductions per installed unit × service life.
Annual carbon reductions per installed unit = annual power generation per installed unit × (Coal consumption for subcritical power generation – Coal consumption for supercritical power generation) × 2.71.

6. Methodology for calculating carbon reduction capacities of relevant technologies that improve the performance of building envelope or improve energy efficiency of heating supply system:

Relevant technologies that improve the performance of building envelope or improve energy efficiency of heating supply system: energy-saving capacities of these two technologies are mainly achieved by complying with building energy efficiency design standards. When calculating energy-saving capacities of relevant technologies such as building envelope, we assume energy-saving capacity of the technology in southern China is zero. Energy-saving capacities in Northern China are calculated in reference to building efficiency design standards in the specific region:

Energy saving capacity of building envelope technology = Area of energy-efficiency buildings in Northern China × energy savings per unit area × 60%

Energy saving capacity of district heating = Area of energy-efficiency buildings in Northern China × energy savings per unit area × 40%

Energy consumption for district heating per unit area in urban buildings in Northern China is approximately 20 kgce/(m^2.a). We assume that approximate energy saving per unit area is 10 kgce/(m^2.a); according to building efficiency design standards, building envelope accounts for 60% of energy savings, and district heating system and equipment account for the rest.

7. Energy-saving capacity of energy-saving lighting fixtures mainly refers to the ability of energy-saving lights to replace ordinary lighting sources such as incandescent lights with equivalent luminous flux, yet the corresponding power of energy-saving lighting fixtures is significantly lower than that of incandescent lights. Luminous flux of an 11 W energy-saving light is the same as that of a 40 W incandescent light. The corresponding energy-saving capacity is calculated as follows:

Energy-saving capacity of energy-saving lights = Additional usage of light × energy-saving capacity of a single light, where energy-saving capacity of a single light = (power of replaced incandescent light-power of energy-saving light) × annual service hours × coal consumption for power generation.

We assume that average power of incandescent lights is 40 W, average power of energy-saving lights is 11 W, annual service hours of a light is 2 × 365, and coal consumptions for power generations are annually recorded measurements for thermal power generation in 2006–2010.

8. Methodology for calculating energy savings by energy-saving air conditioners:

Energy savings by energy-saving air conditioners = Service hours of energy-saving air conditioners × energy saving capacity per air conditioner

Energy saving capacity per air conditioner = Energy intensity of air conditioners × % increase in energy efficiency of air conditioners. According to “energy efficiency limits and energy efficiency standards of household air conditioner adjuster” (GB12021.3), COP of energy-saving air conditioners is 15% higher than regular air conditioners, i.e., using energy-saving air conditioners can save ~15% of energy. In addition, survey results by Tsinghua University Building Efficiency Research Center (THUBERC) show that average electricity consumption for residential air conditioners in China is 135 kWh/household, and air conditioner ownership is approximately 107/100 household, i.e., one air conditioner per household, therefore we can take energy intensity per air conditioner as 135 kWh.

9. Methodology for calculating energy saving capacity of renewable energy sources (solar water heaters, biogas, and other renewable energy sources):

> Energy-saving capacity of a particular renewable energy source = Increase in amount of the renewable energy source × energy-saving capacity per unit of renewable energy.

Take solar water heater as an example. Energy saving capacity of solar water heaters = increase in households that have installed solar water heater × energy-saving capacity per household.

During the 11th FYP, cumulative increase in solar water heater installations in China amounted to 930 million m^2; average area of solar collector per household is approximately 2 m^2, equivalent to additional 46.5 million households using solar water heaters. Assume every household take baths 300 times a year, and use 60L of hot water for each bath, which requires heating water from 15°C to 40°C, then every household needs 64.3 kgce of energy for water heating for baths. If this energy demand can be provided by solar water heaters, then energy-saving capacity per household is 64.3 kgce/household. In fact, due to improvement in quality of life, many households take more than 300 baths a year. Moreover, limited by weather condition, solar energy cannot meet 100% of the hot water demand. As a result, actual energy savings by solar water heaters are undoubtedly different from results in this report.

10. Methodology for calculating carbon reduction costs of soil testing-based fertilizer prescription technology:

a. Determine reduced fertilizer consumptions due to promotion of soil testing-based fertilizer prescription in 2005 and 2010, respectively:

> According to "China Agricultural Yearbook" from 2006 to 2010, total area that has adopted soil testing-based fertilizer prescription technology in 2005–2010 was 5.33 million hectares, 17.33 million hectares, 42.67 million hectares, 60.00 million hectares, and 66.67 million hectares, respectively.

Accordingly, reduced improper fertilizer consumptions (pure fertilizer consumption) in 2006–2010 were 500 thousand tons, 1.1 million tons, 1.2 million tons, and 1.5 million tons, respectively. Linear correlation coefficient of area adopting the technology vs. reduced fertilizer consumption is as high as 0.96. In 2010, soil-testing based fertilizer prescription has been applied to 73.33 million hectors. By linear extrapolation, reduced improper chemical fertilizer consumptions in 2005 and 2010 due to soil testing-based fertilizer prescriptions were 310 thousand tons and 1.56 million tons, respectively. Based on the structure of chemical fertilizer applications in corresponding years, nitrogen, phosphorus and potassium fertilizer consumptions were reduced by 200 thousand tons, 67 thousand tons, and 45 thousand tons in 2005, and 1 million tons, 340 thousand tons, and 220 thousand tons in 2010. Compared to 2005, additional reductions in fertilizer applications in 2010 due to soil testing-based fertilizer prescription were 1.25 million tons.

b. Greenhouse gas emissions reductions:

Fertilizer-related greenhouse gas emissions include embodied carbon emissions in fertilizer production, and N_2O emissions in applying nitrogen fertilizer. Carbon emission factors of nitrogen, phosphorus, and potassium fertilizers are 7.2 tCO_2/tN, 1.8 tCO_2/tP_2O_5, and 1.5 tCO_2/tK_2O, respectively. N_2Oe emission factor in applying nitrogen fertilizer is $0.01tN_2O\text{-}N/tN$, global warming potential (GWP) of N_2O during heat absorption is 295. Please refer to (Climate Policy Initiative at Tsinghua, 2011) for calculation methodology. Estimated greenhouse gas emissions reductions due to soil testing-based fertilizer prescription in 2005 and 2010 were 2.6 $MtCO_2$-eq and 12.8 $MtCO_2$-eq, respectively. Compared to 2005, carbon reduction capacity of soil testing-based fertilizer prescription in 2010 was 10.2 $MtCO_2$-eq.

c. Rough estimation of carbon reduction costs:

According to chemical fertilizer prices in 2010 (www.3456.tv/, 2010-10-28), assume average price of chemical fertilizer is 2400 RMB/ton, and conversion-to-pure-amount factor is 50%, then estimated cost savings from reduced improper application of chemical fertilizers were 600 thousand RMB in 2010, and CO_2 emissions reduction costs were approximately 587 RMB/CO_2-eq.

d. Promotion potential and penetration rate of soil testing-based fertilizer prescription:

Total arable land in China is 1.82 billion mu (121.33 million hectares); promotion potential of soil testing-based fertilizer prescriptions is total arable land. Area where this technology has been adopted was 73.33 million hectares in 2010, so penetration rate was 61%. (In fact, the areas given in agricultural yearbooks were total areas where soil-testing-based fertilizer prescriptions have been made, but it remains a question whether farmers actually fertilized their land in accordance with the prescriptions, plus statistical data for the areas were missing, thus the 61% penetration rate should be much higher than reality.)

11. Methodology for calculating carbon reduction costs of afforestation/reforestation:

a. Calculation of carbon sequestration of afforestation/reforestation:

Average carbon sequestration potential of juvenile wood is 3.26 tCO_2/(ha·a).[38] Compared to 2005, additional reforestation area in 2010 was 24.16 million hectares, equivalent to carbon sequestration of 101.4 $MtCO_2$.

b. Estimation of carbon sequestration costs:

According to "Reforestation project for reclamation of Pearl River basin in Guangxi," construction cycle of the project is 30 years, total CO_2 sequestration is 774 thousand tons, investment 22.7 million USD,[39] estimated carbon sequestration cost is 29.3 USD/tCO_2, i.e., 197 RMB/tCO_2.

c. Promotion potential of afforestation/reforestation and penetration rate:

"11th FYP forestry development and medium-to-long term guideline" proposes a forest coverage of 26% in 2050, thus we take 26% as the promotion potential of afforestation/reforestation. According to forest coverage in 2005 and 2010 (18.21% and 20.36%, respectively), estimated penetration rate was 27.6%.

---

[38] Climate Policy Initiative at Tsinghua (2011).

[39] Zhang Xiaoquan, Wu Shuhong, *Theory and practice in forest carbon sequestration* (Beijing: China Forest Press, 2010).

CHAPTER THREE

ADJUSTMENT OF THE ECONOMIC STRUCTURE

Abstract: The adjustment of China's economic structure during the 11th FYP period realized 143 Mtce in energy savings, about 23% of its overall energy savings. However, the contributions in energy savings are rather different for the sectoral, industrial and product structure. Although the sectoral structure has accomplished a comprehensive adjustment, it has caused a negative contribution towards energy savings (–50.8 Mtce) due to the fact that China's sectoral structure remains oriented towards the heavy industry. The tertiary sector's proportion has improved, but in the meantime, the proportion of the secondary sector also continued to rise annually. On the other hand, the adjustment of the industrial structure conserved 77 Mtce in energy. The industrial sector achieved noteworthy results, mainly due to the adjustment of the ratio of value-added products within energy-intensive industries. The product structure adjustment achieved the most prominent contribution—117 Mtce was conserved.

Under the energy saving target constraint in the 11th FYP period, enterprises improved production efficiency and reduced energy consumption through upgrading equipment, merger and reorganization. The production equipment shows a clear trend towards large-scale and the industry concentration ratio has improved quickly.

The spatial reallocation of industries in the 11th FYP period has caused significant changes in the regional economic development pattern in China. Economic growth rate in western and northeastern China with relatively high energy intensity was faster than the east, thereby increased approximately 0.5% of national energy consumption per unit of GDP. Meanwhile, energy saving capability of these regions has improved, which encouraged the realization of energy conservation target. Industry reallocation improved the industry structure in the east, but the expansion of production capacity has brought greater challenge to national economic structure adjustment.

In the 11th FYP period, China saved 630 Mtce annually through a reduction in energy intensity per unit of GDP. The adjustment of its economic structure realized 143 Mtce in energy savings, which is about 23% of total energy savings (Figure 3.1).

Methods used in this study are similar to those of many domestic institutions and researchers.[1] This report regards the economic structure as a

[1] Feng and Wang (2009) analyze structural adjustments for the first half of the 11th FYP period by starting from the sectoral level and high energy-consuming industries; they concluded that the structural energy saved was negative. See: Feng Fei, Wang Jinzhao 冯飞, 王金照, "Shiyiwu jienengjianpai zhongqi pinggu '十一五'" 节能减排中期评估 [11th FYP

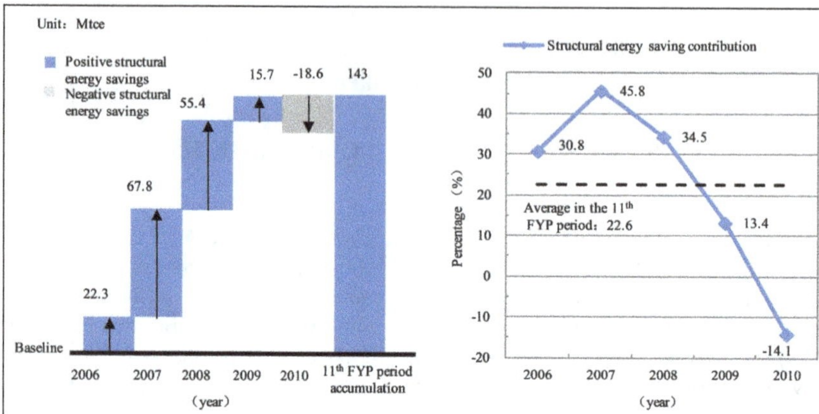

(a) Overall Structural Energy Savings (b) Structural Energy Savings Contributions

Data Source: 2006–2011 China Statistical Yearbook; 2009–2010 China Energy Statistical Yearbook

Calculation method: Appendix 3.1

Figure 3.1. Annual structural energy savings and its contribution in the 11th FYP period.

Mid-Term Estimates on Energy Saving and Emissions Reduction]," *Lüye* 4 (2009): 113–118. Xiong (2011) considers the sector, industry and product structure and concludes that all structures contributed about 30% of energy saving in the 11th FYP period: Xiong Huawen 熊华文, "Jiyu danwei GDP nenghao de jieneng qianli fenxi fangfa yu shizheng yanjiu. 基于单位 GDP 能耗的节能潜力分析方法与实证研究 [Analysis Method and Empirical Research Based on Energy Consumption Potential Per Unit of GDP]," *Zhongguo nengyuan* 33, no. 4 (2011): 21–25. Zhang et al. (2011), Yao et al. (2007), Chen et al. (2011), Zhao et al. (2011), Qi et al. (2010) and other researchers have considered sector and industry or both in studies of energy saving contribution from economic structure adjustment: Zhang Lei, Li Yanmei, Huang Yuanxi, Wu Yangmei 张雷, 李艳梅, 黄园淅, 吴映梅, "Zhongguo jiegou jieneng jianpai de qianli fenxi 中国结构节能减排的潜力分析 [Analysis of China's Structural Energy Saving Potential]," *Zhongguo ruankexue* 2 (2011): 42–51; Yao Yufang, Chen Jie, Li Hujua 姚愉芳, 陈杰, 李花菊, "Jiegou bianhua de jieneng qianli jisuan de fangfa yanjiu 结构变化的节能潜力计算的方法论研究 [Structural Changes In Energy-saving Potential Calculation Methodology]," *Shuliang jingji jishujingji* yanjiu 4 (2007): 115–123; Chen Shiyi 陈诗一, *Jienengjianpai*、*jiegou tiaozheng yu gongye fazhan fangshi zhuanbian yanjiu* 节能减排、结构调整与工业发展方式转变研究 [An Analysis of Energy Saving, Emissions Reduction, and Restructuring; And on the Transformation of China's Economic Development Pattern]. (Beijing: Beijing Daxue Chubanshe, 2007): 21–31; Zhao Xiaoli 赵晓丽, *Chanye jiegou tiaozheng yu jieneng jianpai* 产业结构调整与节能减排 [Industrial Restructuring, Energy Saving, and Carbon Emissions Reduction] (Beijing, Zhishi chanquan chubanshe, 2011): 31–39; Qi Yuping, Mu Hailin, Tang Xudong, Wang Peng 齐宝平, 穆海林, 康旭东, 汪鹏, "Zhongguo 1990–2007 nian jieneng yinsu fenjie fenxi 中国 1990~2007 年节能因素分解分析 [Decomposition Analysis of Factors in China's Energy Saving, 1990–2007]," *Nengyuan yanjiu yu guanli* 2 (2010): 1–4. Zhou and Liu (2008) pointed out that structure energy saving should consider sectoral structure adjustment, structural adjustment within the industry and product structure adjustment: Zhou Fuqiu, Liu Zhiping 周伏秋, 刘志平, "Cujin jieneng rongzi de zhengce jianyi 促进节能融资的政策建议 [Policy Proposals to Promote Energy-saving Financing]," *Zhongguo negyuan* 30, no. 8 (2008): 5–8.

multi-level and multi-factor complex composed of several systems. It has the property of top-down and level-by-level decomposition. In each level, energy can be saved through structural adjustments between components within that level. Only when comprehensive consideration is given to the hierarchical systems of economic structure can a contribution to energy saving of all structures be obtained. Products are the smallest production unit to reflect energy consumption as well as the smallest units constituting GDP. Based on this understanding, the contribution to energy savings resulting from structural changes in China's economy is analyzed from three levels: sector, industry, and product. A detailed analysis of how the three levels contribute to energy savings is also given.

Since the 9th FYP, China has proposed to fundamentally change its economic growth pattern away from extensive growth towards acceleration of scientific and technological progress. China set forth guidelines in the 10th FYP document of "adhering to the development as the pivotal theme," "adhering to structural adjustment as the main line" and "persisting in promoting economic structure adjustment in development and maintaining a rapid growth in economic structure adjustment." Meanwhile, China issued a special plan for structural adjustment—The "Outline of the Program of Industrial Structure Adjustment" in the 10th FYP. In the 11th FYP, China put forward the principle "to maintain a stable and rapid economic development" and "to accelerate the transformation of our economic growth mode." For over a decade, the Chinese government has been committed to adjust and optimize China's economic structure, but China's economic development still maintained a development path that relied on extensive resources and energy consumption and high pollution emissions. Economic growth of the heavy industry was lower than that of light industry in the 9th FYP period, but the average growth rate of the heavy industry was 2.6% higher than that of light industry during the 10th FYP and 11th FYP periods, highlighting a clear tendency towards the heavy industry; consequently, the heavy industry fuels economic growth but restricts low-carbon economic development in China.

3.1 GDP Structure

To understand the processes and calculation methods involved in structural energy saving, is helpful in understanding how economic structure adjustment contributes to energy saving.

Xiong Huawen defines structural energy-saving as energy saved due to proportional changes of each subsystem (industry or product, etc.) within a structural level. Taking the sectoral level as an example, structural energy saving can be expressed thusly:[2]

Energy savings of sectoral structure$_{target\ year}$ = *GDP*$_{target\ year}$ × Σ *[(average energy intensity of all sectors*$_{base\ year}$ − *energy intensity of j sector*$_{base\ year}$*)* × *(j sector proportion of value-added*$_{base\ year}$ − *j sector proportion of value-added* $_{target\ year}$*)]*

Energy savings of sectoral structure$_{target\ year}$ = *GDP*$_{target\ year}$ × Σ *[(average energy intensity of all sectors*$_{base\ year}$ − *energy intensity of j sector*$_{base\ year}$*)* × *(j sector proportion of added value* $_{base\ year}$ − *j sector proportion of added value* $_{target\ year}$*)]*

Energy intensity and the added value proportion of each sector are the key factors that affect structural energy saving. Structural energy saving only exists when there is a difference in energy intensity between sectors. If the energy intensity of a sector is higher than the average energy intensity of all sectors,[3] then reducing the share of that sector is an effective measure to achieve structural energy savings; if the energy intensity of a sector is lower than the average energy intensity of all sectors, then raising the share of that sector would facilitate structural energy savings as well. (Specific calculation method is shown in Appendix 3.1).

Whether structural energy saving can be achieved is a tug-of-war between high- and low-energy consumption intensity sectors (industries and products).

According to accessible data and based on the principle of matching energy statistics and economic data statistics in China, the economic structure is divided into a sectoral structure, industry structure, and product structure (Figure 3.2).

- Sectoral structure includes the primary industry, industrial industry, construction industry, and tertiary sector.[4]

---

[2] Xiong Huawen 熊华文, "Jiyu danwei GDP nenghao de jieneng qianli fenxi fangfa yu shizheng yanjiu. 基于单位 GDP 能耗的节能潜力分析方法与实证研究 [Analysis Method and Empirical Research Based on Energy Consumption Potential Per Unit of GDP]," *Zhongguo nengyuan* 33, no. 4 (2011): 21–25.

[3] All sectoral value-added energy intensity = (national energy consumption − household energy consumption) / GDP.

[4] The reasons to separate the measure of industrial and construction within the secondary industry are: First, energy intensity of the construction industry is lower than that of the tertiary sector and much lower than that of the industrial. If the construction industry and the industrial industry are combined as secondary industry for analysis, it will blur the focus of examine the industrial industry as the main energy consumption subject; second, because the construction industry shapes a large proportion of the economy, if it is classified within the level of industry structure, it will be mismatched with other industries within the industrial sector. Some studies analyzed industrial and construction industry as

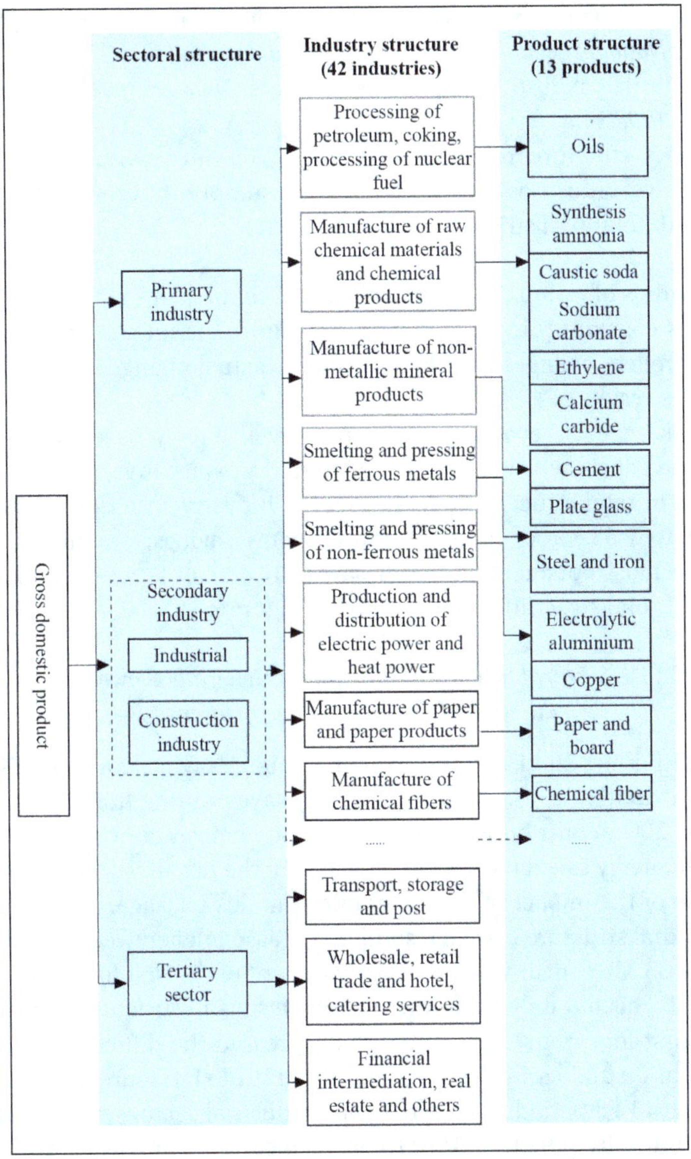

Note: "Industries" in double lined borders are "six major energy-consuming industries" approved by the National Bureau of Statistics.

Figure 3.2. The economic structure levels classification.

---

separate levels (Xiong 2011: 21–25), but from the perspective of policy analysis, the proportional change between industrial and construction industries does not play an actual role in policy guidance. Therefore, this report separates industrial and construction industries and attributes them to the sectoral level.

- Industry structure consists of two parts: 39 industries with two-digit codes in the *National Economy Industry Classification Standard* (GB/T4754-2002); three tertiary industries divided based on the *China Energy Statistical Yearbook 2010.*
- Product structure refers to 13 types of products within six energy-intensive industries and its energy consumption accounts for about 45% of total industrial energy consumption.

The analysis of economic structural change in the 11th FYP in this chapter assumes constant prices in 2005, and excludes the impact of price variation to reflect changes in volume, namely, actual changes in production activities results.

Based on each economic structure level, energy saved by sectoral structure adjustment in the 11th FYP period was negative, accounting for –35.6% of total structural energy savings; Industry structure adjustment contributed 53.8% of total structural energy savings; product structure was the most outstanding one, it saved 117 Mtce of energy, making up 81.8% of total structural energy savings (Figure 3.3).

3.1.1 *Sectoral Structure Adjustment Generally Slowed Down the Rate of Decline in Energy Intensity*

In the 11th FYP period, sectoral structural energy savings totaled—51 Mtce and contributed to –8.0% of total energy saved during this period. From 2007 to 2009, contributions from structural changes grew from –9.8% to 7.2% of energy saved nation-wide, reversing the negative trend. However, because of the impact of the economic stimulus package, the changes in the sectoral structure caused a sudden increase in energy consumption in 2010 by 33 Mtce, nearly double the increment of the first four years of the 11th FYP. This not only offset the positive results of structural adjustment in the first four years, but also greatly increased the difficulty for energy conservation and carbon reduction in the 12th FYP (Figure 3.4).

In the 11th FYP period, the share of the industrial sector with the highest energy intensity (1.92 tce/10^4 RMB on average) accumulatively rose by 0.9 percentage points from 41.8% to 42.7%, and the share of the tertiary sector (0.38 tce/10^4 RMB on average) accumulatively rose by 1.3 percentage points from 40.5% to 41.8%.

This change has had different degrees of impact on energy intensity per unit of GDP. An increase in the share of the energy-intensive industrial

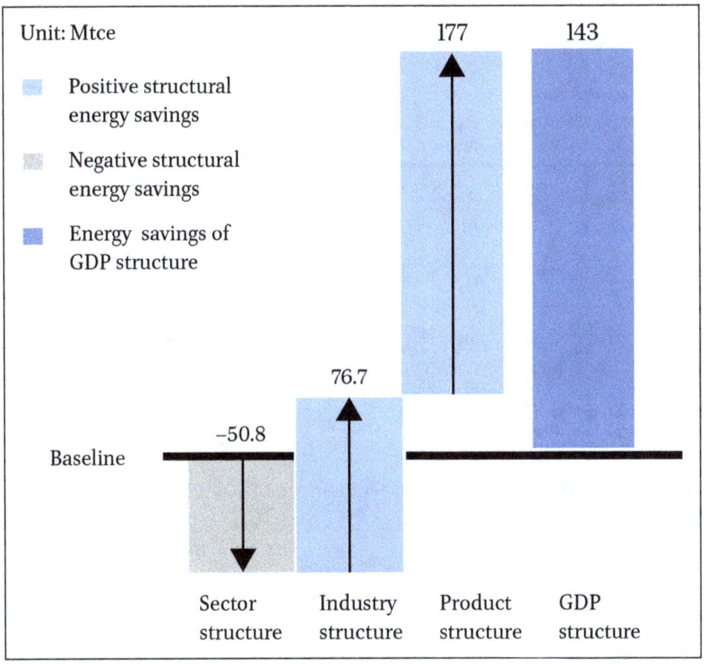

Data Source: 2006–2011 China Statistical Yearbook, and 2009–2010 China Energy Statistical Yearbook.

Calculation method: Appendix 3.1.

Figure 3.3. Structural energy savings of each economic structural level in the 11th FYP period.

sector would increase energy intensity per unit of GDP (Scenario 1), while a proportional decrease will lower energy intensity per unit of GDP (Scenario 3); if the share of the tertiary sector increases relatively to the industrial sector, it would decrease energy intensity per unit of GDP (Scenario 3); if the share of the tertiary sector increases relatively to the primary industry, it would lead to an increase in energy intensity per unit of GDP (Scenario 2). Comparing the year 2010 to the year 2005, the proportional changes of each sector increased energy intensity per unit of GDP by 1.64%. The tertiary and the industrial sector seized a share of the primary industry, which has the lowest energy intensity (0.24 tce/10^4 RMB on average in the 11th FYP period), thereby, the sectoral structure had an overall negative contribution to national energy saving (Table 3.1).

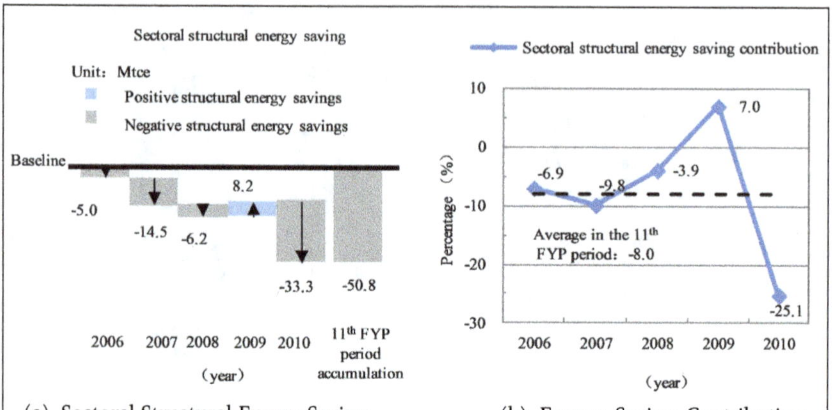

(a) Sectoral Structural Energy Saving

(b) Energy Saving Contribution of Sectoral Structure

Data Source: 2006–2011 China Statistical Yearbook and 2009–2010 China Energy Statistical Yearbook.

Calculation method: Appendix 3.1.

Figure 3.4. Sectoral structural energy savings and its contribution in the 11th FYP period.

Compared with the 10th FYP period, China's economic development fluctuations were larger in the 11th FYP period (Figure 3.5).

From 2005 to 2007, prior to the outbreak of the global financial crisis, the industrial growth rate increased annually. Fortunately, industrial growth was only slightly higher than that of the GDP, thus reversing the situation of the 10th FYP where industrial growth had been much higher than that of GDP; from 2005 to 2007, the growth of the tertiary sector increased rapidly and drove GDP growth along with the industrial sector, thereby alleviating the negative contribution to energy savings in the sectoral structure.

The financial crisis of 2008 slowed down economic growth, affecting most heavily the industrial sector that suffered a faster decline in growth. In 2009, its growth continued to decline and its proportion decreased by 0.2%, which resulted in a positive energy saving balance, the first time since 1991 that industrial growth had been slower than GDP growth.

In an effort to restrain the adverse effects of the financial crisis, China launched "The four trillion RMB economic stimulus package" in November 2008, of which 81% flowed into industrial-related infrastructure and

Table 3.1. Impacts of proportional change in sectors on energy intensity per unit of GDP.

	Change of Sectoral structure Scenario	Resulted Change in Energy Intensity per unit of GDP
Scenario 1	The primary industry dropped 1% and the industrial rose 1%	+1.49%
Scenario 2	The primary industry dropped 1% and the tertiary sector rose 1%	+0.12%
Scenario 3	The industrial dropped 1% and the tertiary sector rose 1%	-1.37%
Actual situation in the 11th FYP	The primary industry dropped 3.2%, the industry rose 0.9%, the construction industry rose 1.0% and the tertiary sector rose 1.3%	+1.64%

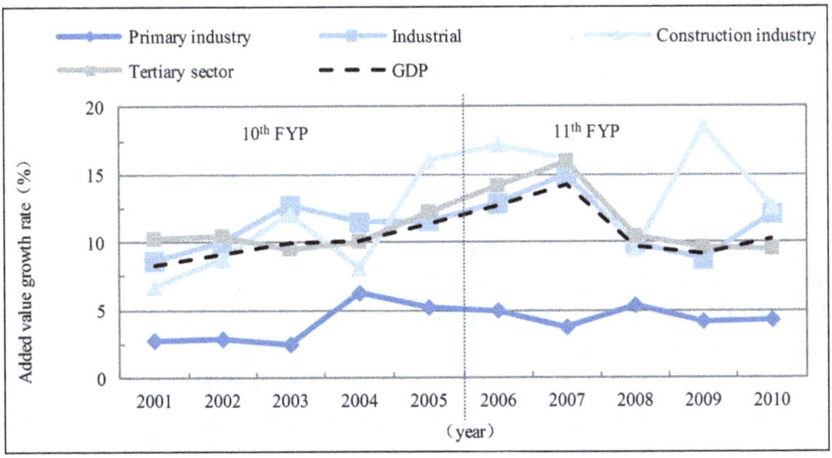

Source: 2010 China Statistical Yearbook and 2011 China Statistical Yearbook.

Figure 3.5. 2001–2010 annual growth rate of sectoral added values.

residential housing projects.[5] In 2010, the industrial growth rate rapidly increased and the proportion of the industrial sector surged by 0.7%, resulting in the largest negative contribution to energy savings in this sector.

The global financial crisis and China's countermeasures have once again highlighted the key role of the industrial sector for China's economy. While China is undergoing a process of industrialization and urbanization, the industrial sector remains the main engine that drives China's economic growth as well as the main carrier for the transformation of China's economic model. Thus, adjustments to China's industrial structure remain crucial at the present stage, and it has still a long way to go in order to achieve its goal to save energy from these adjustments.

3.1.2 *Industry Structure Significantly Improved with Substantial Energy Saving Effects*

In the 11th FYP period, structural changes to 42 industries saved 77 Mtce of energy. The structural change of 39 industries within the industrial sector saved 45 Mtce and presented a positive and negative fluctuation in contributions to energy saving annually. The structural change of three industries within the tertiary sector saved 32 Mtce and made a positive contribution to energy saving. In the 11th FYP period, structural changes of the 42 industries accounted for 12.2% of China's total energy saving (Figure 3.6).

3.1.2.1 *The Proportion of Six Major High Energy-Consuming Industries Declined and High Value-added Mechanical Equipment Manufacture Industry Developed Rapidly*

Thirty-nine industries within the industrial sector have had a rather complex impact on energy saving within industry structure changes. For the sake of clarity, the 39 industries analyzed here are classified by their energy intensity and their economic development characteristics. Figure 3.7 illustrates the average energy intensity and growth rate of the 39 industries in the 11th FYP period. The origin of the figure represents average energy intensity and average growth rate of the industrial sector, the horizontal and vertical axis are composed of four quadrants: industries in quadrant

[5] World Wildlife Fund WWF, *Zhongguo jingji ciji jihua dui qihou he nengyuan de yingxiang* 中国经济刺激计划对气候和能源的影响 [The effect of the stimulus plan for the Chinese economy on climate and energy resources]: 4 (http://www.csrglobal.cn/reports_detail.jsp?fid=306079) [accessed: 2010-11-08].

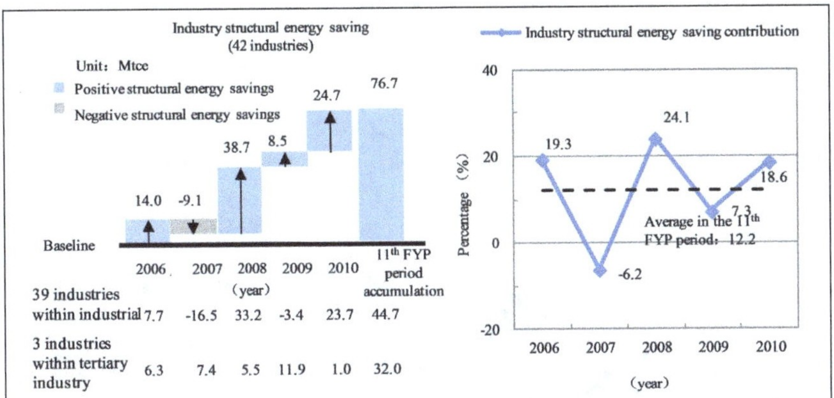

(a) Industry Structural Energy-Saving

(b) Energy Saving Contribution of Industry Structure

Data Source: 2006–2011 China Statistical Yearbook and 2009–2010 China Energy Statistical Yearbook.

Calculation method: Appendix 3.1.

Figure 3.6. Industrial structural energy savings and its contribution in the 11th FYP period.

II and IV generated positive effects on structural energy saving and industries in quadrant I and III generated negative effects on structural energy saving. According to Figure 3.7, the 39 industries are classified into three categories for the analysis of structural energy saving contributions:

- Six major energy-intensive industries: accounted for more than 70% of industrial energy consumption, but only about 30% of industrial added value. Average energy intensity was 4.50 tce/10^4 RMB, more than double that of the industrial sector;
- Five mechanical equipment manufacture industries:[6] Accounted for 25% of industrial added value, but only 5.6% of industrial energy consumption. Average energy intensity was 0.39 tce/10^4 RMB, less than 1/10 of the six energy-intensive industries;
- Other 28 industries: relatively moderate energy intensity (1.01 tce/10^4 RMB) and proportion of added value.

[6] Includes: manufacture of general purpose machinery, manufacture of special purpose machinery, manufacture of transport equipment, manufacture of electrical machinery and equipment, manufacture of communication equipment, computers and other electronic equipment.

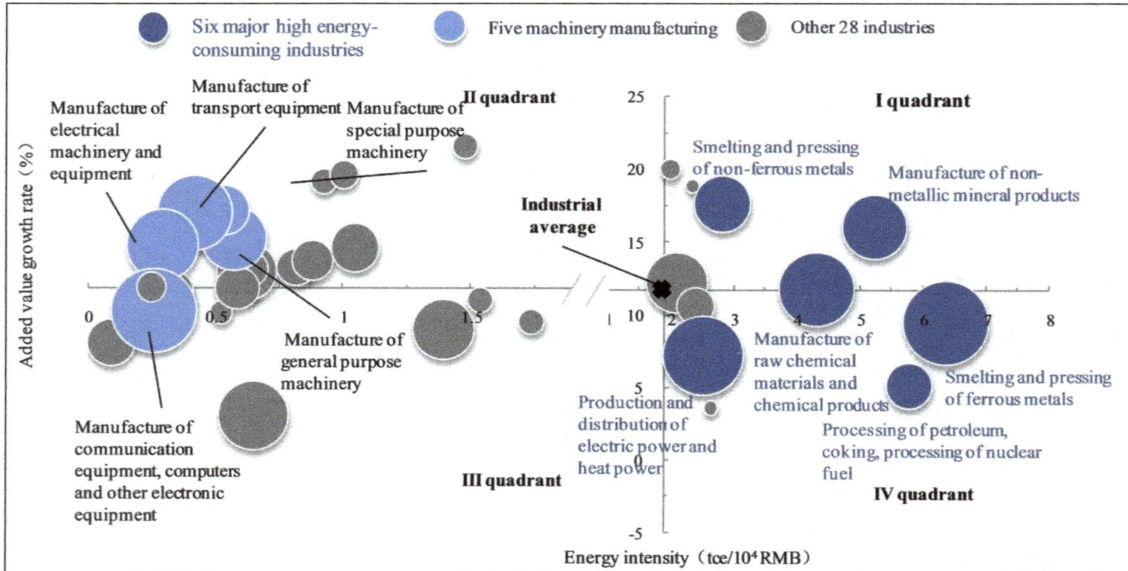

Note: The horizontal axis indicates average energy intensity of each industry in the 11th FYP period; the vertical axis indicates average growth rate of added value of each industry in the 11th FYP period; the origin represents average industrial energy intensity and added value growth rate (1.92, 11.7%) in the 11th FYP period; the circular area of each industry indicates the seize of added value proportion of each industry in total industrial added value.

Source: 2006–2011 China Statistical Yearbook and 2009–2010 China Energy Statistical Yearbook.

Figure 3.7. Average energy intensity and added value growth rate of industries within the industrial sector in the 11th FYP period.

Table 3.2. Impacts of Proportional Changes of Industries' Added Value within Industrial on Energy Intensity per Unit of GDP.

	Structural Change of Industry in the Industrial	Change of Energy Intensity per unit of GDP
Scenario 1	1% decline in added value proportion of six major energy-consuming industries and 1% rise in the proportion of five mechanical equipment manufacture industries	–1.54%
Scenario 2	The added value proportion of five mechanical equipment manufacture industries rose by 1% and the proportion of other 28 industries dropped by 1%	–0.23%
Actual situation in the 11th FYP	The added value proportion of six major energy-consuming industries dropped by 1.3%, the proportion of five mechanical equipment manufacture industries rose by 2.9% and the proportion of other 28 industries dropped by 1.6%	–2.31%

In the 11th FYP period, the proportion of added value of the six major energy-intensive industries fell by 1.3 percentage points from 31.3% to 30.0%, and the proportion of five mechanical equipment manufacturing industries with relatively high added value and low energy intensity accumulatively grew by 2.9 percentage points from 24.6% to 27.5%.

An analysis of the impact of the proportional change of these industries' added value on energy intensity per unit of GDP (Table 3.2) led to the following conclusions: an increase in the proportion of five mechanical equipment manufacture industries with low energy intensity reduces energy intensity per unit of GDP (Scenario 1 and 2); the impact of the six major energy-intensive industries was largest, a 1% decrease in the proportion of six major energy-intensive industries and a 1% increase in the proportion of five mechanical equipment manufacturing industries will drive a decline in energy intensity per unit of GDP by 1.53%. Comparing 2010 with 2005, the five mechanical equipment manufacture industries developed swiftly, which forced the decline in the share of the six major energy-intensive industries and other 28 industries, as well as a decline in energy intensity per unit of GDP by 2.31%.

In the 11th FYP period, the six major energy-intensive industries contributed positively to energy savings in the industrial structure. But growth rates for industries manufacturing non-metallic mineral products and/or smelting and pressing of non-ferrous metals have remained high, as quadrant I in Figure 3.7 indicates, and this structural change is not conducive to energy conservation.

The added value of the six major energy-intensive industries declined proportionally over the 11th FYP period, but their growth rate still increased by 0.2 percentage points more than that of the 10th FYP period (Figure 3.8). This can mainly be attributed to the following reasons:

The inertia effect caused by the rapid development of energy-intensive industries during the 10th FYP period was still in effect at the beginning of the 11th FYP. With the average industrial growth rate at 10.9% during the 10th FYP period, the average growth rate of the six major energy-intensive industries was stable at 10.7%, while their average added value proportion stayed at 31.6%. In the 11th FYP period, it was this inertia effect that drove the continuous rise of the growth rate of the six major energy-intensive industries that peaked at 15.1% in 2007, causing a negative contribution to energy savings in this industrial structure.

China is currently experiencing a process of rapid urbanization which brings about rising demand for energy-intensive products. China's urbanization is also accompanied by intensive infrastructure construction. The impact of changes in statistical calibers excluded, the total building area from 2000 to 2010 has more than doubled (Figure 9.4). In the 11th FYP period, road and railway construction advanced rapidly and the mileage of newly-constructed railways and roads were 2.1 and 4.3 times that of the 10th FYP period, respectively.[7] Infrastructure construction, including housing and railways, consumes large amount of iron, steel, cement, glass, electrolytic aluminum, and other high energy-consuming materials. Its steep demand has exerted a strong push on the growth of China's energy-intensive industries.

In the 11th FYP period, the Chinese government adopted a series of policy instruments geared towards structural adjustment in an effort to slow down the rapid expansion of the six major energy-intensive industries to a certain extent. In June 2007, as energy-intensive industries were still expanding rapidly, the National Development and Reform Commission issued *The Notification on Publishing and Issuing Comprehensive Industrial Plan on Energy Conservation and Emissions Reduction (Guo Fa [2007] No. 15)*. It clearly defined as its goal the elimination of outdated production

[7] Compiled according to "Statistical Communique of National Economic and Social Development" in past years.

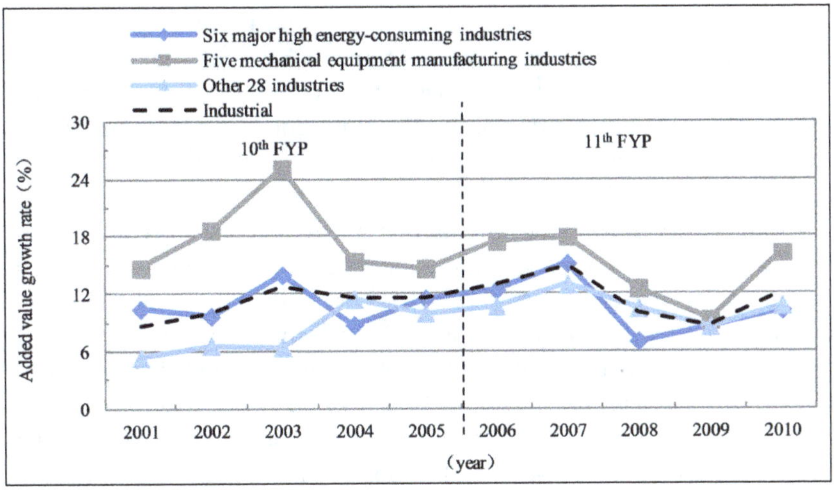

Data Source: 2010 China Statistical Yearbook and 2011 China Statistical Yearbook.

Figure 3.8. The added value growth rate of industries within industrial in the 11th FYP period.

capacity for energy-intensive products, including electricity, iron and steel, cement, glass, electrolytic aluminum and calcium carbide (Table 1.4). Additionally, it proposes a series of policies including the elimination of outdated production capacity by administrative order, strict control of market access, loan restrictions for "energy-intensive and high-pollution" industries, revocation of export refunds for energy-intensive products, as well as a differentiated and punitive pricing system for electricity in order to curb the rapid growth of the six major energy-intensive industries. In addition, under the impact of the financial crisis in 2008, the growth rate of the six major energy-intensive industries dropped by 50% (Figure 3.8), which in turn caused energy savings within this sector to approach 33 Mtce.

At the end of the 11th FYP period, a "four trillion RMB economic stimulus package" led to a rebound in the growth rate of energy-intensive industries. Funding was allocated mostly into infrastructure and residential housing construction projects which caused additional steel and iron consumption of 62 Mt and cement consumption of 349 Mt[8] which accounts for 5.2% and 9.9% of total crude steel and cement outputs from

[8] World Wildlife Fund WWF, *Zhongguo jingji ciji jihua dui qihou he nengyuan de yingxiang* 中国经济刺激计划对气候和能源的影响 [The effect of the stimulus plan for the Chinese economy on climate and energy resources]: 4 (http://www.csrglobal.cn/reports_detail.jsp?fid=306079) [accessed: 2010-11-08].

2009 to 2010, respectively. During 2009 to 2010, the mileage of newly-constructed railways and roads were 4 times to 5.2 times that of the years 2006–2008. It accelerated the growth of energy-intensive industries to a large extent. The growth rate of the six major energy-intensive industries reached the bottom and began to rebound in 2008, and then rose to 8.5% in 2009 and 10.3% in 2010.

Although structural adjustments within the industrial sector contributed positively to energy savings and the proportion of added value within the six major energy-intensive industries has dropped, the phenomenon of excessive production capacity was still widespread during the 11th FYP period: From September 2008 to June 2009, the iron and steel industries ran at about 70% of their overall capacity, while for the aluminum smelting and ferroalloy industries, capacity utilization stood at about 60%.[9] On September 26, 2009, the National Development and Reform Commission, together with several ministries and commissions, issued the *Advice on Curbing Overcapacity and Repetitive Construction of Industries and Guiding Industrial Healthy Development* (Guo Fa [2009] No. 8). It listed energy-intensive industries such as iron and steel, cement, plate glass, and coal chemical industries as those with notable overcapacity issues.

3.1.2.2 *Low-Energy-Consuming Service Industries within the Tertiary Sector Developed Rapidly*

The average energy intensity of the transportation, storage, and postal industries in the tertiary sector was 1.65 tce/10^4 RMB, which is about 4.3 times more than the average level of the entire tertiary sector. Their proportional added value within the tertiary sector accumulatively dropped by 2.1 percentage points from 14.2% to 12.1%; average energy intensity of wholesale, retail and accommodation, and catering services was 0.22 tce/10^4 RMB, which is only 57% of the average level of the tertiary sector, and their proportion in the tertiary sector accumulatively grew by 3.5 percentage points from 24.2% to 26.7%; average energy intensity of the financial, real estate, and other industries was 0.18 tce/10^4 RMB and their share in the tertiary sector dropped by 1.3 percentage points.

Added value of the wholesale and retail, real estate, and financial industries sustained rapid growth from 2006 to 2007 with an average growth rate of more than 20% in 2007; they are the major industries

[9] Li Xiaojian 李小建, "Channeng: Cong 'shengyan' dao 'shengyan' 产能: 从'盛宴'到'剩宴' [Production Capacity: From 'Grand Banquet' to 'Leftovers']," *Zhongguo renda* (2009): 12–14.

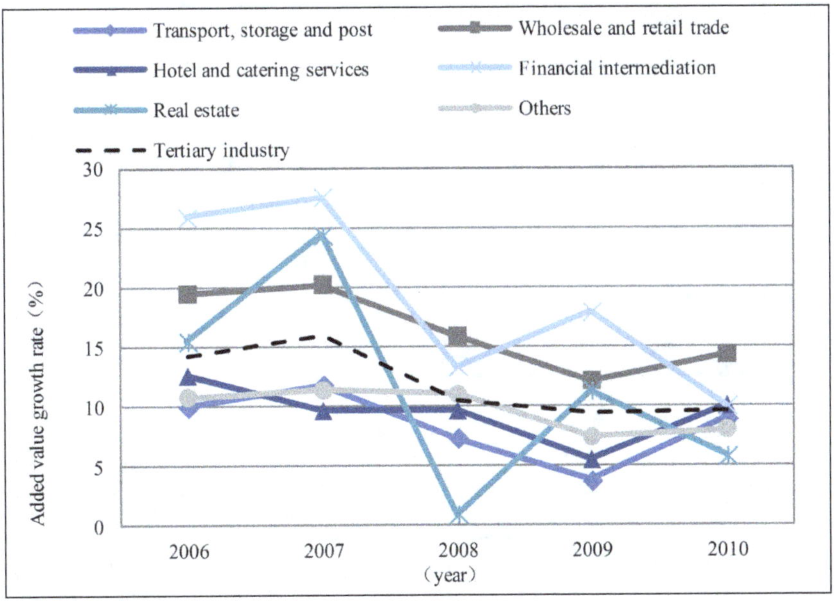

Data Source: 2010 China Statistical Yearbook and 2011 China Statistical Yearbook.

Figure 3.9. The added value growth rate of the industries within the tertiary sector in the 11th FYP period.

that pushed the growth of the tertiary sector. Although the growth rate slowed down due to the financial crisis in 2008, the financial as well as the wholesale and retail industries still experienced stronger growth than the tertiary sector as a whole. Under the "Four trillion economic stimulus package" of 2010, the growth of the transportation, storage, and postal industries with relatively high energy intensity accelerated (Figure 3.9). Compared to 2005, the industries with relatively low energy intensity in the tertiary sector developed rapidly in 2010, thereby causing the proportional added value of the transportation, storage, and postal industries to decline, while energy intensity per unit of GDP dropped by 0.98%, saving 32 Mtce.

3.1.3 *Product Structure Optimization Resulted in Outstanding Contributions to Energy Saving*

Product structure adjustment was the most outstanding contribution to structural energy savings in the 11th FYP period and by estimation has saved 117 Mtce in energy, accounting for 18.5% of China's total energy savings (Figure 3.10).

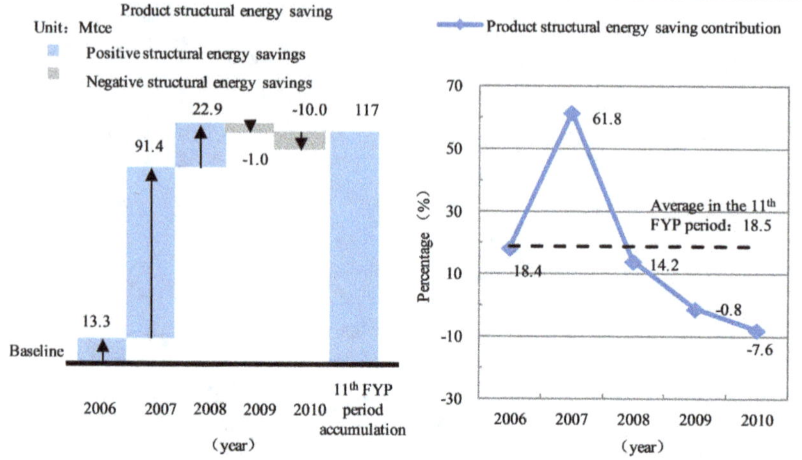

(a) Product Structural Energy Saving

(b) Energy Saving Contribution of Product Structure Adjustment

Data Source: Sampled from 2006–2011 China Statistical Yearbook and 2009–2010 China Statistical Yearbook, researcher Wang Qingyi from China Coal Information Institute compiled and provided the data of output and energy intensity of 13 kinds of products.

Note: The calculation method is shown in Appendix 3.1.

Figure 3.10. Product structural energy savings and its contribution in the 11th FYP period.

During the 11th FYP period, energy-intensive products accounted for 44.5% of industrial energy consumption; 17.7% coming from crude steel; 10.4% from cement, 4.1% from synthetic ammonia, 2.8% from electrolytic aluminum, and 2.2% from calcium carbide. The production of energy-intensive products still showed a rising trajectory (Figure 3.11).

At the beginning of the 11th FYP period, under the momentum of the rapid development of the 10th FYP period, output of energy-intensive products still maintained a relatively high growth rate. In 2006, the output of calcium carbide grew by 30.7% and crude steel and electrolytic aluminum outputs grew by about 20%. Moreover, in 2007, the output growth rates of some products still showed an upward trend: for example, output growth of electrolytic aluminum was 33.1%.

The Chinese government began to devote more efforts to eliminate obsolete production capacities in the second half of 2007. In the same year, the output growth rate of calcium carbide, cement, plate glass, and synthetic ammonia dropped. In 2008, policies to eliminate obsolete production capacities, as well as the financial crisis, caused a steep drop in the output growth of energy-intensive products, while some industries (such

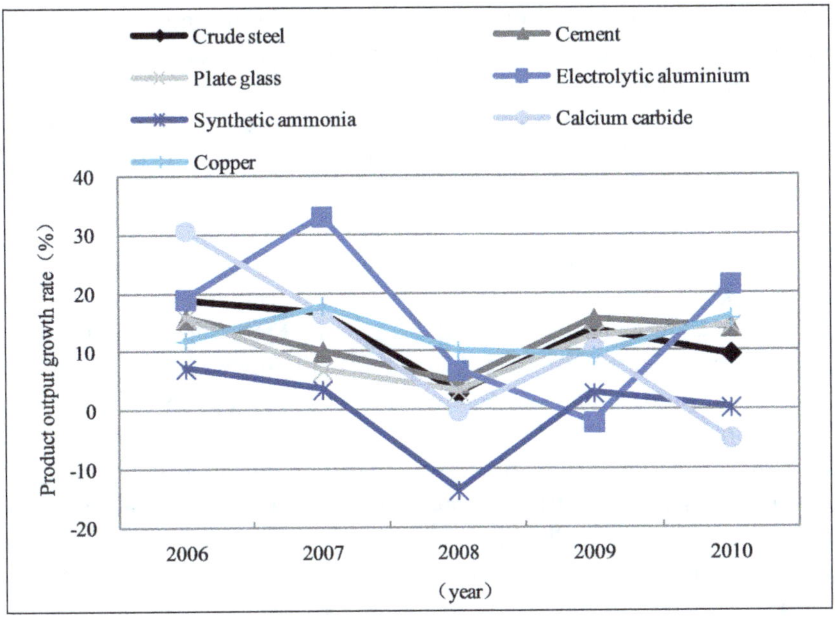

Data Source: Output of Products Sampled from 2006–2011 China Statistical Yearbook.

Figure 3.11. The output growth rate of several high energy-consuming products in 11th FYP period.

as calcium carbide and synthetic ammonia) even experienced negative growth. During this period, the Chinese government's structural adjustment policies aimed at suppressing the expansion of energy-intensive products achieved significant results.

The "Four trillion RMB economic stimulus package" played a significant role as a moving force behind the production output of building materials, which stimulated output growth for cement, crude steel, and electrolytic aluminum.

In addition to these factors, China's export volume increased. From 2005 to 2007, the net exports of goods and services contributed 19.1% on average in three major regards (final consumption expenditure, capital formation, and imports and exports of goods and services).[10] Due to the financial crisis, this contribution was merely 9% in 2008 and 2010, and it was negative in 2009. However, in such difficult times, China's export of energy-intensive products still maintained a high volume. From

[10] Data Source: 2011 China Statistical Yearbook.

2006 to 2007, exports of copper, aluminum, and other non-ferrous metals accounted for about 25% of domestic output; exports of steel products accounted for about 10% of total output of crude steel.[11] The share of energy-intensive products in overall exports showed a downward trend for which strict energy saving policies, restricted export refunds for energy-intensive products, and the effects of the financial crisis were mainly responsible. The share of steel products in exports fell to about 5% in 2009, but rebounded again in 2010.

Due to the large volume of exports, China consumed vast amounts of energy, causing substantial carbon emissions. Embedded carbon emissions of China's net exports began to grow rapidly since 2004, contributing to 29.3% of China's total carbon emissions in 2006[12] and 29.1% in 2007.[13] Thus, carbon emissions from net exports accounted for about 30% of China's total carbon emissions. Even after the financial crisis, embedded carbon emissions from China's net exports are estimated to account for more than 15% of overall emissions.

While energy-intensive product manufacturing continued to grow in the 11th FYP period, energy saving in the product structure adjustment made progress, mainly due to the improvement of products' added value and the gradual expansion of the share of high value-added products.[14] The extension of processing chains, deep processing, and the rising share of high-value-added products were important reflections of the optimization of the industrial product structure. Deep processing of raw materials into a variety of products expanded the use value of products and improved technological contents, thereby increasing added value of products and reducing energy intensity and carbon emissions per unit of product added value.

---

[11] Aluminum includes unwrought aluminum and aluminum alloys; copper includes unwrought copper and copper alloys. Data source: 2006–2010 China Statistical Yearbook and 2010 Statistical Communique of National Economic and Social Development.

[12] Qi Ye, Li Huimin, Xu Ming 齐晔, 李惠民, 徐明, "Zhongguo jinchukou maoyi zhongde yinhantan gusuan 中国进出口贸易中的隐含碳估算 [Estimate On Embedded Carbon Emissions Of China's Export Trade]," *Zhongguo renkou, maoyi yu huanjing* 18, no. 3 (2008): 8–13.

[13] Liu Hongguang, Liu Weidong, Tang Zhipeng 刘红光, 刘卫东, 唐志鹏, "Zhongguo chanye nengyuan xiaofei tan paifang jiegou ji qi jianpai minganxingfenxi 中国产业能源消费碳排放结构及其减排敏感性分析 [Analysis of the Structure of the Chinese Industry's Energy Consumption and Carbon Emissions, and the Sensitivities of Emissions Reduction]," *Dili kexue jinzhan* 29, no.6 (2010): 670–676.

[14] High value-added product refers to products with high "input-output," its technical contents and cultural values are richer than ordinary products, therefore the market appreciation is large and the profit is high.

Table 3.3. High added value of major energy-consuming products in the 11th FYP period.

Industry	Index describing added value	2006	2010
Iron and Steel	Ratio of steel plates and strips (%)	41.78	45.35
Building Materials	Deep processing rate of glass (%)	25	36.8
	Proportion of new-type wall materials (%)	46.1	58
Chemicals	Proportion of high-concentration urea (%)	60.0	66.5
	Proportion of low-concentration ammonium bicarbonate (%)	21.7	11.7

Note: The ratio of steel plates and strips refers to the proportion of plates and strips in total output of steel products.

Source: China Iron and Steel Industry Association, "2006 nian 12 yue quanguo gangcai fen pinzhong chanliang 2006 年 12 月全国钢材分品种产量 [Dec. 2006 Steel Products Vary in Output Capacity]," *Zhongguo gangtieye* 1 (2007a): 45; Xinhua, "Tuiguang xinxing qiangti cailiao 20 nian jiancheng jieneng jianzhu 11 yi pingfangmi 推广新型墙体材料 20 年 建成节能建筑 11 亿平方米 [20 Years of Promoting new-style Building Material Has Created 1.1 billion Square Meters of Energy-Saving Building Space]" (2007-09-24), URL: http://www.gov.cn/jrzg/2007-09/24/content_760239.htm [accessed: 2011-08-24]; Zhang Baiheng, Zhang Jingtao 张佰恒, 张景焘, "Woguo boli hangye jinqi fazhan huangjing ji qushi zoushi 我国玻璃行业近期发展环境及趋势走势 [Recent Environmental Developments and Trends in China's Glass Industry]," *Zhongguo jiancai* 4 (2008): 47–51; China Fertilizer Industry Association, "'Shiyiwu' nongyong huaxuepin fazhan tufeimengjin '十一五'农用化学品发展突飞猛进 [Use of Agricultural Chemicals Grows by Leaps and Bounds During the 11th FYP]" (2011-01-11), URL: http://www.baiinfo.com/article/default/4273/4923112.html.

Products in major energy-intensive industries experienced significant optimization in the 11th FYP period (Table 3.3): output of rolled steel plates and strips that had been in relatively short supply for a long time grew faster than that of long products and the proportion of rolled strips increased progressively; the application of new types of glass such as energy-saving glass and solar glass expanded rapidly and the deep-processing rate of glass improved; production of new types of wall materials with high energy saving performance increased accordingly; the proportion of high-concentration nitrogenous fertilizers that are derivative products of synthetic ammonia in total output of nitrogenous fertilizers increased constantly.

In the 11th FYP period, China's product structure adjustment achieved certain successes, but many indicators are still lagging behind when compared with international levels. The ratio of steel plates and strips in major steel-producing countries reached over 60% already in 2005, 71.35% in the United States, 64.27% in Japan and 63.59% in

Germany.[15] The global deep processing rate of glass was about 55% on average in 2009, compared with 65% to 85% in developed countries, and the value added to processed glass in relation to raw glass in developed countries was five times higher, compared with only 2.5 times currently in China. Therefore, product structure adjustment has great potential for further energy savings by improving added value through extending products processing chains and improving deep-processing rates.

3.1.4 *Impact of Multiple Factors on Structure Adjustment's Contribution to Energy Saving*

In the 11th FYP period, energy saving contributions from sectoral, industry, and product structure adjustments was affected by a series of complex factors. The result of 23% in energy savings was achieved by careful consideration of all of these factors.

3.1.4.1 *The Inertia Effect of Heavy Chemical and Industrial Development in the 10th FYP Period was Significant*

The development objective of both the 10th and the 11th FYP was to achieve structural adjustments while simultaneously ensuring GDP growth. China is in the process of rapid industrialization, the main driving force of its economic growth is the industrial sector. In the first two years of the 11th FYP period, GDP growth, growth of the industrial sector, and of energy-intensive industries and products all maintained the upwards trajectory that had characterized China's economic development trend at the end of the 10th FYP.

3.1.4.2 *China's Rapid Urbanization Process Strongly Contributed to Rising Demand in the 11th FYP Period*

China's urbanization rate increased from 43% in 2005 to 49.95% in 2010, an increase of seven points over five years. Developed countries with urbanization rates of more than 70% such as Canada, the United States, Germany, Japan, and South Korea only displayed an increase of 1% in urbanization rate in the 11th FYP period; Some developing countries with urbanization rates of about 30%, such as Thailand and India noted increases of 1%; the urbanization rate in South Africa increased by 2%, reaching 62% at the end of 2010. China's rapid urbanization in the 11th

---

[15] Weng Yuqing 翁宇庆, "Zuijin wunian (2002–2006) woguo gangcai pinzhongde bianhua qushi zuijin wunian (2002–2006) 我国钢材品种的变化趋势最近五年 [Changing Trends in China's Iron and Steel Products Assortment Over the Last 5 Years]," *Zhongguo gangtie nianhui* (2007-11-15). URL: http://wenku.baidu.com/view/22fe5434eefdc8d376ee32d9.html [accessed: 2011-08-09].

FYP period entailed large-scale infrastructure construction: from 2005 to 2010, investments in urban infrastructure totaled 22.0648 trillion RMB, an average annual increase of 21.8%.[16] China consumed large amounts of iron and steel, cement, and other building materials, the high demand for which led to the growth of energy-intensive industries.

3.1.4.3 *High Exports Volume in the 11th FYP Period Brought Large Volumes of Embedded Energy Consumption*

At the beginning of the 11th FYP period, embedded energy consumption and embedded carbon emissions in exports accounted for nearly 30% of China's total energy consumption and carbon emissions. Although the financial crisis had a negative impact on international trade, in 2009, with trade deficit,[17] China's export of some energy-intensive products still accounted for more than 10%.

The three factors mentioned above meant that China remained in a developmental pattern dominated by heavy industry and energy-intensive consumption. This had far-reaching consequences for China's economic adjustment. Nevertheless, the Chinese government continued to do its utmost in adjusting and optimizing China's economic structure and adopted a series of unique and innovative policies and systems.

3.1.4.4 *The Brake Effect of China's Structural Adjustment Policies was Significant in the 11th FYP*

From 2006 to 2007, little progress was made in China's structural adjustment, and energy-intensive industries showed a trend towards expansion. As a countermeasure, the National Development and Reform Commission issued the *Notification on Publishing and Issuing Comprehensive Industrial Plans on Energy Conservation and Emission Reduction (Guo Fa [2007] No. 15)*, in order to advance a series of policies; this included a loan restriction for "energy-intensive and high-pollution" industries by eliminating outdated production capacities, a strict control of markets access, revocation of export refunds of high-energy-consuming products, and the implementation of a differentiated and punitive pricing system for electricity. These policies had a considerable braking effect that effectively

[16] National Bureau of Statistics of China, "'Shiyuíwu' jingji shehui fazhan chengjiu xilie baogao zhi shiliu '十一五' 经济社会发展成就系列报告之十六 [Report No. 16 on Economic and Social Developments and Successes During the 11th FYP]" (2011-03-11), URL: http://www.stats.gov.cn/tjfx/ztfx/sywcj/t20110311_402709772.htm [accessed: 2011-07-05].

[17] Trade deficit means that the total export value of a country is lower than total import value in a given period (such as a year, half a year, a quarter and a month).

slowed down the expansion of energy-intensive industries and products. The elimination of outdated production capacities saved 110 Mtce of energy and reduced emissions by 400 $MtCO_2$ (see Chapter 4).

The development of China's economy was relatively volatile in the 11th FYP period. The financial crisis caused economic growth to decline rapidly. In response, China adopted an interim economic policy, the "Four trillion RMB economic stimulus package," which had an immense impact on the structural adjustment during the 11th FYP period.

3.1.4.5 *Negative Impacts of the "4 Trillion RMB Economic Stimulus Package" on Structural Adjustments in the 11th FYP Period*

Of the overall four trillion RMB allocated, 81% flowed into the construction of infrastructure and residential housing projects, causing a rapid growth of energy-intensive industries and products. Other than a growth in production capacity and product value, it may also have contributed to overcapacities, resulting in permanent precipitation of social costs, and with negative impacts on energy saving in the 11th FYP period.

3.2 Scale of Production and Organizational Structure

The establishment of binding energy-saving targets, along with the policies to eliminate outdated production capacities and the energy savings accountability system put forward by the central government in the 11th FYP resulted in great external pressure on Chinese enterprises, especially energy-intensive enterprises. Under these conditions, enterprises not only must meet energy-saving targets assigned by the government, but must also cope with intense competition. Enterprises also moved to improve their production efficiency and reduce energy intensity. In addition, replacement and upgrades of internal equipment, expansion of production scale, conducting of mergers and acquisitions, and collaboration with big enterprises were all strategies by which enterprises attempted to proactively shape their own development. Two aspects stand out in this regard: The trend towards upsizing production equipment, and the increasing concentration of industries.

3.2.1 *Scale of Production: A Clear Trend towards Large-scale Equipment*

With the economy growing fast during the 11th FYP period, China's industrial structure was optimized and upgraded in order to accomplish its goals in terms of energy savings and emissions reduction. Production capacities of major energy-intensive industries have significantly improved; the introduction of large-scale and efficient equipment for iron and steel, electricity, cement, and glass production was strongly promoted and implemented (Table 3.4).

Table 3.4. Large-scale trends of the production equipment of major energy-consuming industries in the 11th FYP period.

Industry	Main production equipment indicators		2005	2006	2007	2008	2009	2010
Electricity	Proportion of above 300,000 kW thermal power units (%)		43.4	50.4	61.5	65.2	69.4	71.3
	Proportion of above 600,000 kW thermal power units (%)		11.7	—	25.6	31.3	—	36.4
Iron and steel	Proportion of above 1000 m³ blast furnaces (%)		20.6	20.8	24.8	30.0	—	39
	Proportion of number of 300–999 m³ furnaces (%)		59.5	65.7	64.0	64.7	—	—
	Proportion of number of 300 m³ furnaces (%)		19.9	13.5	11.2	5.3		
Cement	Proportion of 4,000 ton and above production lines in full-year operation	By number (%)	—	—	17.2	26.4	28.4	35.3
		By capacity (%)	—	—	36.8	49.0	—	56.9
	Proportion of newly-added 4,000 ton and above production lines	By number (%)	—	—	37.5	56.1	46.0	57.3
		By capacity (%)	—	—	55.9	73.0	64.1	71.7
Glass	Proportion of 600 tons and above melting furnaces by capacity (%)		32	—	—	—	54.5	60
	Proportion of 400 tons and below melting furnaces by capacity (%)		46	—	—	—	8.1	—
	Average production capacity of float glass per line (t/d)		—	460	480	500	510	521

Note: The proportion of full-year operating 4,000 ton and above production lines in 2007 was that of 5,000 ton and above production lines.[18]

[18] Sources: Power data compiled by Wu Yin 吴吟, "Woguo nengyuan xingshi he 'shiyiwu' nengyuan fazhan silu 我国能源形势和 '十二五' 能源发展思路 [Reflections on China's

Energy Situation and Development During the 11th FYP]" (2011–4-24), http://www.docin.com/p-213335728.html [accessed: 2011-07-19]; Xie Jun 谢军, "Dianli hangye: Guanzhu zichan zhuru 电力行业: 关注资产注入 推荐 4 只龙头股 [Electricity Industry: Pay Close Attention to Assets Input]" (2007-12-21), http://finance.eastmoney.com/071221,748944.html [accessed: 2011-07-19], China Electricity Council, "Quanwei xinxi: Zhongdianlian fabu 2007 nian dianli gongye tongji nianbao shuju 权威信息:中电联发布2007年电力工业统计数据 [Authoritative Information: CEC releases 2007 Electric Power Industry Statistical Yearbook Data]" (2008-09-03), http://www.cpnn.com.cn/dljspd/djxw/200809/t20080903_256392.html [accessed: 2011-07-19]; China Electricity Council, "Zhongdianlian fabu 2008 nian dianli gongye tongji nianbao shuju 中电联发布 2008 年电力工业统计年报数据 [CEC releases 2008 Electric Power Industry Statistical Yearbook Data]" (2007-07-08), http://www.chinapower.com.cn/newsarticle/1097/new1097570.asp [accessed: 2011-07-19]; China Electricity Council, "Zhongdianlian fabu 2009 nian dianli gongye tongji nianbao shuju 中电联发布 2009 年全国电力年报数据 [CEC releases 2009 Electric Power Industry Statistical Yearbook Data]" (2010-07-15), http://qol.qdc.com.cn/xwkd_out.asp?code=19488 [accessed: 2011-07-19].

Iron and steel data are based on data compiled by 2009 China Iron and Steel Statistics Yearbook and Shanghai Coal Trade Net, "Dalian chanpinjiaoyisuo jiaotan xiaoge zhiliang biaozhun sheji shuoming 大连商品交易所焦炭交割质量标准设计说明" (2011-02-23), http://www.shcce.com/Futures/showinfo_20581_10.html [accessed: 2011-09-25].

Cement data: Kong Xiangzhong 孔祥忠, "2007 nian Zhongguo shuini fazhan yu 2008 nian shichang zhanwang 2007 年中国水泥发展与 2008 年市场展望 [2007 Development of Cement in China, and Survey for 2008]" (2008-03-27), http://www.dcement.com/Article/200803/64062.html [accessed: 2011-07-26]; SNSQW, "Zhongguo shuini xiehui gongbu 2007 nian xinxing ganfa shuini shengchanxian touchan shuju 中国水泥协会公布 2007 年新型干法水泥生产线投产数据 [Cement Association Publishes Data on 2007 New-Type Cement Production Line]" (2008-2-22), http://www.snsqw.com/news/ShowArticle.asp?ArticleID=32126 [accessed: 2011-07-26]; China Building Materials Federation, "Xinxing ganfa de fazhan cujinli shuini chanye jiegou youhua. 新型干法的发展促进了水泥产业结构优化 [] (2009-08-05), http://www.jiancai365.cn/news/27936.htm [accessed: 2011-07-26]; China Cement Association, "Zhongguo shuini xiehui gongbu 2009 nian quanguo xinxing ganfa shuini shengchanxian shuju 中国水泥协会公布 2009 年全国新型干法水泥生产线数据 [China Cement Association Announces 2009 National New Dry Cement Production Line Data]" (2010-02-02), http://www.ccc.com.cn/newsshow.asp?id=640 [accessed: 2011-07-26]; China Building Materials Federation, Information Department, "Shuini hangye jiegou tiaozheng jieguo xianzhu 水泥行业结构调整成果显著 [Remarkable Outcomes in the Structural Adjustment of China's Cement Industry]," *Zhongguo Jiancai* 3 (2011): 51; Zhou Hongjin 周鸿锦, "'Shiyiwu' shiqi shuini gongye de bianhua tezheng '十一五' 时期水泥工业的变化特征 [Changes in Characteristics of the Cement Industry During the 11th FYP]," *Zhongguo shuini* 4 (2011): 16–19.

Glass data: Han Yongqi 韩永奇, "Woguo pingbanboli gongye mianlin de xin wenti ji qi duice xuanze 我国平板玻璃工业面临的新问题及其对策选择 [New Problems Facing China's Glass Place Industry, and Measures to Address Them]." *Jiancai fazhan lingxiang* 5 (2007): 24–27; China Architectural and Industrial Glass Association, "Gaigekaifang tuidongle woguo boli gongye kuaisu fazhan 改革开放推动了我国玻璃工业快速发展 [The Reform and Opening Up Policies Promoted the Rapid Development of China's Glass Industry]" (2008-12-24), http://www.cbminfo.com/tabid/234/InfoID/284177/Default.aspx; CBMD, "Woguo boli gongye fazhan zhong zhide guanzhu de wenti 我国玻璃工业发展中值得关注的问题 [Issues Worth Paying Attention to in the Development of China's Glass Industry]" (2009-07-09), http://ccn.mofcom.gov.cn/spbg/show.php?id=9482 [accessed: 2011-07-10]; Peng Shou 彭寿, "Zhongguo boli gongye jiegou tiaozheng de sikao 中国玻璃工业结构调整的思考 [Reflections on the Structural Adjustment of China's Glass Industry]," *Zhongguo qiye bao* 4 (2010-03-12); CBMD, "Woguo boli gongye fazhan zhong zhide guanzhu de wenti 我国玻璃工业发展中值得关注的问题 [Issues Worth Paying

(1) Electricity industry: The policy of "adopting large thermal power generator units and shutting down small thermal power generator units"[19] achieved great success and created direct benefits in optimizing the structure of thermal power units, while at the same time reducing the average power coal consumption in China. In the 11th FYP period, thermal power units with a total capacity of 76.825 million kilowatts were shut down across the country, so that the proportion of thermal power units with an annual capacity of 60 kilowatts and more has more than doubled. The average power coal consumption fell from 343 gce/kWh in 2005 to 312 gce/kWh in 2010.[20]

(2) Iron and steel industry: Elimination of outdated steel production capacity reached 178.55 Mt in the 11th FYP period, which promoted major upgrades in technical equipment in the iron and steel industry and significantly optimized scale structure. Energy consumption per ton of steel produced accounted for over 60% of overall energy consumption,[21] and large-scale blast furnaces can effectively reduce energy consumption, increase production efficiency, and save costs. In the 11th FYP period, the share of blast furnaces with a volume of more than 1,000 m³ increased by about 19% and their production capacity accounted for 62% of total capacity.[22] Energy consumption in the iron making process measured per

---

Attention to in the Development of China's Glass Industry]" (2009-07-09), http://ccn.mofcom.gov.cn/spbg/show.php?id=9482. [accessed: 2011-07-10]; CBMD, "Zhongguo fufa boli gongye fazhan taishi fenxi 中国浮法玻璃工业发展态势浅析 [Analysis of Developmental Trends in China's Float Glass Industry]" (2011b-05-19), http://www.glass.org.cn/hyzx/201105/19/613.html [accessed: 2011-08-03]; Xu Meijun 徐美君, "Zhongguo pingban boli channeng xianzhuang pouxi 中国平板玻璃产能现状剖析 [Analysis of the Present State of China's Plate Glass Production Capacity]," *Boli yu tangci* 38, no. 3 (2010): 29–33.

19 "Adopting large thermal power generator units and shutting down small thermal power generator units" means the construction of large-capacity, high-parameter and high-efficient thermal power units to eliminate and replace old, small-capacity, low-parameter and inefficient ones. It requires gradual termination of: conventional thermal power units with single capacity below 50,000 kilowatts, conventional thermal power units with single capacity below 100,000 kilowatts and running over 20 years, and various power units with single capacity below 200,000 kilowatts and have expired the designed service life, including coal-fired (oil) units, within the coverage of large grid.

20 China Electricity Council, "Dianli hangye 2010 nian fazhan qingkuang zongshu 电力行业 2010 年发展情况综述 [Summary of the Development of the Electricity Industry in 2010]" (2011-6-27), http://tj.cec.org.cn/niandufazhanbaogao/2011-06-27/58873.html [accessed: 2011-8-16].

21 Chen Guanjun 陈冠军, "Liantie xitong nenghao fenxi ji jieneng 炼铁系统能耗分析及节能 [Analysis of Power Consumption in Iron Smelting Systems and Energy Saving]," *Yejin nengyuan* 29, no. 4 (2010): 11–14.

22 Shanghai Coal Trade Net, "Dalian chanpinjiaoyisuo jiaotan xiaoge zhiliang biaozhun sheji shuoming 大连商品交易所焦炭交割质量标准设计说明" (2011-02-23), http://www.shcce.com/Futures/showinfo_20581_10.html [accessed: 2011-09-25].

ton, per capita output, and production costs per ton were 84%, 3.7 times and 96% respectively, compared with blast furnaces with a volume of 300 m^3 or less.[23]

(3) Cement industry: Production of new-type dry-process cement was vigorously promoted in the 11th FYP period. In 2010, the output of new-type dry-process cement accounted for 80% of total cement production. The portion of production lines with a daily production capacity of 4,000 tons and more showed an upward trend. Compared to smaller production lines with a daily capacity of 2,000 tons or less, large-scale production lines (over 4,000 tons daily capacity) have low energy consumption, their overall electricity and clinker burning heat consumption being only 90.6% and 84.7% that of the former.[24] Optimization of production structure has reduced overall energy consumption per unit/product with new-type dry-process technology.

(4) Glass industry: Float glass production was strongly promoted in the 11th FYP period. In 2010, output of float glass accounted for 86%. In the 11th FYP period, the proportion of production capacity of melting furnaces with a daily capacity of 600 tons and more increased by 28%, replacing melting furnaces with a daily capacity of 400 tons and less as the main furnaces used in float glass production. Melting furnaces with a daily capacity of 700 tons and more constituted 34% of overall production capacity.[25] Large-scale float glass production has reduced heat consumption used in the melting process. In 2009, the average heat consumption of melting furnaces with a daily capacity of 600 tons and more reached 6,183 kJ per kg, 15% less than that of smaller melting furnaces with a daily capacity of 500 tons and 22% less than that of melting furnaces with a daily capacity of 400 tons.[26]

---

[23] Liu Wenquan 刘文权, "Taotai luohou, jieneng jianpai, zengqiang liantie liangang jingzhengli 淘汰落后节能减排增强炼铁竞争力 [Elimination of Outdated Capacities, Energy Saving and Emissions Reduction Will Increase Competitiveness of Iron and Steel Sector]" (2011-08-11), http://www.steelplanning.cn/ShowArticle.asp?ArticleID=193 [accessed: 2011-08-11].

[24] Wang Hongbing 王洪兵, "Xinxing ganfa shuini jieneng jianghao jishu 新型干法水泥节能降耗技术" (2007-11-05), http://www.dcement.com/snjishu/ShowArticle.asp?ArticleID=60358 [accessed: 2011-07-23].

[25] CBMD, "Zhongguo fufa boli gongye fazhan taishi fenxi 中国浮法玻璃工业发展态势浅析 [Analysis of Developmental Trends in China's Float Glass Industry]" (2011b-05-19), http://www.glass.org.cn/hyzx/201105/19/613.html [accessed: 2011-08-03].

[26] CBMD, 'Shiyiwu' woguo boli gongye baote jiankang fazhan shitou '十一五'我国玻璃工业保持健康发展势头 [China's Glass Industry Maintained Healthy Development Momentum During the 11th FYP]" (2011a-01-28), http://www.glass.org.cn/hyzx/201101/28/252.html [accessed: 2011-08-02].

3.2.2 *Organizational Structure: Industry Concentration Gathers Rapidly*

Large business groups not only have large production capacities, but also advanced management and internal specialization while utilizing their resources and energy relatively efficiently. They represent the advanced

Table 3.5. The quantity of large enterprises and products output ratios of major energy-consuming industries in the 11th FYP period.

		2005		2010	
Industry	Production Capacity	Quantity of enterprises	Proportion by output (%)	Proportion by number	Proportion by output (%)
Coal	10 Mt and above	31	42.3	50	58.0
	50 Mt and above	—	—	13	41.0
Coal	5 Mt and above	19	51.5	27	73.0
	10 Mt and above	8	29.8	13	54.3
Cement	10 Mt and above	10	15.0	22	45.8
Synthetic ammonia	180,000 tons and above	73(a)	44.4(a)	80	52.0

Note: (a) is based on data in 2007.

Sources: China.com, "2006 nian meitan hangye fazhan tedian ji wenti 2006 年煤炭行业发展特点及问题 [Characteristics and Problems of the Development of the Coal Industry in 2006]" (2007-05-24), http://www.china.com.cn/economic/zhuanti/2007nyfz/2007-05/24/content_8302557.htm [accessed: 2011-09-13]; Xinhua, "Tuiguang xinxing qiangti cailiao 20 nian jiancheng jieneng jianzhu 11 yi pingfangmi 推广新型墙体材料 20 年 建成节能建筑 11 亿平方米 [20 Years of Promoting new-style Building Material Has Created 1.1 billion Square Meters of Energy-Saving Building Space]" (2007-09-24), http://www.gov.cn/jrzg/2007-09/24/content_760239.htm [accessed: 2011-08-24]; National Development and Reform Commission, "Chanye jizhongdu dafu tisheng. Gangtie chanye 'shiyiwu'fazhan zhuyao chengji zhisan 产业集中度大幅提升. 钢铁产业 "十一五" 发展主要成绩之三 [Degree of Industry Concentration Greatly Increases. Three Important Achievements of the 11th FYP in the Iron and Steel Industry]" (2011-08-05), http://gys.ndrc.gov.cn/115125/t20110830_431503.htm [accessed: 2011-09-23]; Ministry of Industry and Information Technology of the PRC, "2009 nian quanguo xin jiancheng touchan daxing shuini shengchanxian 176 tiao 2009 年全国新建成投产大型水泥生产线 176 条 [176 New Large-scale Cement Production Lines Put Into Operation in 2009]" (2010-03-17), http://www.miit.gov.cn/n11293472/n11295125/n11299605/13080335.html [accessed: 2011-07-26]; Visteel.com, "2010 nian Zhongguo gangtie chanye zuzhi jiehou zhuangkuang fenxi 2010 年中国钢铁产业组织结构状况分析 [2010 Analysis of the Organizational Structure of China's Iron and Steel Industry]" (2011-07-01), http://www.visteel.com/modules/article/view.article.php/511 [accessed: 2011-08-09]; ZGNCW 2008-12-24. 氮肥工业——走向自给、自主、自强的历程. http://www.ampcn.com/news/detail/46593.asp [2011-07-23]; China Building Materials Federation, Information Department, "Shuini hangye jiegou tiaozheng jieguo xianzhu 水泥行业结构调整成果显著 [Remarkable Outcomes in the Structural Adjustment of China's Cement Industry]," *Zhongguo Jiancai* 3 (2011): 51. Han Hongmei 韩红梅, "Woguo hecheng'an gongye fazhan pingshu 我国合成氨工业进展评述 [Comment on the Development of China's Synthetic Ammonia Industry]," *Huaxue gongye* 28, no. 9 (2010): 1–5.

productivity of the manufacturing industry, while production lines with high energy efficiency also are concentrated mostly in large enterprise groups. In the 11th FYP period, the proportion of large enterprise groups in the coal, iron and steel, cement, and chemical industries increased both in terms of quantity and output (Table 3.5).

In the 11th FYP period, the Chinese government strongly promoted organizational restructuring by encouraging company mergers and acquisitions; it has also fostered collaboration among big enterprises and integrated operation of upstream and downstream businesses. Thus, the concentration rate within several industries increased.

(1) Coal industry: the government has engaged in a reorganization of small coal mine businesses, shutting down 9,535 small coal mines and eliminating outdated production capacities of 530 Mt which led to an improvement in the concentration ratio within the coal industry (Figure 3.12 (a)). The proportional output of the top four, resp. top eight, coal enterprises increased by 3.7 percentage points and 6.4 percentage points.
(2) Iron and steel industry: Since 2005, the rate of mergers and acquisitions in the iron and steel industry has accelerated due to industrial policies, market forces, and local government encouragement (Figure 3.12 (b)). The proportional output of crude steel by the nation's top-ten iron and steel enterprises (groups) grew from 34.8% in 2006 to 48.4% in 2010.
(3) Cement industry: Proportional cement output from China's top ten cement enterprises (groups) grew from 16.7% in 2005 to 30% in 2010 (Figure 3.12 (c)).
(4) Glass industry: The industry's concentration ratio of top-ten glass enterprises (groups) was 45.8% before 2005 (*China Securities Journal*, 2008), and output of China's top ten plate glass enterprises accounted for 57% of national output in 2010.[27]

Moreover, the number of small and medium-sized enterprises (SMEs) in energy-intensive industries is large and their output accounted for 40% to 50% in total. Obsolete technology, equipment, and management, as well as outdated manufacturing techniques and production processes have caused the energy consumption of SMEs to be 30% to 60% higher than

[27] Cao Kaihu 曹开虎, Channeng guosheng yulianyulie 产能过剩愈演愈烈 [Excess Production Capacity Grows More and More Severe] *Diyi Caijing Ribao* B02 (2011-05-10).

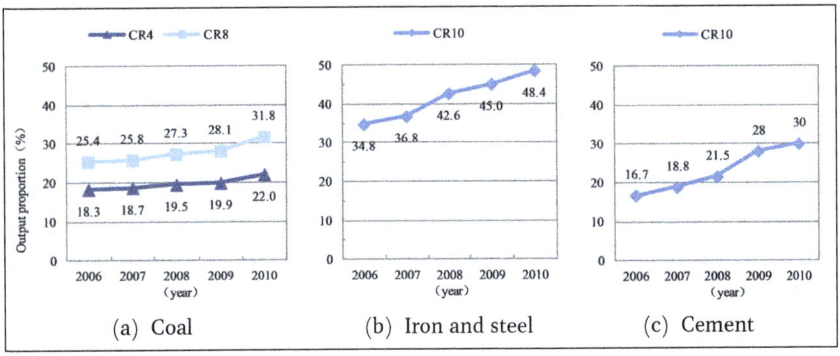

Note: CRn indicates industry concentration indexes and means the proportion of output of top N enterprises in total output.

Sources: Coal data are from: Wu Yin 吴吟, "Woguo nengyuan xingshi he 'shiyiwu' nengyuan fazhan silu 我国能源形势和'十二五'能源发展思路 [Reflections on China's Energy Situation and Development During the 11th FYP]" (2011-4-24), URL: http://www.docin.com/p-213335728.html [accessed: 2011-07-19].

Iron and steel data are based on: China Iron and Steel Industry Association, "2009 nian woguo gangtie hangye yunxing qingkuang fenxi 2009 年我国钢铁行业运行情况分析 [Operational Aspects of China's Iron and Steel Industry in 2009]," *Zhongguo gangtieye* 2 (2010): 15–22; China Iron and Steel Industry Association, "Zhongguo gangtie gongye xiehui 2008 nian diyici hangye xinxi fabuhui 中国钢铁工业协会 2008 年第一次行业信息发布会 [China Steel Industry Association Holds First 2008 Industry Information Conference]" (2008-01-30), URL: http://www.chinaisa.org.cn/news.php?id=2140809 [accessed: 2011-07-28]; China Iron and Steel Industry Association, "Zhongguo gangtie gongye xiehui 2011 nian diyici hangye xinxi fabuhui 中国钢铁工业协会 2011 年第一次行业信息发布会 [China Steel Industry Association Holds First 2011 Industry Information Conference]" (2011-02-24), URL: http://www.chinaisa.org.cn/news.php?id=2170517 [accessed: 2011-07-28]; China Steel Industry Association, "2006 nian gangtie hangye yunxing tezheng fenxi 2006 年钢铁行业运行特征分析(上) [Analysis of the Operational Characteristics of the Steel Industry in 2006 (Part A)]" (2007b-03-21), URL: http://www.projectbidding.cn/info/scfx/16520.html [accessed: 2011-07-28].

Cement data are based on: SNSQW, "Zhongguo shuini xiehui gongbu 2007 nian xinxing ganfa shuini shengchanxian touchan shuju 中国水泥协会公布 2007 年新型干法水泥生产线投产数据 [Cement Association Publishes Data on 2007 New-Type Cement Production Line]" (2008-2-22), URL: http://www.snsqw.com/news/ShowArticle.asp?ArticleID=32126 [accessed: 2011-07-26]; Sun Qian 孙钤, "Shuini gongye yingli nengli de zai tantao 水泥工业盈利能力的再探讨 [More Enquiries Into the Profitability of the Cement Industry]" (2009-04-03), URL: http://www.dcement.com/Article/200904/72968.html [accessed: 2011-08-02].

Figure 3.12. The industry concentration ratio changes of major energy-consuming industries in the 11th FYP period.

that of their larger competitors. That is also the main reason why some Chinese energy-intensive products consume more energy than those produced internationally on a more advanced level.

3.3 Industry Reallocation Led to the Changes of Regional Structure

The 11th FYP's national binding energy-consumption target yielded puzzling results. According to data provided by the National Development and Reform Commission, all provinces with the exception of Xinjiang completed or even over-fulfilled their respective targets, but the national target did not reach 20%.

One reason was the adjustment of energy-saving targets. According to a 2006 agreement, the national energy-saving target was 20%; while 25% were set for Shanxi and Inner Mongolia, and 30% for Jilin. However, in 2008, all targets were re-adjusted to 22% considering the large pressure on the aforesaid provinces to save energy. Based on the share of GDP and energy consumption intensity of all provinces, municipalities and autonomous regions in 2005, this adjustment led the national energy-saving target to fall to 19.6%. In the 11th FYP period, energy intensity per unit of GDP decreased by 19.06%. However, the 0.5% gap is not clearly determined.

The following section explores the reason for this gap by comparing differences of energy intensity and GDP growth rate among all provinces, municipalities, and autonomous regions.

3.3.1 *Rapid Economic Growth of Western and Northeastern Regions Slowed Down National Energy Intensity Reduction*

To better account for structural variations, this analysis divides 31 provinces, municipalities and autonomous regions into four major economic regions: east, central, west and northeast.[28] An examination of these four major economic regions' past GDP proportion (Figure 3.13) reveals that

---

[28] East China: (10 provinces and municipalities) Beijing, Tianjin, Hebei, Shanghai, Jiangsu, Zhejiang, Fujian, Shandong, Guangdong, and Hainan; Central China: (6 provinces) Shanxi, Anhui, Jiangxi, Henan, Hubei, and Hunan; West China: (12 provinces and municipalities) Inner Mongolia, Guangxi, Chongqing, Sichuan, Guizhou, Yunnan, Tibet, Shaanxi, Gansu, Qinghai, Ningxia, and Xinjiang; Northeast China (3 provinces) Liaoning, Jilin, and Heilongjiang.

Source: National Bureau of Statistics, June 13, 2011. The method to divide East, West, Central and Northeast regions: http://www.stats.gov.cn/tjzs/t20110613_402731597.htm [June 13, 2011].

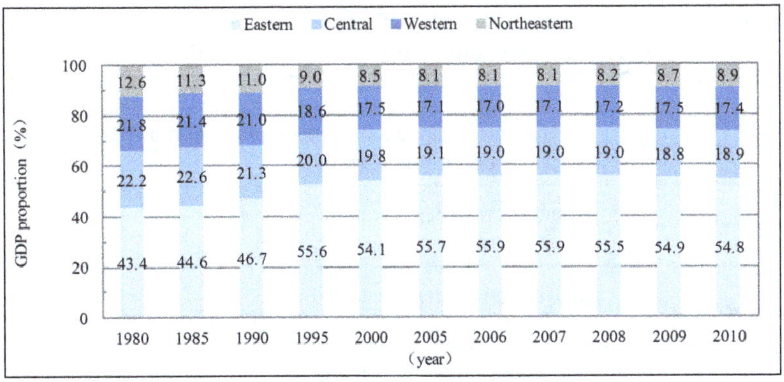

Source: China Statistical Yearbook 2010, and China Statistical Yearbook 2011.

Figure 3.13. The historical GDP proportional changes of the four major economic regions in China.

the 11th FYP was a turning point in inter-regional economic adjustment. The eastern region's share dropped one percentage point from the continuous increase it had enjoyed ever since the Reform and Opening Policy in 1978, while the proportion of the western and northeastern regions increased by 0.3 and 0.8 percentage points, reversing a downwards trend. In the 11th FYP period, average national GDP growth was 11.2%, with 11.65% for the western and 13.16% for the northeastern region, and merely 10.79% for the eastern region that previously had always been developing rapidly.

In the 11th FYP period, China's regional GDP growth pattern changed, creating a significant impact on nationwide energy consumption and energy intensity patterns. The difference in energy intensity among Chinese provinces was substantial, which was reflected in the four major economic regions. In the 11th FYP period, average energy intensity of the western, northeastern, central and eastern regions reached 1.65 tce/10^4 RMB, 1.44 tce/10^4 RMB, 1.34 tce/10^4 RMB, and 0.91 tce/10^4 RMB respectively, showing a downward trend along this line. In accordance with the computational model of energy intensity per unit of GDP affected by changes in the sectoral structure, the proportion of high-energy-intensity regions grew and the proportion of low-energy-intensity regions declined, which would mean an increase the country's energy intensity per unit of GDP (Table 3.6). In the 11th FYP period, the proportional change of the four major economic regions' GDP caused national energy intensity to grow by 0.52%, which explains why all provinces completed their respective binding energy-saving targets, but did not reach the national target of 20%. From this perspective, China should

Table 3.6. Impacts of proportional GDP changes of the four major economic regions on energy Intensity per unit of GDP.

	Structural changes of four major economic regions	Changes on national energy intensity per unit of GDP
Scenario 1	Western region grow by 1% and western region dropped by 1%	0.63%
Scenario 2	Northeastern region grew by 1% and eastern region dropped by 1%	0.45%
Actual situation of the 11th FYP period	Western region grow by 0.3%, northeastern region grew by 0.8%, Eastern region dropped by 1.0%, central region dropped by 0.1%	0.52%

Data Source: 2006–2011 China Statistical Yearbook.

consider factors of regional structure when composing national energy saving targets. Particularly the central and western regions are projected to experience rapid economic development in the future, a factor whose impact should be fully considered by the central government when setting new energy-saving targets.

3.3.2 *The Change in Regional Structure Promoted High Energy Intensity Regions to Achieve Energy-Saving Targets*

Guided by the national macro-economic policies of the 11th FYP period, the change in China's regional economic structure has led to a positive trend with regards to energy intensity per unit of GDP. This may give the intuitive impression that regional development policies have had a negative impact on energy savings. On the contrary, a recognition of the aforementioned 'western—northeastern—central—eastern regions' successive decline pattern in energy intensity is helpful to understand the role regional structural adjustments have played in terms of energy saving.

Since the Opening and Reform period, China has followed a "gradual development strategy of allowing some regions to develop early and thereby stimulating the economic development of other regions." China's initial strategy was "prioritized development of the eastern region." China has since implemented policies that favor its eastern region with regards to taxes, investments, and international trade. Although the eastern region is less rich in energy resources than the central and western regions, capital, labor, and energy are complementary according to the theory of energy substitution.[29] With rising energy prices, factors such as intensive labor and abundant capital that were characteristic for the eastern region made them more attractive for investment, which resulted in rapid economic development and annual expansion of GDP in the eastern region (Figure 3.13). Since the 1990s, the Chinese government put forward several market-oriented policies to encourage energy conservation and make the capital-rich eastern region adopt energy-saving technologies more easily. Furthermore, the east is a coastal area and capable of attracting more foreign investments while introducing up-to-date production equipment. With the promotion of energy-saving policies, more and more energy-saving technologies and management mechanisms were introduced and adopted. Due to these factors, the eastern region became the region in the country with the lowest energy intensity and shaped the regional energy intensity pattern during the 11th FYP period.

Since the beginning of the 11th FYP period, income levels in the eastern, central, western and northeastern regions have gradually widened; this regional pattern needed to be urgently addressed. Campaigns such as "Western Development," "Revitalization of the Northeastern Old Industrial Base" and "Rise of Central China" were put on the agenda, signaling the entry into a stage of "development driven by the eastern region." In the 11th FYP period, China followed the path of prioritized development and implemented favorable policies in taxes and investments for the western, northeastern and central regions. Meanwhile, governments on all levels in all regions had a strong incentive to improve the growth rate of GDP, on the basis of national policies that offered preferential policies to enterprises and attracted domestic and foreign capital with relatively low costs for leases, labor, energy, and transportation in order to strengthen their respective capabilities to adopt energy saving technologies.

On the other hand, the eastern region has experienced more than 30 years of industrialization, effectively limiting its potential and scope for development. Restricted by pressure to save energy, costs and other factors during the 11th FYP period, some enterprises in the eastern region diverted their attention to the western, northeastern and central regions. The transfer of these enterprises brought along both capital and advanced energy

29 E.R. Berndt, D.O. Wood, "Technology, Prices and the Derived Demand for Labor," *Review of Economics and Statistic* 57 (1975): 376–384; and: J.M. Griffin, P.R. Gregory, "An Intercountry Translog Model of Energy Substitution Responses," *American Economic Review* 66, no. 5 (1976): 845–857.

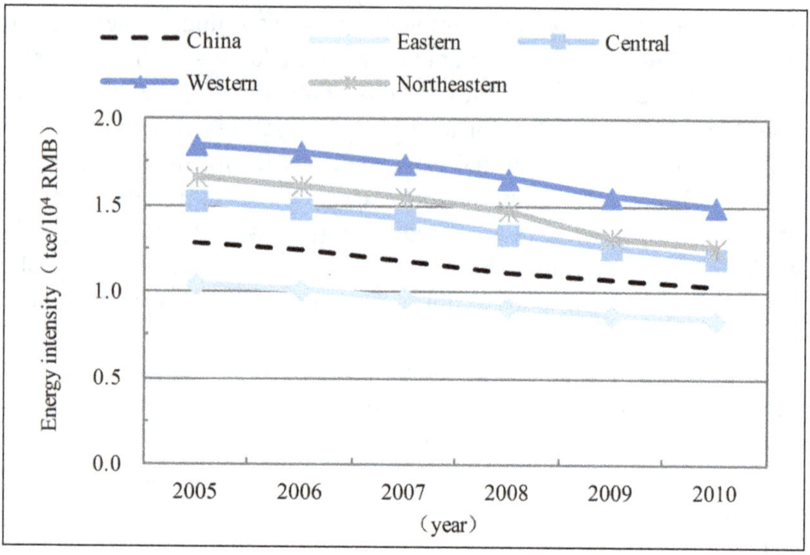

Data Source: 2010 China Statistical Yearbook and 2011 China Statistical Yearbook.

Figure 3.14. The changes in energy intensity of the four major economic regions in the 11th FYP period.

saving technologies, which have improved the capability for energy savings in these regions, and assisted the completion of energy-saving targets.

In the 11th FYP period, energy intensity the in eastern, central, western, and northeastern regions dropped by 19.6%, 21.4%, 18.9%, and 24.5% respectively, or 0.20 tce/10^4 RMB, 0.33 tce/10^4 RMB, 0.35 tce/10^4 RMB, and 0.41 tce/10^4 RMB in quantitative terms (Figure 3.14). An acceleration of the economic development of China's central and western regions and continuing of the industrial transfer from the eastern region is conducive to achieve a rapid decrease in energy intensity.

3.3.3 *Internal Sectoral Structure of the Eastern Region has Improved*

GDP growth rates of different provinces, municipalities, and autonomous regions in the four major economic regions were still relatively diverse during the 11th FYP (Figure 3.15). Among the 12 western provinces, municipalities, and autonomous regions, only the rate of Inner Mongolia, Shanxi, Guizhou, and Guangxi came close to the mid-eastern regions that experienced higher GDP growth rate than the national average; out of the "northeastern three provinces," Jilin and Liaoning ranked first and fourth in terms of GDP growth rate in China; with the exception of Fujian, Tianjin, Jiangsu, and Hainan, six of ten other provinces and municipalities

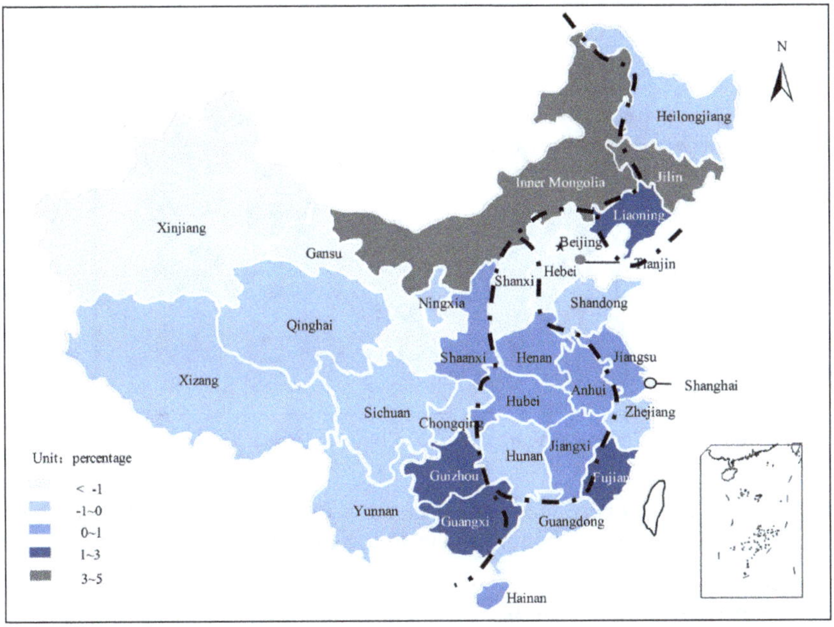

Note: The percentage range represented by each color refers to differences in GDP growth between all provinces and country as whole in the 11th FYP period.

Data Source: 2010 China Statistical Yearbook and 2011 China Statistical Yearbook.

Figure 3.15. The differences of average GDP growth rate of all provinces, autonomous regions and municipalities in the 11th FYP period.

in the eastern region experienced a slower GDP growth rate than the national average; four out of six provinces except Shanxi and Hebei in the central region noted a faster GDP growth rate than the national average, by merely one percentage point.

The existing differences indicate: China's provinces, municipalities and autonomous regions are at different stages of industrialization, and hence their tasks with regards to economic development are different. In 2010, GDP per capita in the eastern region was 2.2, 1.3, and 2.0 times that of the western, northeastern, and central regions respectively. Therefore, the western region put more emphasis on its pursuit of rapid quantitative economic growth, while the eastern region was capable of focusing on the pursuit of rapid qualitative economic growth through optimizing its sectoral structure.

Compared with 2005, the proportion of the national secondary sector increased by 1.1% in 2010, while in the central, western, and northeastern regions, it grew by 4.5%, 5.4% and 3.3% respectively. However, in the eastern region, it shrunk by 0.2 percentage points (Figure 3.16). Thus, the eastern region's sectoral structure was optimized. However, the "Four

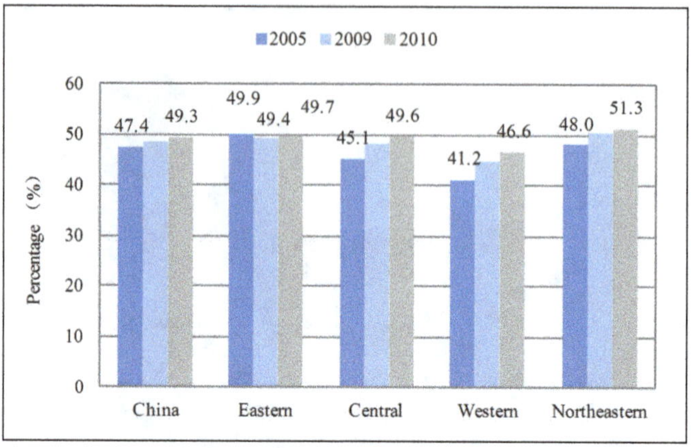

Source: 2006 China Statistical Yearbook, 2010 China Statistical Yearbook, and 2011 China Statistical Yearbook.

Figure 3.16. The proportional change of the secondary sector within the four major economic regions in the 11th FYP period.

trillion RMB economic stimulus package" of 2010 stimulated the development of energy-intensive industries and caused the proportion of the secondary sector to increase by 0.3 percentage points as compared to 2009.

In the 11th FYP period, the eastern region has achieved a certain stage of industrialization characterized by high-tech industries and a modern service industry while the extensive development of heavy industries has slowed down. In contrast, the western region is stimulating the growth of its economy by extensively promoting its heavy industry. In 2001, the output value of large-scale heavy industries (assuming constant prices in 2005)[30] of the eastern region accounted for 57.2% of the nationwide total, which increased to 64.6% in 2005 due to the rapid industrialization process of the eastern region under the 10th FYP. In 2009, the share of heavy industry in the eastern region dropped to 62.6%, while the proportion in the western region increased by 1.1 percentage points.

Moreover, compared with other regions, the eastern region featured the lowest energy intensity, making it relatively difficult to achieve the 20% reduction target in energy intensity per unit of GDP by technological measures in the 11th FYP period. A series of measures such as "adopting

[30] Total output value of heavy industries above the designated size of each region in 2001, 2005 and 2009 stems from 2002, 2006 and 2010 China Industrial Economy Statistical Yearbook, and the Producer Price Index of each region from 2002 to 2009 stems 2010 China Statistical Yearbook.

large thermal power generator units and shutting down small thermal power generator units," "mergers and acquisitions," "eliminating outdated production capacity" and "industry access" have accelerated the optimization in the eastern region's industrial structure.

3.3.4 *Regional Industry Reallocation Posed Great Challenges to Economic Structure Adjustment*

In the 11th FYP period, high-energy-consuming enterprises operating in the eastern region began to relocate to the western, northeastern and central regions, while acceptance on the local level to continue with the industry reallocation policy remained high.

From 2008 to the first half of 2010, 754 projects relocated to Inner Mongolia from the eastern region, with investments totaling 551 billion RMB;[31] Henan province received capital investments of more than 250 billion RMB from other provinces and foreign countries in 2009, ranking first in the central and western regions, making up approximately 25% of the province's urban fixed asset investments with the petrochemical industry as the main focus;[32] from 2004 to 2008, Hunan and the pan-Pearl River Delta implemented 12,600 cooperative projects, with 784.1 billion RMB in contractual investments, including paid-in capital of 238.5 billion RMB; Business investments in Weinan city (Shaanxi province) through industry relocation in 2009 accounted for 30.59% of the city's fixed assets investments, and the projects introduced were mainly metallurgy, energy, chemical and building materials.[33]

The relocation of energy-intensive enterprises accelerated the expansion of national production capacity and aggravated the issue of overcapacity for industries such as iron and steel, cement, plate glass, and coal chemical. In 2009, national cement investments were mainly concentrated in a few provinces and municipalities in the western and central regions, including Sichuan, Chongqing, Henan, Anhui, and Hunan. For the first

---

31 Miao Fang, Chen Jinghui, Liu Xiuling 苗芳, 陈景辉, 刘秀玲, "Chengjie 'ergezhuanyi' tuijin Neimenggu youshi tese chanye shengji de sikao 承接"二个转移"推进内蒙古优势特色产业升级的思考 [Reflections on Continuing the 'Second Transfer' In Order to Advance Inner Mongolia's Industrial Upgrade]," *Tequ jingji* (2011): 265–267.

32 Wu Yan 武验, "Yanhai chanye zhuanyi zhongyuan diqu liangdian pinlü 沿海产业转移中原地区亮点频现 [Highlights in the Transfer of Coastal Industries to Central China]," *Zhongguohua gongbao* 001 (2010-05-14).

33 Sun Zhongdong 孙仲东, "Dongbu diqu xiang xibu diqu zhuanyi channeng guosheng hangye qingkuang diaocha 东部地区向西部地区转移产能过剩行业情况调查 [Survey of the East-West Transfer of Industries with Excess Production Capacity]," *Xibu jinrong* 7 (2010): 74–75.

seven months, cement investments in Sichuan and Chongqing grew by 250% on a comparable basis and accounted for 19.62% of national total cement investments;[34] the cement sales-output ratio in Guizhou Province shifted from 49% in 2009 (shortage) to 150% in 2011 (redundant); Weinan City in Shaanxi province lacked overall planning in the process of integrating reallocated industries which resulted in overcapacities in industries such as iron and steel, cement, and coal chemical (Column 3.1).

In the 11th period, the trend of industry relocation from the eastern to the western, central, and northeastern regions just began to emerge. In the 12th FYP period, the central and western regions will receive industrial reallocation on a large scale, and to some extent will drive the development of energy-intensive industries which will pose great challenges to China's economic structure adjustment.

Column 3.1. Weinan City (Shaanxi) Adoption of Reallocated Industries Results in Redundant Capacity.

Until 2009, output value of enterprises through industry reallocation in Shaanxi Weinan City accounted for 18.03% of local GDP, resulting overcapacities in energy-intensive industries.

Iron and steel industry: output of iron and steel reached 8.13 million tons equivalent to 6.7 billion RMB in 2009. Since the elimination of outdated production capacity in 2008, a total of 481,000 tons has been eliminated. However, newly-built production capacities in 2008 reached 3 million tons. Cement industry: Since 2007, 3.6 million tons of outdated production capacity has been eliminated. From the fourth quarter of 2008, 5 million tons of dry-process cement production line has been build, which doubled the capacity in 2008. Coal chemical industry: 3.14 million tons of outdated coking production capacity has been shut down in 2007. In the fourth quarter of 2008, newly-built production capacity was 6.28 million tons, an increase of 100% over 2008.

The above mentioned industry reallocations have caused a state of overcapacity in certain industries in Weinan City. Investments in excessive capacity industries accounted for 23.43% of total investments in reallocated industries. Output value of reallocated industries with excess capacity in Hancheng, an energy industry concentrated city in Shaanxi Province, accounted for as much as 70% of local GDP.

Source: Sun Zhongdong 孙仲东, "Dongbu diqu xiang xibu diqu zhuanyi channeng guosheng hangye qingkuang diaocha 东部地区向西部地区转移产能过剩行业情况调查 [Survey of the East-West Transfer of Industries with Excess Production Capacity]," *Xibu jinrong* 7 (2010): 74–75.

[34] Digital Cement Net, "Shuini hangye: Quyu channeng guosheng zhuangkuang chayi da 水泥行业：区域产能过剩状况差异大 [Cement Industry: Great Regional Disparities in Excess Production Capacities]" (2009-08-31), URL: http://www.jmsbeifangshuini.com/Article.asp?ID=1595 [accessed: 2011-09-07].

Appendix 3.1 Calculation Method of Structural Energy-Saving

Total structure energy savings: $\Delta E_{str,t} = \Delta E_{strj,t} + \Delta E_{strk,t} + \Delta E_{strl,t}$
Sectoral structure energy savings:

$$\Delta E_{strj,t} = GDP_t * \sum_{j=1}^{n_j} [(p_{j,t} - p_{j,t-1}) * (e_{t-1} - e_{j,t-1})]$$

$n_j = 1, 2, 3, 4$. (1: *Primary industry*; 2: *industrial sector*; 3: *Construction industry*; 4: *Tertiary sector*)

Industry structure energy savings:

$$\Delta E_{strk,t} = G_{j,t} * \sum_{k=1}^{n_k} [(p_{k,t} - p_{k,t-1}) * (e_{j,t-1} - e_{k,t-1})]$$

when $j = 2$, $n_k = 1, 2, \ldots, 39$; *when* $j = 4$, $n_k = 1, 2, 3$.

Product structure energy savings equals the technological energy savings in industry level minus the technological energy savings in product level. Considering available data, a calculation of the technological energy savings of 13 high-energy consuming industrial products using the output and energy intensity per weight data. Estimating the technological energy savings of all the industrial products, according to the share of the energy consumptions of the above 13 products in industrial sector.

$$\Delta E_{strl,t} = \Delta E_{teck,t} - \Delta E_{tecl,t} = \sum_{k=1}^{n_k} G_{k,t} * (e_{k,t-1} - e_{k,t}) - \frac{\left[\sum_{l=1}^{n_l} Q_{l.t} * (e_{l.t-1} - e_{l.t}) / 1000\right]}{\propto_t}$$

$n_l = 1, 2, \ldots 13$.

Where *subscript t* represents target year; *subscript j, k, l* represents j sector, k industry, l product; ΔE_{str} represents structure energy savings, 10^4tce; ΔE_{tec} represents technological energy savings; GDP_t represents gross domestic product at 2005 constant prices, 10^8 RMB; $G_{j,t}$, $G_{k,t}$ represents the added value of j sector, k industry at 2005 constant prices, 10^8 RMB; p_j, p_k represent the added value proportion of j sector to GDP, the added value proportion of k industry to its upper sector; e represents average energy intensity of all sectors, tce/10^4 RMB; e_j, e_k represent the energy intensity of j sector, k industry, tce/10^4 RMB; Q_l represents the output of l product, 10^4t (Millions of weight box); e_i represents the energy intensity of l product, kg ce/t(kg ce/ weight box); $\propto$ represents the share of 13 products' energy consumptions in industrial sector.

CHAPTER FOUR

LOW-CARBON DEVELOPMENT POLICY

Abstract: During the 11th FYP, China established an initial low-carbon development policy framework that aims to boost new and renewable energy development, reduce energy consumption, and increase carbon sink capacity.

Policy instruments such as pricing, funding, and taxation were proactively applied in coordination with government investment. The effects of the 11th FYP's low-carbon policies were significant; administrative measures realized a carbon deduction of 472 $MtCO_2$; incentive measures accomplished 784 $MtCO_2$ in carbon deduction; and market instruments realized 140 $MtCO_2$ in carbon deduction.

Energy conservation has essentially become an excellent field for investment, supported by grants from both the central and local governments and accompanied by dynamic societal participation. Both government-led administrative measures and incentives played critical roles in achieving carbon reductions. Unfortunately, the lack of market instruments has increased policy costs.

During the same period, the establishment of a comprehensive low-carbon development policy depended on an interplay of factors; while the executive concentrated on enforcing energy target responsibility, this policy was supported by relatively lavish government funding, loans, and sectoral adjustments; decision-making processes were led by democratic and scientific considerations, while China's political framework and government structures provided a significant advantage. Nonetheless, further improvement in policy coordination and diversification of economic means is still necessary and urgent.

During the 11th FYP, China enacted a series of laws, legislations, and regulations which preliminarily outlined the legal and political framework for low-carbon development. On the basis of the 2003 *Promotion Law on Cleaner Production* and the 2006 *Renewable Energy Law*, the amendment to the *Energy Conservation Law* from 2008 has for the first time in history explicitly established a legal statute that makes resource conservation a basic national policy. This further emphasizes the strategic importance of energy conservation for China's energy agenda.

The promulgation of the *Recycling Economy Promotion Law* of 2008 specifically aims at transforming China's traditional economic growth model that had been characterized by high energy consumption, high carbon emissions, and low resource utilization. It calls for proactive measures such as reduction, reuse, and recycling in the process of production, circulation, and consumption. As a novel development mode, a recycling economy calls for the dynamic integration of resource conservation, environmental planning, and economic development while aiming to break the structural dilemma that has constrained China's economic development. The *Recycling Economy Promotion Law* insists on prioritizing reduction as a key consideration of economic development, including a series of structural adjustments in China's industrial sectors, and political regulations that promote energy conservation and emissions reductions. The *Recycling Economy Promotion Law* provides political guidance for a sound and fast stimulation of China's economy.

The adoption of a revised version of the *Renewable Energy Law* on April 1, 2010 reiterates the importance of renewable energies within the framework of sustainable development. It stipulates the mutually supporting roles of aspects such as resource exploration, planning, R&D, investment, pricing, and taxation of renewable energy development. The law explicitly defines the roles and responsibilities of end-users, enterprises, and the government during the process of development and utilization of renewable energy. In addition, the law draws up provisions that regulate renewable energy development targets, mandatory grid connections, power tariff classifications, cost sharing, and project funding. The *Renewable Energy Law* establishes various specific policies for the development of China's renewable energy—it provides legal warrants and legal foundation while at the same time overcoming several difficulties with regards to technology, regulation, and market forces.

During the 11th FYP, China carried out energy conservation campaigns such as the "Closure of Small Power Plants," the "Top 1000 Enterprises Program," the "Ten Key Energy-Efficiency Improvement Projects," the "Elimination Of Outdated Facilities Program," the "Subsidizing Consumers for Energy Efficient Products Project" and the "Low-carbon Transportation Project." The implementation of these campaigns contributed to the achieving of the 20% energy conservation target during the 11th FYP. Moreover, by means of green loans, tax incentives, differential electricity pricing mechanisms, and other economic incentives instruments, along with mandatory energy efficiency standards, the whole society was encouraged to participate in the transformation of China's economy towards a low-carbon approach. Meanwhile, China has launched reforms and explorations in the fields of taxation, resource pricing, bank credits, and market cultivation; China's initial steps towards low-carbon development rests on a fiscal, financial, and economic framework.

Low-carbon development is a fundamental strategic imperative for China. China has chosen energy conservation and environmental protection as the elements that will radically alter its economic development path. The experience and practice of policy formulation, the construction of a policy framework towards low-carbon development, and the achievements of the 11th FYP policies have established the foundation on which China aims to cope with climate change in the long run.

4.1 Low-Carbon Development Policy Framework in the 11th FYP

China's main approach towards low-carbon development and climate change mitigation includes the development of new and renewable energy, the promotion of energy conservation, and the increase of carbon sink capacities. To optimize its energy structure, China has proposed a medium and long-term plan that promotes the development of renewable and nuclear energy and expands the scale on which new and renewable energy will be applied throughout China. The supporting policy framework consists of elements facilitating the development of relevant industries, as well as investment, pricing, taxation, and other components. In addition, China has adopted a series of energy conservation policies and campaigns that have effectively increased energy efficiency and mitigated GHG emissions. With regards to carbon sink, China has enhanced its afforestation efforts and forest management, which has led to an increase in forest coverage rate and carbon sink potential.

A sound market environment alongside solid economic incentive instruments is essential for the implementation of low-carbon policies. During the 11th FYP period, China took various actions in the fields of "green" credits, tax incentives, market training, and others. Figure 4.1 illustrates the main policy instruments supporting low-carbon development. Solid lines represent direct policy goals (or direct policy effects) while dotted lines represent side policy goals (or side policy effects).

4.1.1 *New and Renewable Energy Development Promotion Policies*

4.1.1.1 *Renewable Energy*

China initial policy framework regarding renewable energy consists of six parts:

First, the *Renewable Energy Law* has virtually eliminated possible obstacles for the large-scale expansion of renewable energy such as price, market entrance, and vulnerability of the industry.

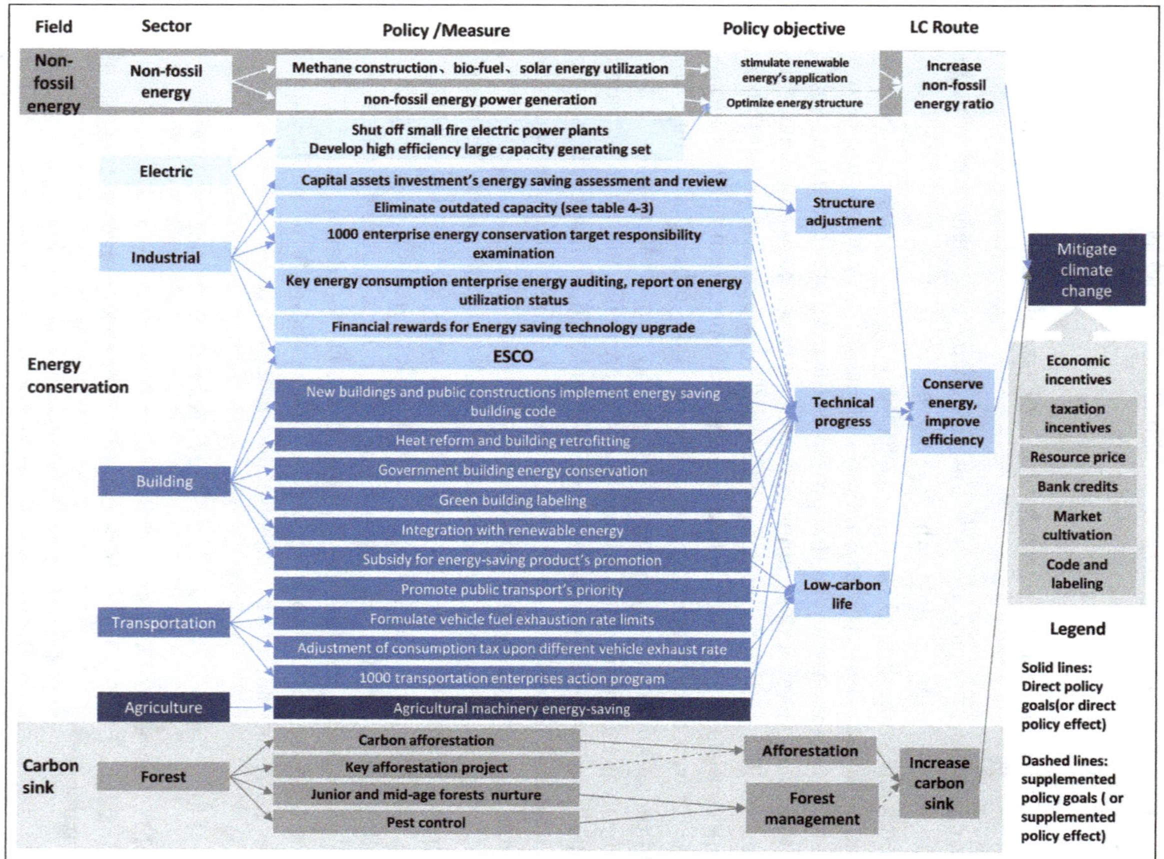

Figure 4.1. China's low-carbon policy map.

Second, the development plan has explicitly stated short-term and long-term total production volume targets for renewable energy. The plan points out a clear direction towards the development of renewable energies.

Third, on-grid power tariffs for wind power, solar photovoltaic power, and biomass power were established. (1) With regards to wind power tariffs, the whole country is categorized into four types of wind power resource areas. Subject to the status of wind energy resource on-site and the construction condition of project sites, local wind energy benchmarks are set forth accordingly. Hence, new on-shore wind farms must all execute grid tariffs according to the local benchmark. (2) With regards to solar photovoltaic power tariffs, the on-grid tariff has been unified across provinces and cities (except Tibet). (3) A biomass power on-grid tariff was established as well.

Fourth, the implementation of regulation guaranteeing full acquisition has alleviated difficulties in regards to power grid admittance: Power grid companies are required to sign agreements with renewable energy businesses that have either obtained administrative permits or filed records and have business plans in accord with the renewable energy utilization plan. The grid firms must then acquire renewable energy in compliance with the technical standards within the grid network coverage at full price. The grid firms also have to provide grid connection services to the renewable power generation companies and adhere to a cost allocation system: Wherever the price of electricity from renewable sources exceeds the benchmark price of local desulfurized coal burning units, the price mismatch is to be shared among electricity sales networks of provincial levels and above.

Fifth, to support the development of renewable energy, China implemented a series of fiscal policies such as establishing funds and granting tax benefits. China has progressively implemented new procedures to manage the funding of renewable energy, including *Renewable Energy Funding Management Interim Measures, Wind Power Generation Equipment Industrialization Fund Management Interim Measures; Straw Energy Conversion Utilization Subsidy Management Interim Measures; Solar Power Building Application Subsidy Management Interim Measures; and Golden Sun Demonstration Projects Subsidy Management Interim Measures*. Since 2008, the income tax regulations for enterprises list wind, solar, and biomass energy businesses as eligible for tax benefits, which include: three years of income tax exemption and half of income tax for the subsequent two years for renewable power generation enterprises; this regulation also applies to manufacturers of technological equipment in this field. These measures have alleviated the financial burden of renewable energy businesses and reduced the utilization cost for renewable energy.

Sixth, the technical norms and regulations of the renewable energy sector are currently under construction. For instance, the *Wind Farm Grid Connection Technology Regulation* has passed its final review and has been submitted to the Standardization Administration of China. The *Large Scale Wind Farm Grid Connection Design Technology Regulation* has also passed final review and it has been submitted to the National Energy Administration. Regarding grid detection regulations, the *Wind Turbine Low Voltage Penetration Capability Testing Procedure*, setting the industry standard, exists as an expository draft. The *Wind Power Dispatching Operation and Management Regulation* and *Wind Power Forecasting Function Regulation* are currently reviewed by experts in the field.

4.1.1.2 *Nuclear Power*

In 2007, the National Development and Reform Committee (NDRC) promulgated a *Nuclear Power Medium-Long Term Development Plan (2005–2020)*. The plan proposed that installed nuclear power units should reach a capacity of 40 million kW by 2020. This is the first time since the endorsement of the pressurized water reactor technology in 1983 that China explicitly put forward a nuclear development target. It treats nuclear power as an important strategic substitution for coal and as crucial in reducing greenhouse gas (GHG) emissions. Nuclear power is included in the national energy development planning, facing unprecedented opportunities. China is currently amending its *Nuclear Power Medium-Long Run Development Plan* of 2007, and has proposed to reach the 40 million kW target by 2015, while the capacity in 2020 should exceed 60 million kW.[1]

China's development of nuclear power adheres to the principal of 'safety first, quality first' and emphasizes a comprehensive mastery of the pressurized water reactor technology. In order to realize its plan and achieve independency in its development of nuclear power, the National Energy Administration published the *Pressurized Water Nuclear Power Plant Standard Listing* and *Nuclear Power Standard Construction Procedures* in 2009. In 2011, the Fukushima nuclear power plant incident caused a global nuclear crisis. China has since launched four important

[1] Hexun, *Xinjian hedian xiangmu shenpi huo mingmian qidong* 新建核电项目审批或明年启动 [New nuclear project approval will start next year] (2011-09-22), http://news.hexun.com/2011-09-22/133626320.htm [accessed: 2011-9-25].

security measures: It (1) organized a full security inspection on all domestic nuclear units; (2) accelerated the establishment of a nuclear power security regime; (3) reviewed and improved the planning for the medium and long-term development of nuclear power; (4) suspended all pending approvals for nuclear power projects before the implementation of a comprehensive nuclear power security regime. The 12th FYP energy plan has not yet been announced, but Japan's nuclear incident will not have significant impacts on China's nuclear development plan in the long run. Nuclear power remains one of the most important choices for energy development in China.

4.1.1.3 *Other Applications of New and Renewable Energy*

In addition to power generation, China is exploring other possible applications for renewable energy, in order to improve rural environmental conditions and to increase the income of rural residents. For instance, to resolve fuel issues in rural areas, the State Council has published *Suggestions to Strengthen Rural Infrastructure Construction and Further Promote Agriculture Development to Increase Rural Income*, which promotes the use of inexpensive small hydro plants, enabling rural residents to replace their wood and coal stoves. Moreover, China is also striving to promote the integration of renewable energy in building construction. For example, free funding, interest-free bank loans, and other financial assistances are available for renewable energy utilization projects in rural and remote areas. Certain amounts of financial subsidies are also given to projects that integrate solar power applications in buildings, such as indoor hot water supplies, heating and cooling systems, photovoltaics, lighting systems, and other important applications. The Ministry of Finance also published *Rural Methane Projects Construction Fund Management Measures, Non-Food Bio-Energy And Biochemical Rewards Funding Management Interim Measures, Straw Energy Conversion Utilization Subsidy Management Interim Measures*, and others. These measures provide subsidies for rural methane and straw solid burning and other projects. Aside from the measures mentioned above, rural elementary and secondary schools, residential buildings, clinics, and other public buildings have become prioritized targets for subsidies: Geothermal heat pump technology is subsidized with 60 RMB/m^2, integrated solar power applications are subsidized with 15 RMB/m^2, and each solar house is subsidized by 60% for newly added investment.

China is also engaged in the development of biomass energy. Despite the fact that China has vast a territory, its rural population still outnumbers its urban population, and inaccessible areas impede the development of biomass energy there. Taking land limitations, agricultural development, production demands, and other impeding factors into account, China has adopted a biomass development strategy that states: "not competing for food with humans, not competing for land with food, not competing for forage with livestock."[2]

4.1.2 *Energy Conservation Policies*

China's energy conservation regime establishes binding conservation targets, target decomposition, and target accountability (including energy statistics, monitoring, and examination) to conduct specific energy conservation campaigns. In September 2006, the State Council determined the schedule of the amount of energy consumption per unit of GDP that each province must exert during the 11th FYP. In May 2007, the State Council promulgated the *Notice from the State Council Regarding Printing Comprehensive Proposal of Energy Conservation and Emissions Reduction* that contains 45 policy measures from ten aspects. An accountability system that treats the status of energy conservation and emissions reduction target completion as part of the comprehensive evaluation for both government officials and enterprise executives was established. In November 2007, the State Council published "The Three Schemes" (this refers to the "Energy consumption per unit of GDP statistical indicator system implementation scheme; energy consumption per unit of GDP monitoring system implementation scheme and energy consumption per unit of GDP examination system implementation scheme").

The "energy conservation data acquisition, monitoring, and verification system" is specifically designed for information management and progress control with regards to the energy conservation efforts of local governments. This system allows the central government to monitor the status of policy implementations and performances regarding the energy conservation target on the provincial level.

The procedures and standards of this system strictly follow the requirements of developing countries established in the Cancun agreement. It ensures the reduction of energy-related GHG emissions and other

---

2 Du Xiangwan, Huang Qili, Li Junfeng, Gao Hu 杜祥琬, 黄其励, 李俊峰, 高虎, "Woguo kezaisheng nengyuan zhanlüe diwei he fazhan luxiantu yanjiu 我国可再生能源战略地位和发展路线图研究 [Research on the strategic status and development road map of renewable energy in China]," *Zhongguo gongcheng kexue* 11 (2009): 6.

mitigation policies of climate change and independent action to engage systematic measurement, reporting, and verification. In August 2008, the state-owned Assets Supervision and Administration Commission formulated the *Central Enterprises Energy Conservation and Emission Reduction Supervision Management Interim Measure*. This further strengthened the energy efficiency management of key enterprises. In the context of specific energy conservation campaigns, China has launched the "Closure of Small Power Plants," the "Top 1000 Enterprises Program," the "Ten Key Energy-Efficiency Improvement Projects," the "Elimination Of Outdated Facilities Program," the "Subsidizing Consumers for Energy Efficient Products Project" and the "Low-carbon Transportation Project." The implementation of these individual campaigns has effectively boosted China's progress in energy conservation.

4.1.2.1 *Closure of Small Power Plants*

In 2007, the promulgation of the *Notice from the State Council Endorsing State Development and Reform Commission and National Energy Administration Regarding Opinions on Acceleration of the Termination of Small Power Plants* clearly stated China's goal to terminate small power plants (Figure 4.2).

The NDRC led the charge to close down small power plants countrywide; it assigned targets for individual provinces (autonomous regions and municipalities) and signed accountability agreements with each

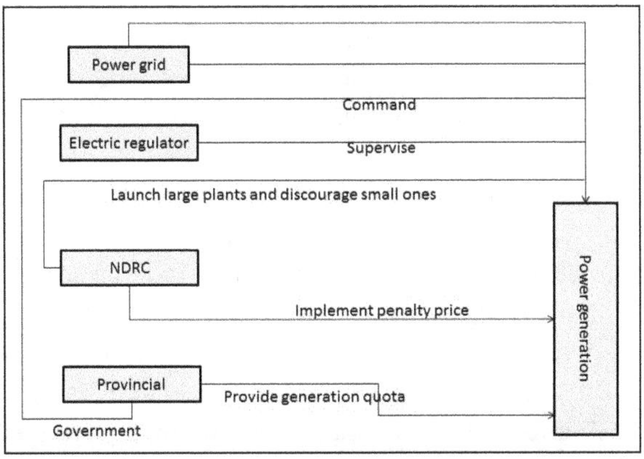

Source: Climate Policy Initiative at Tsinghua University, *2010 nian Zhongguo ditan fazhan baogao* 2010 中国低碳发展报告 [Annual review of low-carbon development in China] (Beijing: Science Press, 2011): 201.

Figure 4.2. Measures of small power plants termination.

provincial government and state-owned large-scale power company. The power grid companies and regulators lent active support by formulating corresponding policy measures and jointly promoting the campaign.

The power grid companies strengthened their supervision of management practices, improved their dispensation methods, prioritized power dispensation of renewable energy and high-efficient clean generators, and limited access for energy-intensive and high-pollutant generators. The regulatory institutions enhanced their supervision of small power plants and established an information system to help in this effort. Small power plants not in compliance with the required regulation would not be issued licenses. At the same time, if a small power plant refused to be closed, it could be ordered to terminate immediately by the responsible departments within the provincial government, which could also temporary suspend the latter's rights to build any new projects until the termination was completed. Projects that terminate small power plants via merger, acquisition, and/or restructuring with the goal to form large-scale power plants were also encouraged. Thus, the NDRC increased the energy construction capacity of each province, autonomous region, and municipality accordingly. Where such transformation was conducted across provinces (and autonomous regions and municipalities), the freed capacity would remain at the respective site, and correspondingly the capacity of the region for new project was reduced.Apart from administrative measures, market instruments played a significant part as well. Each province (and autonomous region and municipality) strengthened the grid tariff management for small power plants; grid tariffs for all small coal burning power generators were lowered to a point not higher than the local standard benchmark tariff, with no subsidy provided. Small power plants with prices lower than the local benchmark tariff will still adhere to the existing tariff.

Moreover, those small power plants that are included in the 11th FYP termination plan and have terminated within 3 years will enjoy power generation quotas; the quotas are transferable to large scale power plants for a certain amount of financial compensation; the scale of the quotas as well as their temporal validity are decreasing progressively after the termination. During the 11th FYP, the energy sector has accumulatively terminated 76.8 million KW of small power plant capacity, exceeding the 50 million KW target.[3]

---

[3] NDRC, *Taotai luohou channeng chengxiao xianzhu* 淘汰落后产能成效显著 [Elimination of outdated production capacities achieved considerable effects] (2011e-3-10), http://www.sdpc.gov.cn/xwfb/t20110310_399045.htm [accessed: 2011-3-10].

4.1.2.2 *1000 Enterprises Energy Conservation*

The "1,000 enterprises" refers to enterprises among nine key industries that had a high energy consumption level of 180,000 tce or more in 2004; these nine key industries are the steel, non-ferrous metal, coal, electricity, petroleum and petrochemical, chemical, building materials, paper, and textile industries. In 2004, these 1,000 enterprises accounted for 33% of the gross energy consumption, 47% of the industrial sector's energy consumption. In 2006, the NDRC and five other commissions jointly issued *The Implementation Plan for the 1000 Enterprises Energy Conservation Campaign* and clearly stated an energy conservation target of 100 million tce between 2006 and 2010.[4] In 2006, 998 enterprises and local governments signed the energy conservation target accountability document (15 of the central enterprises have signed the document directly with the NDRC). The government has unveiled four policy measures to implement this campaign:

First, signature of the target accountability document: The energy conservation target accountability document contains energy conservation targets and an evaluation system. The energy conservation target states the amount of energy savings that enterprises must realize; the evaluation system refers to the related arrangement on evaluated objects, contents, and rewards and punishment mechanism. The evaluation system then classifies results into four degrees: overfulfilled, fulfilled, mostly fulfilled and not fulfilled. The degree of compliance with the energy conservation target is part of the performance assessment of provincial governments and its officials. Enterprises and provinces that do not complete the target will not be allowed to participate in annual awards nor receive any honorary titles; the head personnel of state-owned and state-held enterprises will not be promoted. The whole evaluation system effectively forces government officials to supervise and support enterprises to complete their targets.

Second, implementation of energy audits: In the fourth quarter of 2006, the 1,000 enterprises implemented an energy auditing plan. During the first half of 2007, all energy auditing reports provided by energy companies in each province were closely monitored by governmental experts; any unqualified reports submitted had to be amended. The companies could opt to recruit the services of an independent third party for the

[4] NDRC, *Qianjia qiye jieneng xingdong fang'an 2006* 千家企业节能行动方案 2006 [Action plan of Top 1000 enterprises energy conservation program 2006], http://hzs.ndrc.gov.cn/newzwxx/t2060413_66111.htm [accessed: 2011-9-21].

auditing. Currently, 60–70% of enterprises have chosen to implement energy audits in this way.[5]

Third, implementation of energy utilization status reports: The NDRC has created software, the *Energy Utilization Status Report Fill in System for Key Energy Consumption Enterprises*, which gives every enterprise the option to submit an electronic version of its report online. The environmental department of the NDRC is responsible for the distribution of the software and organizing training workshops. The head of the energy department of each company is in charge of analyzing and editing the reports on the company's energy consumption status, and the report on the previous year must be submitted to the local energy conservation management department by the end of March.[6] Because the submission procedure was only implemented in 2008, all enterprises were asked to submit their 2006 and 2007 reports by the end of September 2008. The final deadline for all these three reports was on the end of March 2009.

Fourth, launch of a campaign for energy-efficiency benchmarks: In 2007, the NDRC issued a *Notice Regarding the Implementation of Key Energy-Consumption Enterprises' Energy Efficiency Benchmark Campaign*. Benchmarking pilot projects were launched for the steel, cement, and caustic soda industries, enterprises are allowed to choose benchmarks of their own, led by the guidance of industry associations, while the state and the local governments hand out compensations and rewards for businesses that adopt relevant technologies.

Various government bodies have different responsibilities within the campaign (Table 4.1). The NDRC is the head commission, and alongside the National Bureau of Statistics (NBS) responsible for public announcements regarding the previous year's energy utilization status. The NBS is the head department for collecting and analyzing data. Through establishing a campaign information system and an official website, it tracks statistics, verifies energy utilization statuses, and regularly analyzes and summarizes the progress report on enterprise energy conservation. The NBS is also responsible for personnel training. The National Bureau of Quality Inspection (NBQI) is responsible for inspecting energy meter

---

[5] Shen Bo, Lynn Price, Lu Hongyou 沈波, 蒲思琳, 鲁虹佑, 2010 *Zhongguo nengyuan shenji de shixian 2010* 中国能源审计的实践 [2010 Report on China energy audit practices], http://china.lbl.gov/publications/energy-audit-practices-china-national-and-local-experiences-and-issues-english-version [accessed: 2011-9-21].

[6] NDRC, "Guojia fazhan gaigewei gongbao 国家发展改革委公告 [Bulletin of NDRC]," 58 (2008).

Table 4.1. Government departments involved in the "1000 enterprises energy conservation campaign."

Government department	Function
NDRC	• Lead department • Every year in April, together with NBS, publicly announce statement on previous year's energy utilization report
NBS	• Establish campaign information system and official website and links to NDRC and individual enterprises' website • Tracking, accounting, verifying energy utilization progress and related data • Related personnel's training • Regularly analyzing and summarizing progress report on enterprise energy conservation • Announce previous year's campaign report together with NDRC
NBQI	• Inspecting enterprises energy meter equipment and its usage • Lead enterprises to establish and improve its measurement management system • Supervise enterprises regular inspection and calibration on all energy meters • Guide enterprises to enhance its application on energy meter data
SASAC	• Improve the evaluation on central enterprises performance
Provincial (autonomous regional and municipal) energy conservation departments	• Responsible for tracking, guidance and supervision of enterprises within the local territory • Organize related industry association and experts to evaluate enterprises' energy auditing report and energy-conservation planning, and to urge immediate implementation • Enhance energy conservation monitoring and inspections, Routine and arbitrary control and spot-check inspection on enterprises' energy utilization progress • Converting the target reduction of energy consumption per unit of GDP to quantity of energy conservation and to decompose to each enterprises with strict evaluation • Actively promotes energy conservation voluntary agreement and other novice energy conservation mechanisms • Hand out rewards to enterprises that have achieved energy conservation target ahead of schedule and international advanced standard

Reference: NDRC (2006).

equipment and its proper use by companies. The State-owned Assets Supervision and Administration Commission (SASAC) is responsible for the energy conservation progress of the central enterprises. The energy conservation departments of each province (autonomous region and municipality) are the main executives of this campaign, they focus on the tracking, guidance, and supervision of the 1,000 enterprises. The campaign itself is in accordance with the principle of territorial management, the energy conservation departments of each province (autonomous region and municipality) are responsible for their local enterprises, the main functions of energy conservation departments are: To supervise and urge enterprises to strengthen their energy conservation management; to submit related information in accordance with work requirements; coordinate related industry associations and experts to evaluate energy auditing reports and energy-conservation plan, and to urge immediate implementation; enhance monitoring and inspections; to include all 1,000 enterprises within the local energy conservation key monitor and inspection listing. Routine and random inspections of the progress of companies' energy utilization; conversion of the target reduction of energy consumption per unit of GDP into the quantity of energy conservation and; active promotion of voluntary conservation agreements and other new conservation mechanisms; rewarding enterprises that have achieved their energy conservation target ahead of schedule and have adhered to advanced international standards.[7]

The performance evaluation of the 1,000 Enterprises Energy Conservation Campaign is divided into several phases (Figure 4.3). First, enterprises submit self-evaluating energy conservation target completion reports to provincial energy conservation departments. Then, the provincial government gathers associated industry experts to form a provincial expert evaluation team to assess and evaluate the report provided by the enterprises. The results are then submitted to the provincial government and the NDRC. The NDRC then gathers a team of national experts to proceed with field evaluations among provinces, and performs random inspections on 10% of the enterprises. This team of experts then draws up an evaluation report it submits to the NDRC. Lastly, depending on the expert evaluation reports and the energy conservation utilization progress of all 1000 enterprises, the NDRC and the NBS jointly announce the previous year's campaign progress report. The NDRC already published the *1000 Enterprises Energy Utilization Progress Public Report* in September 2007; and

[7] NDRC (2006).

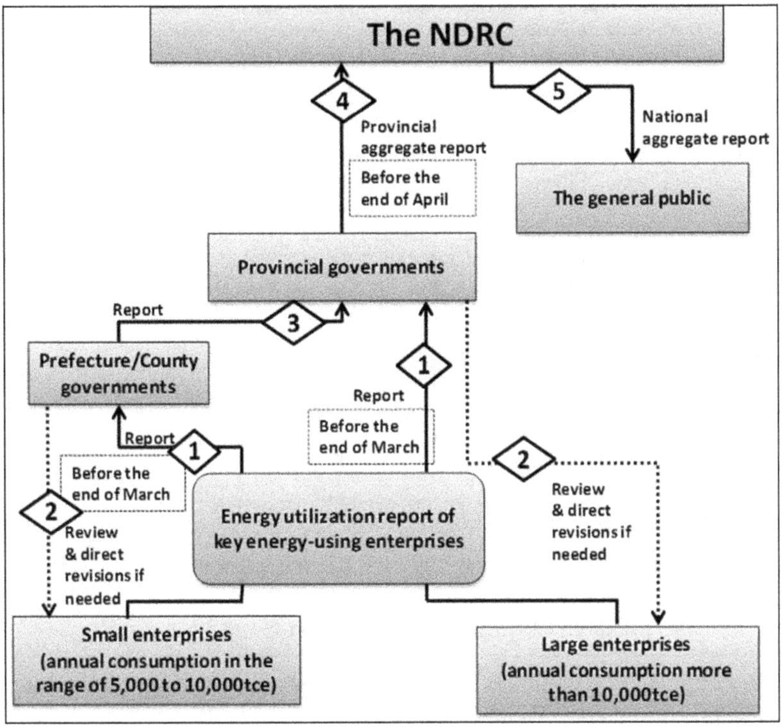

Figure 4.3. 1000 enterprises energy utilization report and evaluation.

published the previous year's evaluation reports in August 2008, November 2009 and June 2010. From 2006 to 2010, this campaign realized 150 Mtce in energy conservation. CO_2 emissions were reduced by 388 Mt.[8]

4.1.2.3 *Elimination of Outdated Capacity*

The acceleration of the efforts to eliminate outdated capacities in energy-intensive industries was one of the most effective energy conservation campaigns during the 11th FYP. This again illustrates that the focus of the 11th FYP was on the optimization of the sectoral structure of China's industry. Figure 4.4 sums up the dynamics of eliminating outdated capacity. The elimination campaign relied on various policy measures; Table 4.2 shows these policy measures as divided into command-control and incentive-based policies.

---

[8] NDRC, *Qianjia qiye chao'e wancheng 'shiyiwu' jieneng renwu* 千家企业超额完成'十一五ь'节能任务 [Top 1000 enterprises exceeded the energy conservation target of the 1th FYP] (2011a-3-14), http://sdpc.gov.cn/xwfb/t20110314_399363.htm [accessed: 2011-03-14].

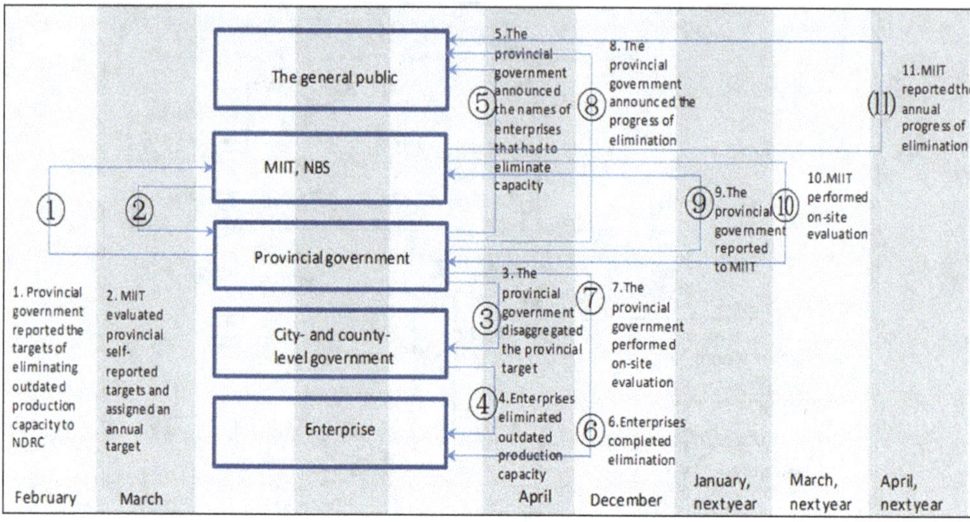

Figure 4.4. Procedures of eliminating outdated production capacity.

Table 4.2. Policy measures of elimination of outdated capacity.

Policy objects	Command-control measures	Incentive-based measures
Enterprises that completed elimination on schedule		• Financial compensation and help settle down employees • Provide technology adaptation funding
Enterprises that did not complete elimination on schedule	• Revoke pollution emissions permits • Restriction on projects and land use approval • Ban register of new production license or withdraw production license that already registered • Revoke business registration • Cut off power supply • Enforce regional planning restrictions	

The campaign to eliminate outdated capacities was mainly implemented by administrative measures; it enforced the closure of energy-inefficient companies not in compliance with industrial policies. Such companies fall mainly into two categories: Those with a legitimate and approved status, but not in compliance with governmental energy policy because of their energy capacity. In these cases, the government grants a certain amount of compensation. The second category comprises enterprises that have not been examined or approved prior to operation. These enterprises face immediate closure, while their owner may also face a fine. Any region, city, or county significantly impacted by the elimination campaign is eligible for financial compensation. On the other hand, any region, city, or county that has not performed well during the elimination campaign will face punitive measures in terms of financial transfer payments. If enterprises delay their elimination, the following measures are applied: revocation of the pollution emissions permit; restrictions on projects and land use approvals; freeze on new production licenses or withdrawal of already registered production licenses; revocation of business registrations, and separation from power supply. If a region has delayed its elimination, regional planning restrictions are enforced.

The implementation of such administrative measures may also mean direct economic losses for regions that are the least developed; for this reason, the government has allocated financial means to lend its support. For instance, the central financial rewards fund and the local specific compensation fund will be able to compensate parties affected negatively in the course of the elimination campaign. Between 2007 and 2009, the central government provided funds of 16.2 billion RMB in this context.

To some extent, market measures are complementary to administrative measures. In this context, this refers to a system of differential energy pricing for energy-intensive enterprises. In 2006, the electrolytic aluminum, ferrous alloys, caustic soda, cement, steel, yellow phosphorous, and zinc smelting industries—all characterized by relatively high energy consumption—all implemented measures towards differential energy pricing. In 2007, the NDRC promulgated four policy documents, and firmly demanded the implementation of a differential energy price policy: (1) *Notice Regarding the Firm Demand to Fully Implement the Differential Price Policy and Forbid Locally Enacted Power Price Discounts*, this further emphasized the implementation of differential power prices and the prohibition of price discounts for energy-intensive enterprises. (2) *Notice Regarding Inspections of the Full Implementation of Differential Power Price Policy and Forbid Locally Enacted Discounted Power Price*, this notice demanded to run a full inspection on the progress of each province's implementation of differential power price policy and banned energy price discounts. (3) *Notice Regarding Some Questions on Further Implementation Of Differential Power Price Policy* and (4) *Notice Regarding Some Questions On Cancellation Of Electrolytic Aluminum And Other High Energy Consumption Industries' Discounted Power Price*. These documents repeatedly confirmed the cancellation of energy price discounts for electrolytic aluminum, ferrous alloys, and chlorinate alkali enterprises. In 2010, the NDRC requested the cancellation of energy price discounts for some energy-intensive enterprises after the outbreak of the global financial crisis, and vouched to enforce the implementation of the differential pricing policy with heightened commitment. Starting from June 2010, enterprises in the eight fields mentioned above have witnessed increasing energy price from 0.05 RMB/kWh to 0.1 RMB/kWh and the additional charge on power prices for enterprises scheduled to be eliminated has gone up from 0.2 RMB/kWh to 0.3 RMB/kWh. If their energy consumption exceeds the national and local standard regarding energy consumption per unit of production, then punitive energy prices are implemented. Should an enterprise exceed the standard by more than double, an additional price charge would be applied.

During the 11th FYP, a total of 72 million kW worth of small thermal power units have been shut off; 122 million tons of outdated iron capacity has been eliminated; 69.69 million tons of steel capacity and 330 Mt of outdated cement capacity.[9] Progress has also been made in shutting down high-polluting enterprises in the paper, chemical, textile, dyestuff, alcohol, MSG, and citric acid industries.

Although the elimination campaign has been effective, it was relatively inefficient in its execution. The problems were mainly due to four factors: economic incentives, rent seeking behavior by local governments, faulty compensation schemes, and over-emphasis on administrative measures. First, given the framework of a market economy, even enterprises with outdated capacities can still find profit margins. While they have a higher consumption of raw materials and energy costs than advanced firms, they also need less investment, have low labor, environmental, and social costs, and can hence often continue to operate profitably. The current elimination mechanism has not considered the internalization of external costs, which may actually give outdated firms a cost advantage; thus, outdated companies cannot be counted on to proactively close themselves. Secondly, the central and local governments have not agreed on common goal; the phenomenon of local governments solely pursuing the maximization of their GDP still exists. Since shutting down outdated capacities means impacting local revenues, some local governments have silently taken a protective attitude towards these outdated capacities. Thirdly,

---

[9] NDRC, *Taotai luohou channeng chengxiao xianzhu* 淘汰落后产能成效显著 [Elimination of outdated production capacities achieved considerable effects] (2011e-3-10), http://www.sdpc.gov.cn/xwfb/t20110310_399045.htm [accessed: 2011-3-10].

the current elimination campaign emphasizes accountability and transparency, while compensation mechanisms alleviating the financial losses of local governments, business owners, and employees are insufficient. Hence, the elimination campaign has been confronted with obstacles coming from interests at the grass-roots level. Lastly, the elimination campaign focused mainly on administrative measures whose overly emphasis on restrictions and binding measures came at the expense of leadership and incentive-based measures. Although administrative measures can suppress the development of outdated capacities in the short run, they cannot optimize the allocation of resources, and thus cannot establish a permanent efficiency mechanism.

4.1.2.4 *Ten Major Energy-Saving Projects*

Ten major energy-saving projects (Column 4.1) are the key compartments of the *Energy Conservation Mid-Long Term Plan*. In 2006, governmental departments such as the NDRC confirmed the energy conservation target of the ten major energy saving project, with a total of 240 Mtce to be achieved during the 11th FYP. The actual performance was outstanding: Between 2006 and 2010, the policy actually conserved 340 Mtce.[10]

The basis of the ten major energy-saving projects is to encourage enterprises to adopt energy conservation technology by means of financial support and rewards, and to foster the capabilities of enterprises with regards to energy conservation and emissions reduction.

The *Implementation Comment on Ten Major Energy-Saving Projects during 11th FYP* calls for an acceleration of the development of energy conservation technologies, particularly demonstration projects of independent intellectual property that could boost independent innovation capability. In 2007, the Ministry of Finance and the NDRC issued a *Notice* of *Energy Conservation Technology's Financial Rewards Fund Management Interim Measure* which pointed out that the central government will provide the necessary guidance funding, "substituting subsidies with rewards" in order to support and reward the ten major energy-saving projects. The notice also states that the central government will financially reward projects that adopt energy conservation technology such as coal-fired industrial boilers and furnaces, fossil oil conservation or replacement, waste heat and waste pressure utilization, energy saving electromechanical system,

10 NDRC, *Shida zhongdian jieneng gongcheng qude jiji jinzhan* 十大重点节能工程取得积极进展 [Ten Key Projects program made great progress] (2011b-3-14), http://www.sdpc.gov.cn/xwfb/t20110314_399214.htm [2011-3-14].

or energy system optimization. The specific approach is as follows: for any project conserving 10,000 tce or more, the projects in the eastern region will receive 200 RMB/tce, while the middle and western region will receive 250 RMB/tce.

The quantity of energy savings will be verified by 28 outside parties appointed by the Ministry of Finance. The procedure is divided into two stages. Stage one is set before the start of the project; enterprises must submit an energy conservation estimate to the city council for an initial assessment, which then reports to the provincial government, which in turn proceeds with a comprehensive review. After that, the provincial government reports to the NDRC and lets a third party examine the initial conservation estimate. Lastly, the government will transfer 60% of the financial rewards to the enterprise before the project starts. The second stage of the procedure is carried out again by the third party after completion of the project. The remaining 40% of the financial reward is given to projects that have fulfilled 100% of their self-estimated quantity. To ensure the transparency and fairness of the procedures, third parties are not allowed to review local enterprises; the quantity in energy conservations can only be reviewed across provinces. Until 2010, more than 5,100 energy conservation projects have completed the rewards evaluation procedure, and the procedure is still ongoing.

Column 4.1. Ten Major Energy-Saving Projects.

(1) Coal-fired industrial boilers and furnaces renovation project: upgrade and replace inefficient industrial boilers, build regional boiler exclusive distribution and processing center; eliminate outdated industrial furnaces, comprehensive energy conservation renovation on current industrial furnaces.

(2) District thermoelectricity cogeneration project: construct thermoelectricity cogeneration plant that focuses on supplying heat and industrial thermoelectricity cogeneration; Distributed thermoelectricity and combined cooling, heating and power; low calorific value of fuel, straws and other comprehensive utilized thermal power plants.

(3) Waste water and waste heat pressure utilization project: in steel, building materials and chemical industries renovate and construct waste hear power generation without additional fuel; differential pressure power generator, by-product combustible gas and low-heat value gas fuel recycling and other waste water, waste heat pressure utilization equipment's and devices.

(4) Conserve and replace fossil oil project: implement fossil oil conservation and substitution renovation in electricity, petroleum, petrochemical, building materials, chemical, transportation industries; develop coal liquefaction fossil product, Alcohol ethers fuel and bio-mass diesel.

Column 4.1 (*cont.*)

(5) Electromechanical system energy-saving project: renovate and upgrade inefficient power generator, speed readjustment of medium-large varying condition power generator system.
(6) Energy system optimization project: systematic energy conservation renovation to refining, alkene, ammonia, and steel enterprises.
(7) Building energy conservation project: new building strictly enforce 50% energy saving standard, Four municipalities and the north severe coldness regions implement 65% energy saving standards for new buildings, the entire process is under strict supervision. Build low energy consumption, super low energy consumption buildings and renewable energy and building integrated demonstration project. City level demonstration renovation for existing residential building and public buildings, promote industrialization of new wall materials and energy-saving building materials.
(8) The green lighting project: to improve product quality, reduce production cost and enhance the ability of independent innovations the focus of technology adoption of energy-saving lights production line; promotion and application of efficient lighting products.
(9) Government body energy conservation project: include both building energy saving adoption and general energy efficiency adoption, new building energy saving evaluation and entire process monitoring and promote government procurement of energy-saving products.
(10) Energy conservation monitor and technology service system construction project: Provincial energy conservation (monitoring) center, renewal of energy conservation monitoring instruments and equipment; organize key energy consumption enterprises' energy auditing, etc.

Source: NDRC website.

4.1.2.5 *Building Energy Conservation*

Building energy conservation includes four aspects:[11] (1) the implementation of energy saving standards for new buildings, (2) energy-saving retrofitting for existing buildings, (3) energy conservation for public and government buildings, (4) promotion of the "green building" concept.

(1) New building energy saving design standard in China has been through phases of 30%, 50%, and 65%. With regards to public building design, the Ministry of Housing and Urban-Rural Development announced a *Public Building Energy Conservation Design Standard* in 2005 (GB50189-2005), the

[11] Detailed briefing about renewable energy building integration and promotion of energy-saving domestic appliances can be found in 4.1.1.2 and 4.1.2.6 respectively.

Table 4.3. Mandatory building design standards for energy conservation.

Classification	Standards				Energy consumption target			
	Type of building	Climate zone	Document Name	Code	Central heating	Decentralized heating	AC	lighting
Building design standards	Residential	Frigid	Energy conservation design standard for new heating residential buildings	JGJ26-2010	50% less than buildings in 1980		—	—
		Hot summer / cold winter	Design standard for energy-efficiency of residential buildings in hot summer and cold winter zone	JGJ134-2010	—	50% lower than before energy conservation		—
		Hot summer / warm winter	Design standard for energy-efficiency of residential buildings in hot summer and cold winter zone	JGJ75-2003	—	50% lower than standard building		—
	Public	Frigid Hot summer / cold winter Hot summer / warm winter	Design standard for energy efficiency of public buildings	GB50189-2005	50% lower than before energy conservation			

Source: CPI at Tsinghua, 2011.

so-called 50% energy conservation design standard.[12] China's current *Code of Acceptance for Energy Efficient Building Construction* (GB50411-2007) was implemented in October 2007. Table 4.3 summarizes the mandatory building design standards.

[12] GB50189-2005 rules, comparing public building in compliance with the design standard with before the implementation of energy saving measures, heating, ventilation, air condition and lighting total energy consumption decreased by 50%.

Beginning with 2004, the Ministry of Housing and Urban-Rural Development initiated a countrywide special review of energy-saving building designs, a program which evaluates the implementation of energy-saving building codes during the course of design, construction, approval, and market transaction. In this process, the implementation of energy conservation standards for new buildings and the adoption of heat insulation schemes for buildings in China's northern cities are the main points of concern. In support of this special review, some local construction departments have also taken measures such as close supervision during the process of building design, construction, and approval. Tangshan City's management program for energy-efficient building construction is one example in this context.

(2) Retrofitting of existing buildings includes: Improvements in the thermal insulation of building envelopes, reduction of air conditioning and heating, improvements in the energy consumption system and equipment efficiency of public buildings, encouragement of passive energy usage designs, and reduction of the energy use in ventilation and lighting systems. Furthermore, the reform of the heat supply system for northern China is an important part of the retrofitting campaign for existing residential buildings.

In 2006, the Ministry of Housing and Urban-Rural Development issued *Implementation Advice on Promoting Heat-Supply Thermal Measuring* which suggested establishing and improving a heating charge mechanism based on the thermal usage measurement, starting with administration and public buildings. In 2007, the NDRC and the Ministry of Housing and Urban-Rural Development issued the *Urban Heating Price Management Interim Measure* to improve the pricing mechanism for heating and to provide a basis for heating price regulation. In 2007, the Ministry of Finance determined incentive measures according to climate zones: frigid regions would receive 55 RMB/m^2, cold region 45 RMB/m^2.

From 2006 onwards, China began to promote the mandatory installation of heating meters. The installation of heating meters is necessary in order to charge heating supply according to the amount of heat used: from the perspective of the end-user, such a charging system incentivizes conservative heat usage, and also the retrofitting of building envelops in order to reduce heat consumption. As far as heat supply enterprises are concerned, this measure provides incentive to improve the performance of heating networks, and moreover rewards attention to heating efficiency, to lower production, and to supply costs. Thus, the installation of heating meters is directly or indirectly promoting energy conservation and emissions reductions for the central heating system of northern China.

The objective of installing heating meters is to encourage heating conservation through a charging mechanism based on quantity of heat used. However, the current heating price in China cannot reflect the operation cost of supplying heat, nor can it formulate economic incentives for heat providers. On the other hand, the base price unit for heat is per household, which cannot reflect the differing thermal characteristics of individual buildings; therefore it cannot effectively motivate end users. Because the heat pricing mechanism does not provide sufficient incentives yet, the installment of heating meters has had very limited effects on end-user behavior. We can see that the installation rate is relatively low: From 2006 to 2009, heating meter installation amounted to 360 million m^2, a merely 4.5% of the total central heating areas of northern cities.[13]

(3) The intensive promotion of energy conservation measures for government office buildings and large-scale public buildings is driven by three aspects. First, the legal, regulatory, and national planning dimension: The State Council's *Civil Building Energy Conservation Regulation* and *Public Building Energy Conservation Regulation* states that governmental office buildings and large-scale public buildings must establish a measurement mechanism for their energy consumption and conduct a complete energy auditing. Local governments at the county level or above shall formulate energy consumption limits for local public buildings and propose reduced reduction in energy consumption by 20% during the 11th FYP. In 2007, the State Council's *Work Program for Energy Conservation and Emissions Reduction* ruled that twenty-five trial provinces and cities should conduct large-scale energy-consumption accounting for public buildings, energy auditing, issue public notifications on energy efficiency and energy-consumption quotas, and realize a 12.5 Mtce energy conservation target.

Secondly, issue of relevant standard regulations. Regulation of relevant technology is a key factor in building energy consumption system operation management and complementary energy consumption accounting system. The Ministry of Construction promulgated *Air Condition and Ventilation System Operational Management Standards* (GB50365-2005) to enhance the operational management of public buildings. It also promulgated *Government Office Building And Large-Scale Public Building Energy Auditing Guidelines* and *Notice Regarding Issue of Government Office Building*

[13] CPI at Tsinghua 2011: 191.

And Large-Scale Public Building Energy Consumption Monitoring system's relevant technology Guidelines.

Third, issue-relevant management measures and incentive-based polices. The Ministry of Finance from 2007 onwards has conducted energy accounting trials for public buildings in 23 cities. Additionally, the Ministry of Finance has established a fund specifically designed to support the process of establishing a public building energy conservation monitoring and supervision system. The fund subsidizes spending such as setting up an energy consumption monitoring platform, conducting energy consumption accounting, energy auditing, and public notifications on energy efficiency. In 2009, the Ministry of Construction and the NBS jointly published a *Civil Building Energy Consumption and Conservation Statistical Report System*, and in 2010, issued a *Notice Regarding Print and Distribution of Civil Building Energy Consumption and Conservation Statistical Report System*. Across the country, a full-scale energy consumption accounting of all types of buildings was being conducted.

(4) China has devoted itself to disseminate a "green building" concept and to encourage energy-efficiency labeling. In 2006, the Ministry of Construction and the Ministry of Science and Technology (MOST) jointly published China's very first standard regulation regarding green buildings, the *Green Building Evaluation Standards* (GB/T50378-2006). This standard classifies green buildings into one star, two stars, and three stars. This standard is voluntary in nature; the evaluation is conducted by the government. Soon after, the Ministry of Construction and the MOST issued *Green Building Evaluation labeling regulations Green Building Evaluation's Technology Detailed Rules, Additional Remarks (Operation Section) on Green Building Evaluation's Technology Detailed Rules* and *Additional Remarks (Planning and Design Section) on Green Building Evaluation's Technology Detailed Rules* and more. These publications are used in guiding projects to evaluate labeling of green buildings and green building designs. Until 2010, 112 projects countrywide had earned the "green building" label, totaling an area of over 13 million m^2.

4.1.2.6 *Energy Saving Products Benefit People Project*

The promotion and application of domestic appliances and lighting with high energy efficiency is critical in reducing building energy consumption. The current policy system takes a two-pronged approach: one is the establishment of standards and norms for energy-efficient products, a measure that will shape the market environment and thus improve the energy efficiency grades of such products; the second is the launch of related construction projects, which complements incentive-based policies to promote energy saving products.

With regards to lighting, the Ministry of Finance and the NDRC in 2007 ruled that the central government will subsidize 30% to 50% of the price for highly energy-efficient lighting products of the bid-winning applicant, the ultimate beneficiary of which would be the general public. As far as domestic appliances are concerned, the Ministry of Finance in 2007 proposed that the central government would subsidize manufactures of high energy efficiency products, by subsidizing market prices for said products, leaving consumers as the ultimate beneficiaries. In 2009, the Ministry of Finance and other departments promoted air conditioning units and ten other products with energy efficiency grades of one or two; they also suggested that starting from June 1, 2009, the standard subsidy for grade-one air conditioning systems would be between 500 and 850 RMB per unit, for grade-two between 300 and 650 RMB.

Aside from the allocation of product subsidies, China also encourages scrappage schemes for automobiles and home appliances. For automobiles: early scrappage and trade–ins while purchasing a new car with medium, small and light trucks and some medium-sized passenger vehicles that have met the condition of certain service years; or early scrappage and trade in purchase of petrol vehicles that have not met national standard I pollution emissions and diesel vehicles that have not met national standard III. The subsidy should not exceed the purchase tax of a similar model. For home appliances: Beijing, Shanghai, and other cities have launched trials for scrappage schemes involving television sets, refrigerators, washing machines, air conditioning units, and computers. For all participating consumers, 10% of the new products' sales price is subsidized; the maximum amount of subsidy depends on the category the product belongs to. The transportation cost for home appliances delivered to recycling companies is also subsidized with a fixed amount.

Since June 2009, the central government has allocated 11.54 billion RMB to promote more than 34 million units of high-energy-efficiency air conditioning, an annual realization of 10 billion kWh.[14] One million vehicles with a mileage of 1.6 liters or below were supported with 3.4 billion RMB, which led to annual fuel savings of 300,000 tons of petrol.[15] An additional

---

[14] NDRC, *Jieneng chanpin huimin gongcheng qude mingxian chengxiao* 节能产品惠民工程取得明显成效 [The program "Benefitting the general public of utilizing energy efficiency products" made noticeable impacts] (2011c-3-14), http://www.sdpc.gov.cn/xwfb/t20110314_399362.htm [accessed: 2011c-3-14].

[15] Ibid.

1.62 billion RMB has supported the promotion of 360 million units of energy-saving lighting bulbs, which conserved 12.5 billion kWh annually.[16]

4.1.2.7 *Low-Carbon Transportation Campaign*

In 2010, China launched the "1,000 Enterprises Low-Carbon Transportation Campaign." This campaign treats 1,000 transportation enterprises (including automobiles, ships, road and port) as vehicles to promote energy conservation and emissions reduction. Automobiles: promote energy-saving driving experiences; enhance the evaluation on operating vehicles' petrol quotas, strictly implement vehicle fuel consumption limits, eliminate vehicles with high fuel consumption, promote new energy and clean fuel vehicles and drop-pull transport. Ships: promote ship model standardization, ships at port use shore power. Road: promote high-speed roads without toll stations, optimize transportation organizations, promote drop-pull transport, road and tunnel energy conservation, apply renewable technology on pavement materials and promote solar power's application on road system. Port: promote rubber-tiered containers, gantry cranes, oil-to-electricity conversion, and shore power construction.

The transportation sector has already formulated and implemented several technological standards and detailed industry guidelines concerning vehicle fuel consumption limits. In 2004, the General Administration of Quality Supervision Inspection and Quarantine (AQSIQ) and the National Technical Committee of Auto Standardization approved and issued a mandatory national standard, the *Standard on Limits of Fuel Consumption for Passenger Vehicle* (GB19578-2004). The standard measurement unit is fuel consumption per 100 kilometers, the heavier the vehicle the stricter the standard applies. Thus, all vehicles are divided into 16 mass segments respectively, and each segment applies its corresponding standard.

Phase one of the fuel consumption limit was implemented on July 1, 2005, the second phase started on January 1, 2008. The relevant department is currently preparing the implementation of the third phase. Likewise, the mandatory national standard light vehicle fuel consumption level labeling (GB22757-2008) has been in effect since January 1, 2010, the label requires manufactures importers of vehicles to examine and declare levels of fuel consumption by a standardized measure, newly manufactured and imported vehicles must display the fuel consumption level label at a visible place before sale. The Ministry of Transportation has issued a series of documents so that unqualified vehicles are not allowed

16 Ibid.

to enter the market and that vehicles that are above the fuel consumption limit will gradually be phased out of the market in given time.

Improvement of road transportation efficiency is also a key approach towards reducing energy consumption in transportation. To encourage the development of road transportation towards large-scale, multi-axis, and boxcar carriage vehicles, the Ministry of Transportation has printed and issued a *Notice Regarding Promoting the Development of Drop-Pull Transport*. Drop-Pull transport is an approach for organizing transportation, it means that the tractor unit operates according to a preset operating schedule, delivers its trailer compartment at the cargo handling station and takes up another trailer compartment to continue its operation. Compared to traditional transportation approaches, drop-pull transport can reduce empty runs and inefficient transport, thus reducing energy consumption and emissions. China has also dedicated itself to establish a road freight transportation information service network platform, which improves the utilization rate of cargo transport vehicles in terms of kilometers and tonnage, as well as the load rate of passenger vehicles.

Furthermore, China has taken several measures to reducing overall vehicle activity. To be more specific, these include: An increase in the share of public transportation within the overall transportation structure; traffic restrictions based on the last digit of license plates, introduction of a fuel tax, and encouragement of other means of transportation (i.e. rail transport). For a detailed analysis of low-carbon transportation please refer to Chapter 9 of this report.

4.1.2.8 *Standard Labeling*

The National Standardization Administration formulated and amended 46 standards that are complementary to *Energy Conservation Law*, in that, 37 of them are newly formulated national standards and 9 of them have been amended. 36 are national mandatory standards, and most of the standards had already been implemented on June 1, 2008. Overall, the 46 national standards include: 22 standards regarding limits on energy-intensive products per unit of product, five standards for fuel economy, 11 standards regarding end products' energy efficiency,[17] 8 standards regarding energy accounting, energy consumption calculation, economic

[17] These 11 end products covers residential gas water heater, water storage electronic water heater, induction cooker, air conditioner, multi-connected air conditioner, computer monitor, copy machine, medium-small three phase asynchronous motor, AC contactor, external power supplies, rinsing centrifugal pump and other products. The standard

operation and other basic energy conservation. In 2006, the Ministry of Construction and the Ministry of Science and Technology jointly issued China's first green building standard regulation: *Green Building Evaluation Standards* (GB/T50378-2006).

On March 1, 2005, the *Energy Efficiency Labeling Regulation* was implemented; Until October 2010, China had issued seven batches of catalogues on products featuring energy efficiency labels, including flat screen television sets, microwaves, refrigerators, and 23 other products.[18]

4.1.3 *Carbon Sink Policies*

In June 2009, the Chinese government convened a national forestry meeting. It clearly pointed out the unique importance of forestry in dealing with climate change, and it emphasized that China must treat development of forestry as a national strategic priority to cope with climate change. In November 2009, the State Bureau of Forestry published a *Forestry Campaign Plan To Address Climate Change* that established a three-phased target, 15 mitigation actions (Column 4.2) and 7 adaptation actions.[19]

The three targets are: (1) To increase the national forest's function as carbon sink, an annual afforestation average of 4 million hectares or more, a countrywide forest coverage rate of 20%, forest stock of 13.2 billion m^3, to be achieved by the year 2010. (2) An annual afforestation average of 5 million hectares or more by the year 2020, a countrywide forest coverage rate of 23%; and a forest stock of up to 14 billion m^3 that would further improve the carbon sink function of national forests. (3) Until 2050, forest area should experience a net increase of 47 million hectares compared to 2020, forest coverage rate would stabilize at over 26%, making the carbon sink function of China's forests relatively stable. The main approach for China to increase carbon sink capacity is afforestation. Apart from directly

---

has specifically determined these products' energy efficiency limits, energy efficiency level and energy conservation evaluation value.

[18] The 23 products are: flat screen television, microwave, power transformer, ventilator, room air conditioner, automated rice cooker, AC electric fan, AC contactor, displacement air compressor, refrigerator, multi-connected air conditioner (heat pump) units, water storage electronic water heater, induction cooker, computer monitor, copy machine, self-ballasted lamp, high pressure sodium lamp, medium-small three phase asynchronous motor, chillers, residential gas water heater, gas fired heater and boiler, washing machine, unitary air conditioner.

[19] The 7 climate change adaptation actions include: improve the adaptability of planted forest ecosystem, establish representative forest species nature reserve, enhance key species' protection, improve wild animals' epidemic sources and disease monitoring and early warning system, enhance desertification region's vegetation protection, enhance wetland's protection's foundation work, establish and improve wetlands nature reserve network.

planting carbon sink forests, several other important forestry projects deserve mentioning; projects dedicated to curb soil erosion, prevent and control desertification, and protect natural forestry resource, all of which have a synergistic effect for the increase of carbon sink capacity. Meanwhile, forest carbon sink will also be increased through enhanced forestry protection and management to improve forest quality and stock volume.

Column 4.2. Forestry 15 Climate Change Mitigation Actions.

Promote voluntary planting
Implement key afforestation projects
Speed up rare commercial breeding tree species
Implement energy source forestry fostering and processing and utilization integration projects
Implement sustainable national forestry management
Expand areas of closed hillsides to facilitate afforestation
Enhance forest resource harvesting management
Enhance forestland requisition and occupation management
Improve forestry law enforcement capability
Improve forest fire prevention and control capability
Improve forest disaster (disease, pest, mice, rabbit) prevention and control capability
Rational develop and utilize bio-mass materials
Enhance high efficiency recycle utilization of timber
Launch rescue protection and restoration of important wetlands
Launch demonstration of sustainable utilization in agriculture, livestock and fishery

4.1.3.1 *Afforestation*

Afforestation is a vital approach for China to mitigate climate change. There are three mechanisms to implement afforestation: (1) key forestry projects with synergistic effects to increase carbon sink, (2) promotion of clean development mechanism projects, (3) The China Green Carbon Sink Foundation guides enterprises and public to reduce emissions voluntarily.

(1) During the 11th FYP, China continued to implement projects aimed at natural forest resource protections: projects converting cultivated land into forests, projects aiming to control the causes of sandstorms in Beijing, and project to build a fast-growing and high-yielding timber forest base in key regions such as "the three norths" and the Yangtze River area, and other wildlife protection and nature reserve construction projects. In 2009, the State Council in addition to one-million hectares of newly-planted forest demanded to build large-scale and concentrated planted forest bases in three northern regions in order to create a point-line

surface of green ecological barriers. In converting cultivated land into forests, the focus has shifted from large-scale operations to integrated projects that aim to alleviate poverty through development and reservoir resettlement. This shift constitutes a great effort towards a long-term improvement of the livelihood of farmers who lost their farm lands. To consolidate the results of the conversion, the government has enhanced its policy guidance, encouraged farmers to transform their production and living style, and made efforts to alleviate poverty. Specific measures include: build basic grain ration cropland, construction of rural power supplies, resettlement in accordance with ecological consideration; development of follow-up industries, reforestation, and other projects. In 2009, the State Council office tasked the National Bureau of Forestry and other 7 departments to draw up an *Evaluation Method of Provincial Level Government Desertification Prevention and Control Target Accountability* which ruled that from 2010 onwards, the administration of 13 key desert provinces and autonomous regions alongside the Xinjiang Production and Construction Corps proceed with a desertification prevention and target accountability evaluation. This is the first time that China has performed such quantified evaluations of desertification prevention and target accountability.

(2) China already has several ongoing CDM carbon sink afforestation projects. In 2006, China and the World Bank collaborated on the Guangxi Pearl river valley reforestation project, which is the first forestry carbon sink project in the world that has acquired a UN CDM executive board permission. This project has already entered the second stage of its implementation. Another collaborative afforestation project on carbon sink, this time with the Ministry of Environment of Italy, to prevent desertification in Aohan (Chifeng region) was approved in 2007. In 2009, the world-first carbon sink project based on climate, community, and bio diversity standards' CDM (CDM-AR) in Northeastern Sichuan land degradation afforestation and reforestation project successfully registered. In 2010, the Ministry of Finance and the National Bureau of Forestry issued the *Advice Regarding Implementing Afforestation Subsidy Pilot Program 2010* which has led 20 provinces to launch pilot program afforestation subsidies.

(3) The China Green Carbon Sink Foundation is China's first national public foundation that aims to increase carbon sink capacity, reduce emissions, and cope with climate change. It was registered in July 2010 under the Ministry of Civil affairs and approved by the State Council. This foundation offers assistance to enterprises and the general public to reduce emissions voluntarily; it established a platform for enterprises and the public to show social responsibility through support of forestry measures that store carbon credit. Currently, the foundation has started

5 green funding projects, obtained 300 million RMB in public donations, and accomplished afforestation of more than 1 million acres.[20]

4.1.3.2 *Forestry Management and Protection*

In 2006, the National Bureau of Forestry reaffirmed its active commitment to sustainable forestry management and to optimize the structure of forestry resources. Two carbon-sink related forestry management and protection measures deserve mentioning in this context: (1) younger and middle-aged plantations, and (2) pest prevention and control measures.

(1) Focus on junior and middle-aged plantations is the key approach to improve forest quality and to increase forest stock volume. In 2010, the central government had allocated 2 billion RMB for the forest plantation subsidy pilot program, and plans to complete forest plantations of 78.75 million acres.[21]

(2) In 2009, the National Bureau of Forestry and twenty provincial (autonomous regions and municipalities) governments signed the *2008–2010 Pine Wood Nematode Disease Prevention and Treatment Target Accountability Agreement.* 16 provinces (autonomous regions and municipalities) have set up provincial prevention and treatment command centers; this established the 'dual-track' target accountability system among governments and forestry departments, with differing results across regions and provinces. Zhejiang and other provinces have issued specific laws regarding the prevention and treatment of incidents of nematode infestation in pine woods. Anhui, Henan, and more than ten other provinces have issued specific documents to further strengthen the prevention and treatment of pine wood nematodes. In 2011, the National Bureau of Forestry organized an inspection team which inspected and evaluated the target accountability completion status of twenty provinces (autonomous regions and municipalities). From 2008 to 2010, 28 counties eradicated epidemics, as did Yunan province; Moreover, the three years in county-level eradication efforts account for 45.61% of total eradication since the occurrence of pine wood nematode infestations, this is the first time a positive trend occurred at the county level compared to the past.

---

[20] NDRC, *Zhongguo yingdui qihou bianhua de zhengce yu xingdong. 2010 niandu baogao* 中国应对气候变化的政策与行动—2010 年度报告 [China's policies and actions for addressing climate change: 2010 progress report].

[21] Ibid.

4.1.4 *Economic Incentives Policy Instruments*

4.1.4.1 *Tax Incentives and Restrains*

Business income tax. Income tax benefits encourage enterprises to implement energy conservation projects or to purchase energy conservation equipment. Moreover, renewable power generation enterprises enjoy preferential treatment on income tax as well.

For instance, energy conservation projects, such as public waste disposal, methane utilization, or energy conservation and emissions reduction technology enjoys a three-year income tax exemption and a subsequent two years of halved levy; 10% of the cost of purchasing energy conservation equipment is tax-deductible within the same year or can be carried forward to the subsequent five years. Renewable power generation firms enjoy a three-year income tax exemption and halved levy for the subsequent two years; manufactures of equipment used in renewable power generation can enjoy income tax benefits as well.

Value Added Tax (VAT). VAT preferences apply to comprehensive resource utilization, nuclear power companies, energy conservation service enterprises, and others. For instance, from 2008 onwards, nuclear reactors are eligible for VAT reimbursement schemes within 15 years of their official start of operation, the reimbursement diminishes gradually according to three phases, and in the first 5 years, 75% of VAT is reimbursed.

VAT preferential treatments for comprehensive resource utilization include four different aspects: exemption, immediate refund, 50% immediate refund and refund after collection. Table 4.4 summarizes preferential VAT policies related to energy conservation.

Import tax: Lower import tax to encourage the import of products beneficial to technology innovation and resource commodities products. For instance, since November 1, 2006, the import tax on key components used in the manufacture of energy conservation equipment was lowered from 1–7% to 0–3%. This applies to computers and plate machines, textile machinery components, semi-conductor models with variable flow function, and seven other key manufacturing components that foster technological innovation and energy conservation; the import tax on coal, petroleum products, alumina, and 26 other resource commodities was lowered from 3–6% to 0–3%. From January 1, 2008, imports of gasoline, diesel, and aviation kerosene have been taxed with 1% instead of 2%.

Export tax: China has increased its export taxes to control the exportation of products that are either resource-intensive, or are characterized

Table 4.4. Resource comprehensive utilization VAT preferential policies.

Type	Utilized product	Implementation scale	Implementation time
Exemption	Rubber powder made of waste tires	Rubber powder should meet GB/T19208-2008 performance criteria	From January 1, 2009
	Retreaded tire	Retreaded tire should meet GB7037-2007, GB14646–2007 or HG/T3979-2007 performance criteria and retreaded tire matrix is 100% from waste tires	
	Specific building materials' product is produced with waste residue's mixture ratio of no less than 30%	Specific building materials include: brick(exclude fired brick), blocks, ceramsite, wallboard, pipe, concrete, mortar, manhole cover, road fence, fireproof material, refractories, heat insulate materials, mineral wool	
Immediate refund	Electricity or heating produced by burning waste	Waste dosage accounts for no less than 80% of the power generation fuel and production emissions meets GB1322-2003 first period standard or GB18485-2001 related regulation. Waste is defined as household waste, crop straws, bark waste residue, sludge and medical waste	From July 1, 2008
	Cement with waste residue mixture no lower than 30%	Adopt rotary kiln cement production (include cement clinker) and raw material's waste residue mixture should be no less than 30%	
	Renewable asphalt concrete	Waste asphalt concrete as raw material, waste asphalt concrete ratio to total raw material should be not less than 30%	
	Shale oil	Abandoned shale during coal mining process as raw material	
50% immediate refund	Waste distiller and bottom pot water as raw material to produce steam, active carbon, white carbon black, lactic acid, calcium lactate, methane	Waste distiller and bottom pot water as raw material ratio should be no less than 80%	From July 1, 2008

Table 4.4 (*cont.*)

Type	Utilized product	Implementation scale	Implementation time
	Coal gangue, coal slime, stone coal, oil shale as burning fuel to produce power and heat.	Coal gangue, coal slime, stone coal and oil shale as burning fuel ratio should be no less than 60%	
	Utilize wind power		
	Some new wall materials	Include some bricks, blocks and plates	
Refund after collection	Home grown comprehensive bio-diesel	Disposed animal oil and vegetable oil as raw material of diesel, the ratio of disposed oil should be no less than 70%	From July 1, 2008

by high energy intensity or high levels of pollution. For instance, from November 1, 2006 onward, China has imposed export taxes on 110 resource products. Among them, the export tariff for apatite, ores of rare-earth metals, metallic mineral ore, and 44 other minerals was set at 10%; for coal, coke, crude oil, and four other energy resource products, the tax was 5%; 15% for copper, nickel, electrolytic aluminum, and eleven other nonferrous metal products; 10% for iron alloy, pig iron, billet steel, and thirty other steel products; and 10% for rare earth compounds, floor timber, disposable chopsticks and 21 other products. From January 1, 2008 onwards, exports of billet steel and coke were taxed with 25% (previously 15%), while for certain other steel products, export tariffs increased from 5% to 15%.

Export tax refunds: To prevent energy consumption from developing even faster, China has cancelled or lowered export tax refunds for products manufactured by high-emissions, high-pollution industries. From July 1, 2007 onwards, China has canceled export tax refunds for 553 resource products.

However, in order to cope with the financial crisis, China on April 1, 2009 opted to increase export tax refunds: Refunds for steel strips and other steel products were increased to 9%, for aluminum, other nonferrous metal materials, PVC, and other plastic products to 13%. This has proven to be a temporary measure, as China again cancelled export tax refunds for 406 products (including steel and nonferrous metal processing materials) on July 15, 2010.

Refined oil consumption tax. The refined oil consumption tax reform started on the January 1, 2009, replacing the imposed road maintenance fee, waterway maintenance fee, transportation management fee, road passenger and freight surcharge, waterway transportation management fee, waterway passenger and freight surcharge and 6 other fees. Gradual cancellation of government loans for secondary road tolls; at the same time, an increase in the gasoline consumption tax, which is included in the price for refined oil, from 0.2 RMB per liter to 1 RMB per liter; diesel consumption tax increased from 0.1 RMB to 0.8 RMB; Other refined oil products' consumption tax have increased accordingly, the reformed tax framework means that drivers pay less when they drive less and vice versa.

Resource tax: China's resource tax collecting mechanism takes two forms: the amount of volume and the rate of value. The amount of volume means imposing a fixed amount of tax for every unit of product and the rate on value means imposing a fee proportional to the sales price. Coal products have implemented the amount on volume approach, from February 1, 2007, coke resources tax's applicable tax standard increased from 3.2 RMB/t to 8RMB/t. and from May 2007, 22 coal production province's coal resources tax increased from 0.5–1.2 RMB/t to 2.5–4.2 RMB/t.

The current resources tax for oil and gas is also amount on volume approach, crude oil tax range from 8 RMB–30 RMB/t, natural gas tax range from 2 RMB–15 RMB/thousand m^3. But Xinjiang crude oil and natural gas resources tax has reformed since June 1, 2010, its resource tax on crude oil and natural gas is levied as rate on value; the tax rate is currently at 5%. Payable tax amount = sales × tax rate. In 2010, Xinjiang collected resource tax amounted to 3.578 billion RMB.[22] This oil and gas resources tax reform plan set to expand to Inner Mongolia, Gansu and 12 other provinces from December 1, 2010.

4.1.4.2 *Resource Prices*

China has implemented a state-controlled price management system for energy, water, gas, and direct heating. While the government intends to protect the interest of China's low-income population by controlling resource prices, it makes resource allocation via market mechanisms inefficient due to price inflexibility. This in turn fails to provide the right signal

---

[22] Pang Dongmei 庞东梅, *Cong ziyuanshui shidian kuangda dao zengzhishui kuoda* 从资源税试点扩大到增值税扩围 [Expanding tax pilots from resource tax to value-added tax] (2011-9-21), URL: http://finance.sina.com.cn/roll/20110609/07169963307.shtml [accessed: 2011-9-21].

to both consumers and suppliers in the energy market, causing resources to be consumed at low cost. To solve this problem, China has launched pilot programs such as a graduated power tariff, a graduated water tariff, and a peak/off-peak power tariff.

Since 2009, China has executed a petrol price reform, adjusted prices for petrol several times, and improved its pricing mechanism. In June 2010, China increased the benchmark price for natural gas factories, cancelled the "dual track" pricing system and expanded the price range. Since 2007, the heating supply system reform has provided a basis on which to improve price formation for heating oil. A differentiated pricing strategy for energy was enacted to restrict and eliminate groups of enterprises, such as electrolytic aluminum, iron alloy, calcium carbide, cement, steel, yellow phosphorus, zinc smelting, and other restricted and eliminated products groups. Power surcharges were increased to 0.1 RMB/kWh and 0.3 RMB/kWh.

4.1.4.3 *Green Credit*

In 2007, the China Banking Regulatory Commission (CBRC), the State Environmental Protection Administration (SEPA) and the People's Bank of China called on financial institutions to maintain pressure on and exercise differential treatment with regards to corporate lending: they should provide active credit support to projects the government wanted to encourage and implement credit control for enterprises and projects that were not in compliance with the government's industrial policy or violated environmental law, thus suppressing the reckless expansion of energy-intensive and high-polluting industries. This credit policy was named the "green credit." In 2010, the CBRC further strengthened credit control for energy-intensive industries. With the green credit in place, the share of loans to energy conservation and environmental protection projects increased from 1.87% in 2005 to 8.93% in 2009 of total lending.[23] In 2009, 2,348 customers quit the "two-high" industry loan, an increase of 823 as compared to 1,525 in 2008.[24] Table 4.5 summarizes the energy conservation and environment protection projects supported by Chinese financial institutions in between 2005 and 2009.

---

23 China Bank Association, *2009 niandu Zhongguo yinhangye shehui zeren baogao* 2009 年度中国银行业社会责任报告 [Social responsibility report of China's Banks: 2009], http://www.china-cba.net/do/bencandy.php? Fid=65fid=5299.

24 Ibid.

Table 4.5. Energy conservation and environment protection credit loans.

Years	Projects' total Loan (billion RMB)	Ratio of energy conservation and environment protection projects' loans to total loans (%)	Number of projects that are approved with loans	Number of enterprises that are involved
2005	132.306	1.87	1334	1847
2006	202.894	2.65	1999	2649
2007	341.1	2.70	2715	3505
2008	371.016	3.11	2983	3615
2009	856.046	8.93	6412	4099
Total	**1903.362**			

Source: CBRC (2009), (2010).

4.1.4.4 *Market Cultivation*

As the Chinese government is actively transforming itself in its intervention approach towards economic management, improving market mechanisms regarding energy conservation has become the new development trend. In April 2010, the State Council Office issued the *Advice Regarding Further Boost of Energy Management Contract to Promote Energy Conservation Service Industry's Development*[25] by the NDRC and other departments, to initiate further support for energy management contract projects and energy conservation service enterprises from the side of investment, government spending, taxation, finance, and other aspects. In June 2006, The Ministry of Finance and the NDRC decided to reward projects that had conserved between 500 tce/a and 10,000 tce/a, with 240 RMB/tce on the national level, while rewards on the provincial level were no less than 60 RMB/tce.

---

[25] NDRC, *Zhongguo yingdui qihou bianhua de zhengce yu xingdong. 2010 niandu baogao* 中国应对气候变化的政策与行动—2010 年度报告 [China's policies and actions for addressing climate change: 2010 progress report].

4.2 Policy Review

4.2.1 *Policy Implementation Results*

Because policy measures and policy effects do not have a one-to-one corresponding relationship, it is rather difficult to quantify the effects of the implementation of a particular policy. To have a certain impact, most policy measures must synergize with other policy measures. For instance, during the 11th FYP, the energy conservation campaigns applied a command-control mechanism as their main approach (e.g. the Top 1,000 Enterprises Program and the Campaign to Eliminate Outdated Capacities) alongside a certain amount of support from government subsidies. Therefore, it is difficult to isolate precisely the actual effect of one policy measure out of a set of policies. Nevertheless, quantifying policy results is of significant importance. Not only does quantification allow decision makers to track the success of a certain policy, but quantifiable results provide decision makers with accessible information regarding a policy's effectiveness as far as importance, enforceability, penetration, and other factors are concerned. This report's attempt at quantification aims to analyze the individual effects of administrative measures, incentive-based measures, and market measures, and then evaluate the effect of the 11th FYP policies on China's long-term efforts towards energy conservation and carbon reduction.

The main policy campaigns during the 11th FYP achieved significant reductions in the amount of energy-related CO_2 emissions. Our research results differ from past understandings and judgments, and revise past conclusions. If we divide the main policies into individual policy measures, then administrative measures realized 473 $MtCO_2$ in reductions; incentive-based measures realized 777 $MtCO_2$, and market-based measures realized 15 $MtCO_2$ (Figure 4.5). This indicates that during the 11th FYP period, administrative measures and incentive-based measures were the key factors. Comparing the two, incentive-based measures seem to surpass administrative measures, although in the past the general intuitive feeling regarding China's achievements in energy conservation and carbon reduction had mainly pointed towards the efficiency of administrative measures. One must note, however, that administrative and incentive-based measures frequently interact and overlap; our approach to isolate individual effects has to be understood as an initial attempt. Detailed calculation methods are available in Appendix 4.1.

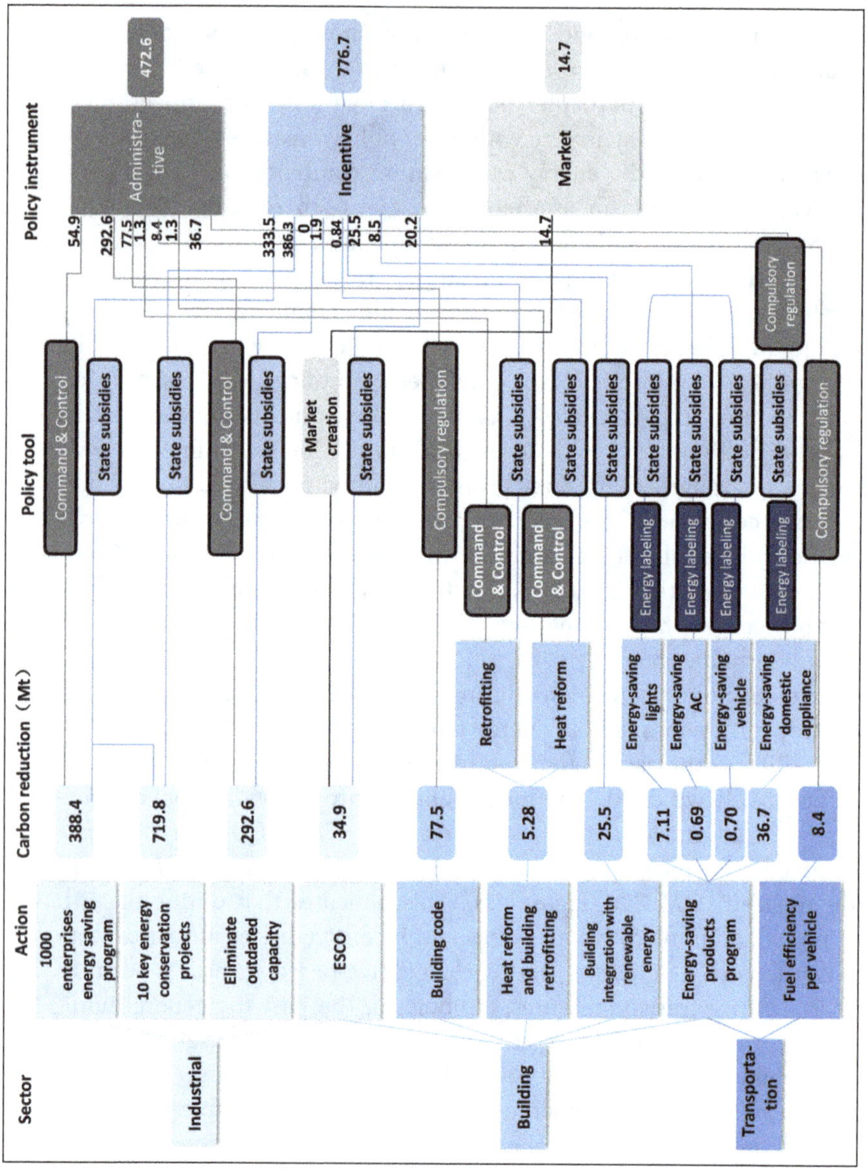

Figure 4.5. Policy Campaigns and Achievements.

4.2.1.1 *Administrative Measures have Reduced Carbon Emissions by 473 MtCO$_2$*

Because of the particular framework of China's political institutions, the government has strong administrative and political power, while at the same time being in possession of enormous resource allocation capability. Political pressure in favor of energy conservation and carbon reduction can hence be passed from the central to the provincial, from the provincial to the local authorities, so that the task of energy conservation can be compartmentalized and eventually realized. The 11th FYP energy conservation campaign, in summary, was characterized by: high targets, tight schedule, and challenging tasks. The factor market being still in its initial build-up phase, administrative measures were the most effective means to achieve energy conservation in the short run. Administrative measures during the 11th FYP took on two forms: command-control and mandatory standards.

Command-control measures are represented by campaigns such as the "1,000 Enterprises Energy Conservation Campaign, "Elimination of Outdated Capacity Campaign," or the "Existing Building Heat Meter and Energy Conservation Retrofitting." The specific measures adopted included a clearly-defined energy conservation target; decomposition of the energy target from central to local levels; and evaluation of target accountability.

The command-control approach resulted in considerable amounts of energy conservation and carbon reductions: The "1,000 Enterprises Energy Conservation Campaign" conserved 150 Mtce,[26] reduced carbon emissions by 388 MtCO$_2$; the "Elimination of Outdated Capacity Campaign" conserved 110 Mtce, and reduced carbon emissions by 293 MtCO$_2$; the "Existing Building Heat Meter and Energy Conservation Retrofitting Campaign" conserved 1.95 Mtce and reduced carbon emissions by 5.28 MtCO$_2$.

Further energy conservation and emissions reduction was realized through the establishing of national mandatory standards, in supplement to energy conservation monitoring and supervision. During the 11th FYP, significant results in this regard were achieved by the Ministry of Construction's new mandatory standard on building energy conservation, and the Ministry of Transportation's standard on fuel economy & end product

[26] NDRC, *Qianjia qiye chao'e wancheng 'shiyiwu' jieneng renwu* 千家企业超额完成'十一五'节能任务 [Top 1000 enterprises exceeded the energy conservation target of the 11th FYP] (2011a-3-14), http://sdpc.gov.cn/xwfb/t20110314_399363.htm [accessed: 2011-3-14].

energy efficiency. Until the end of 2010, a new urban building design phase was implemented with mandatory energy conservation standards that reached 99.5% countrywide for new buildings in their construction phase. During the 11th FYP, the housing area total in which energy conservation schemes were implemented reached 4.857 billion m^2, for which the execution of the new mandatory energy conservation standard for housing achieved a 28.61 Mtce in carbon emissions reductions, 77.53 $MtCO_2$. In the transportation sector, the results from imposing the new standard on fuel economy have been significant. During the 11th FYP, it realized 3.79 Mtce in carbon emissions, 8.37 $MtCO_2$. Energy efficiency standards of end products such as air conditioners, monitors, microwaves, or copying machines realized reductions as high as 13.5 Mtce, and reduced carbon emissions by 37.35 $MtCO_2$ in the 11th FYP. It is worth to point out that despite the fact there was a trade-in scheme for home appliances in place during the 11th FYP, this subsidy was only a pilot program limited to Beijing, Shanghai, and a few other cities, and was only applicable to five categories of products (television sets, refrigerators, washing machines, air conditioners, and computers). The result of this subsidy is not considered here.

It is also worth to point out that these campaigns were not implemented by any single one policy measure, but were the joint efforts of several policy measures. For instance, the "1,000 Enterprises" were able to apply for subsidies from the "Ten major energy conservation projects," enterprises in underdeveloped regions that had successfully eliminated outdated capacities were able to receive special funding from the central and local government to compensate employees. Those participating in the "Existing Building Heat Meter and Energy Conservation Retrofitting Campaign" were able to receive subsidies according to climate zone, i.e. 55 RMB/m^2 in frigid zones or 45 RMB/m^2 at cold zones.

In order to calculate the individual amount of carbon reductions resulting from administrative and incentive-based measures, this section has broken down the policy effects of these mixed policy measures. Because the 1,000 Enterprises make up 47% of the total energy consumption of the industrial sector, it can be assumed that the same 1,000 enterprises have equivalently contributed 47% of the Ten Major Industrial Projects' energy conservation. The Ten Major Energy Conservation Projects Campaign realized a total of 274 Mtce, and reduced carbon emissions by 720 $MtCO_2$. On the basis of this assumption, the 1,000 enterprises realized 129 Mtce, and reduced carbon emissions by 334 $MtCO_2$. Because the Ten Major Energy Conservation Projects Campaign implemented governmental subsidy measures, the 1,000 Enterprises Energy Conservation Campaign's must have experienced related effects which can be seen as effects attributable to incentive-based measures. To calculate the effects of administrative measures, one must deduct the incentive-based measures from the total carbon reduction off the 1,000 Enterprises Energy Conservation Campaign. In other words, given the total of 388 $MtCO_2$ in reductions it realized, minus the incentive-based measures' realization of 334 $MtCO_2$, this would mean that administrative measures amounted to only 54 $MtCO_2$ in realized reductions.

As with regards to the Campaign to Eliminate Outdated Capacities, because the campaign subsidy was only given after enterprises had completed their elimination task and the aim of the subsidy is to compensate employees, it can be assumed that this campaign's subsidy does contribute to any energy conservation. In other words, all carbon emissions reduced by this campaign (293 $MtCO_2$) have to be treated as a contribution from the administrative measures.

As far as the "Existing Building Heat Meter and Energy Conservation Retrofitting Campaign" is concerned, this report assumes that 40% of the energy conserved from the heat meters comes from government subsidies, and that the other 60% comes from command-control policy measures; 60% energy conserved from retrofitting comes from government subsidies and the remaining 40% are contributions caused by command-control policy measures. This assumption is based on the *Northern Heating Region Existing Residential Building's Heat Meters and Energy Conservation Retrofitting Fund Management Interim Measures*, which has confirmed the corresponding coefficient weights of subsidies: subsidy of heating supply system meters, heating source, and heat supply network balance reformation accounts for 40% of the total subsidy, subsidy of building envelope energy conservation retrofitting accounts for 60%. Although this assumption may seem to be not exactly precise because it is derived from confirmation of subsidy proportionality, the subsidy's interim measures has clearly stated that retrofitting received a higher subsidy than that of heat meters. During the 11th FYP, reform of the heating supply system has realized a reduction of 2.11 $MtCO_2$; according to the above assumption, 40% (840,000 tCO_2) comes from government subsidy, 60% (1.27 $MtCO_2$) comes from administrative measures. Retrofitting of existing building reduced 3.17 $MtCO_2$, 60% (1.90 $MtCO_2$) of which came from government subsidy, 40% (1.27 $MtCO_2$) from administrative measures.

4.2.1.2 *Incentive-Based Policy Measures have Realized Reduction of 784 $MtCO_2$*

The "Ten Major Energy Conservation Projects" campaign derived its capability for energy conservation and carbon reduction through government subsidy, making investment in energy conservation technology attractive for businesses. During the 11th FYP, ten major energy conservation projects (in the industrial and construction sector) conserved 340 Mtce,[27] 66.01 Mtce of which were conserved by building projects, while industrial projects conserved 274 Mtce. Therefore, the industrial projects involved in the campaign conserved 274 Mtce, and reduced 720 $MtCO_2$, 334 $MtCO_2$ coming from the 1,000 Enterprises, 386 $MtCO_2$ from other industrial enterprises.

In promoting the utilization of renewable energy, China's main policy measure has been incentive-based subsidies. For instance, rural elementary and secondary schools, residential buildings, clinics, and other public buildings were key targets for subsidies. Subsidy measures included a subsidy of 60 RMB/m^2 for ground source heat pump technology, and 15 RMB/m^2 for solar power integration. In 2010, solar-powered water heaters, methane, and other applications of renewable energy technologies in the building industry were announced to have reached an energy conservation capacity of 12.01 Mtce, and carbon reduction capacity 33.23 $MtCO_2$.

Moreover, to promote energy conservation products, China has allocated subsidies to alleviate the cost mismatch between energy-saving products and normal products. This subsidy has had significant impact on market prices for energy-saving products. For instance, since the beginning of this subsidy in June 2009, prices for highly energy-efficient air conditioners have dropped from 3,000–4,000 RMB per unit to about 2,000 RMB; some models with first-grade energy efficiency have decreased to 1,000 RMB, saving consumers a total of up to 30 billion RMB.[28] With the support of this policy measure, the market share of highly efficient air conditioners has gone up from 5% to a current share of 70% or more, this has in turn lead to the successful implementation of a new energy efficiency standard. Conventional air conditioners with low efficiency grades of three, four, and five have all terminated production, while the industry's overall energy efficiency level increased by 24%, reaching the

---

27 NDRC, *Shida zhongdian jieneng gongcheng qude jiji jinzhan* 十大重点节能工程取得积极进展 [Ten Key Projects program made great progress] (2011b-3-14), http://www.sdpc.gov.cn/xwfb/t20110314_399214.htm [2011-3-14].

28 NDRC, *Jieneng chanpin huimin gongcheng qude mingxian chengxiao* 节能产品惠民工程取得明显成效 [The program "Benefitting the general public of utilizing energy efficiency products" made noticeable impacts] (2011c-3-14), http://www.sdpc.gov.cn/xwfb/t20110314_399362.htm [accessed: 2011c-3-14].

advanced international level.[29] Moreover, the number of energy-efficient passenger vehicles with a mileage of 1.6 liters or less has increased from 101 to 341 models. The monthly sales figure used to be just 50,000 before the promotion, and stands now at approximately 300,000.[30] The market price for energy-efficient light bulbs decreased by 40% after the promotion, 360 million units of low-level mercury light bulbs have been sold, realizing an annual energy savings total of 12.5 billion kW/h, while a total of 62.7 billion kW/h will be saved within the product's life cycle.[31] In June 2010, China launched its promotion of efficient engines; the first batch included 13 enterprises and more than 1,000 models. The second batch was released in March 2011, including 50 enterprises and close to 10,000 models that are in compliance with the promotion policy.[32]

From June 2009 to 2010, subsidized energy-efficient lighting reduced 7.11 Mt in CO_2 emissions, while air conditioners amounted to an annual saving capacity of 0.69 $MtCO_2$. Light 1.6 liter vehicles reduced 0.7 $MtCO_2$. Although the energy efficiency label does help consumers in their decision making process during purchase, this report argues that the large amount of subsidy is the main factor influencing the consumer's purchase decision, therefore this part of carbon reductions should be attributed to incentives.

The most important market-based measure is guidance of energy pricing. Although during the 11th FYP, China has implemented series of energy price reforms, the amount of carbon reductions due to changes in energy price is not measurable. Hence, our calculation has underestimated the amount of carbon reduction realized by market-based measures.

As the Chinese government actively transforms its interventional approach towards economic management, the improvement of energy conservation market mechanisms has become the new development trend. For instance, energy contract management promoted by the industrial and building departments has an annual energy conservation potential

---

29 NDRC, *Jieneng chanpin huimin gongcheng qude mingxian chengxiao* 节能产品惠民工程取得明显成效 [The program "Benefitting the general public of utilizing energy efficiency products" made noticeable impacts] (2011c-3-14), http://www.sdpc.gov.cn/xwfb/t20110314_399362.htm [accessed: 2011c-3-14].

30 Ibid.

31 Ibid.

32 NDRC, *Jieneng fuwu chanye kuaisu fazhan* 节能服务产业快速发展 [Rapid development of energy conservation service industry] (2011d-3-14), http://www.sdpc.gov.cn/xwfb/t20110311_399215.htm [accessed: 2011-3-14].

of 13 Mtce and a carbon reduction potential of 34.85 $MtCO_2$.[33] It is worth to point out that although energy contract management is a market-based measure, its potential for carbon reduction is realized by more than just market-based measures. In fact, a substantial amount of carbon emissions reduction resulted from the support provided by subsidies, thus a decomposed analysis is necessary.

Energy contract management in China is still in its initial phase, and the same applies to the market mechanism of energy conservation service. To promote such a market-driven energy conservation mechanism, the government needs to further enhance fiscal support to stimulate this emerging market, particularly at the initial stage. In 2010, the central government allocated 2 billion RMB in support of energy conservation service enterprises that drive the adoption of energy conservation through energy contract management.[34] The 2 billion RMB subsidy has resulted in a 20.16 $MtCO_2$ carbon reduction potential (with subsidy standard at 240 RMB/tce). Thus, energy contract management has a total carbon reduction potential of 34.85 $MtCO_2$, 20.16 $MtCO_2$ of which were achieved via government subsidy.

4.2.2 *Policy Effects and Input Costs*

4.2.2.1 *Government Capital Investment and Social Capital Investment's Leverage Effect is 1:4.4*

Government capital investment is defined as monetary input in forms such as subsidy and rewards by the central government and/or local governments (provinces and prefecture cities) during the implementation of policy measures. Table 4.6 summarizes funding by the central government and local governments of each policy campaign alongside the carbon reduction capacity for 2010. The 2010 carbon reduction capacity is defined as new energy conservation potential on the basis of 2005.

Table 4.6 shows that the monetary input by the government on average is 167 RMB/tCO_2, in other words the central and local governments spent 167 RMB for every ton of carbon reduced.

Social capital investment is defined as investment other than the government's investment during the implementation of policy measures. It includes mainly business investment and social investment in the process of energy conservation. Table 4.7 lists the social capital investment and the 2010 carbon reduction capacity of the main policy campaigns.

---

[33] Ibid.

[34] Ibid.

Table 4.6. Government capital investment.

Campaign Actions	Central funding (billion RMB)	Local funding (billion RMB)	Carbon reduced ($MtCO_2$)	Cost (RMB/tCO_2)
Major energy conservation projects	31.5	15.8	377.03	125
Elimination of outdated capacity	20.2	10.2	292.6	104
Building energy conservation	15.2	13.4	42.63	671
Energy contract management	2	1.6	32.85	110
Energy saving lights subsidy	1.62	0	7.11	228
Energy saving air conditioner subsidy	11.54	0	0.69	16 725
Energy saving vehicle subsidy	3.04	0	0.696	4368
Total	**85.1**	**41**	**753.61**	**167**

Note:
1. Major energy conservation projects as in the industrial energy conservation projects in the "ten major energy conservation projects campaign."
2. Building energy conservation includes Northern China existing building heat meters and energy conservation retrofitting, renewable energy building integration, government administration building and large-scale building energy conservation supervision system.
3. For detailed calculation of carbon reduction capacity see Appendix 4.1.

Table 4.7. Social capital investment.

Policy	Public funding (Billion RMB)	Carbon reduction ($MtCO_2$)	Cost (RMB/tCO_2)
Ten key projects	186	481.02	387
Top 1000 enterprises	287.4	388.41	740
Building codes	194.3	77.53	2506
Building retrofitting	36.4	5.28	6894
Total	85.1	75.567	739

Note:
1. Major energy conservation projects as in the industrial energy conservation projects in the "ten major energy conservation projects campaign."
2. For detailed calculation of carbon reduction capacity see Appendix 4.1.

Average cost for social funding is 739 RMB/tCO_2, which means that for every ton in carbon reductions, 739 RMB of social investment is needed. Comparing the average cost ratio of government investment and social investment, the leverage effect between government capital investment and social capital investment's is 1:4.4. The Energy Research Institute of the NDRC has come to a similar result;[35] the leverage effect of government capital investment vs. social capital investment was stated with 1:4.7.

4.2.2.2 *For Every Ton of Carbon Reduction Capacity, 906 RMB Investment Needed*

For every ton of CO_2 reduction in 2010, 906 RMB of investment was spent during the 11th FYP. One has to note that this report defines cost as in the amount of investment input for every ton in carbon emissions reduction. This concept is not the same as the average carbon price of 100–200 RMB/tCO_2 (10 Euros–20 Euros/tCO_2) on the international carbon trade market. The carbon price indicates the transaction price for one ton in carbon reduction. In order to compare, one must convert this report's carbon reduction capacity into the life cycle of carbon reduction capacity, for instance, one must consider that the life cycle of industrial projects is 15–30 years, the life cycle of a building is set at 15–30 years, and the life cycle of energy-efficient products is set at 10 years.

4.2.2.3 *Building Energy Conservation Investment Cost is Five Times as Much as Industrial Sector Energy Conservation Cost*

The average investment cost for energy conservation in the building sector (2,786 RMB/tCO_2) is five times the average investment cost for energy conservation in the industrial sector (545 RMB/tCO_2). Social investment cost in the context of the "Ten major energy conservation projects" campaign was 387 RMB/tCO_2 while for the "1,000 Enterprises Energy Conservation" campaign was 740RMB/tCO_2. The latter's investment cost was 1.9 times that of the former, the reason being that the latter campaign, apart from implementing projects that reduced carbon emissions directly, was also heavily engaged in fields that did not yield direct carbon

[35] Energy Research Institute of NDRC, "Zhongguo nengyuan yanjiuhui nengxiao yu touzi pinggu zhuanjia weiyuanhui 中国能源研究会能效与投资评估专业委员会 [China energy research community professional commission on energy efficiency and investment evaluation]," *Zhongneng shitong* (2010).

reduction effects, e.g. energy management staff training, and research and development regarding energy conservation technology. Taking the average of total investment and the carbon reduction capacity of both campaigns, we have derived the average investment cost of industrial energy conservation: 545 RMB/tCO_2.

Compared to the average investment cost for industrial energy conservation, the investment cost in the building sector was substantially higher: total investment costs for new buildings amounted to 2,506 RMB/tCO_2, retrofitting of existing buildings to 6,894 RMB/tCO_2. Thus on average, investment cost in the building sector in general was 2,786 RMB/tCO_2, about 5 times higher than that of industrial energy conservation. This result implies that as the scope of industrial energy conservation is diminishing, energy conservation costs will increase considerably.

Another reason for such high costs for energy conservation in the building sector lies in the fact that some products actual capacity for energy conservation is insignificant, but requires substantial investment. The current approach takes the efficiency of products involved in building energy conservation product's as standard (that is, specific regulation on thermal insulation performance, heating, lighting, air conditioning, equipment performance, etc.), which cannot truly reflect the building operation's energy consumption, whereas, using consumption standard measurement such as energy consumption/m^2 will be able to reflect the energy consumption level of office buildings. Standard consumption measurement is currently applicable as energy consumption behavior between buildings is rather diverse; it is difficult to set a standard value from a technological perspective.

4.2.2.4 *Subsidy Cost for Energy-Saving Air Conditioners is 73 Times Higher than the Cost of Subsidizing Energy Saving Lighting Systems*

The cost of subsidizing energy-efficient air conditioning is 16,725 RMB/tCO_2 (Table 4.6), which is 73 times higher than the cost of subsidizing energy-saving lighting systems (228 RMB/tCO_2). The main reason for this is that the efficiency of energy-saving light is 70% that of incandescent light, whereas this ratio is only about 5%–15% when it comes to air conditioners. Secondly, energy-saving air conditioning units must theoretically operate more than 2,000 hours annually to become cost-effective (see Chapter Two of this report). But the actual operating hours of a standard air conditioning unit are noticeably lower than 2,000 hours; in Beijing, for instance, air conditioning systems on average operate no more than

500 hours yearly.[36] In contrast, energy-saving light bulbs normally fully exhaust their life cycles.

Subsidy for vehicles with mileage of 1.6 liters or less comes at a high cost of 4368 RMB/tCO_2.The reason is that although small engine vehicles generally save more fuel than large engine vehicles, 1.6l–2.0l vehicles might actually save more fuel than vehicles with 1.6l or less. The subsidy currently in place is set according to discharge volume of the engine, not according to the vehicle's actual fuel consumption. In other words, some subsidized small displacement vehicles are actually not fuel efficient; this is another reason for the high investment cost.

During the 11th FYP, the subsidy for promoting energy-saving products was substantial. For instance, the average subsidy standard for energy-saving air conditioning was 340 RMB/unit (the subsidy standard for units with grade 1 energy efficiency was between 500 and 850 RMB/unit, grade 2 between 300 and 650 RMB/unit), which is 10% or more than the sales price. Beijing, Shanghai, and some other cities have additionally implemented 10% trade-in schemes for television sets, refrigerators, washing machines, air conditioning units, and computers. In other words, customers are able to save as much as 20% upon purchase of an energy-saving air conditioner with the trade-in scheme and the subsidy on energy-saving products in place. To give another example, every energy-saving light bulb received on average 4.5 RMB in subsidy; the purchase price for one Philips 11W energy-saving light bulb is 16.9 RMB, making the subsidy ratio as high as 27%. The standard subsidy for a 1.6l or less vehicle is 3,000 RMB per car, 4% of the average sales price.

4.2.2.5 *Negative-Cost Technology Exhausts Government Capital Investment, Government Funding's Efficiency of Usage is Relatively Low*

The conclusion of chapter two indicates: terminal energy utilization technology with negative cost has gained wide circulation in the industrial and building sector. But during the actual promotion process, negative-cost technology has exhausted large amounts of government investment, reducing the efficiency of government funding. For instance, technology costs for energy-efficient light bulbs was –500 RMB/tCO_2 (see Chapter Two), meaning that when incandescent lights are substituted by energy-saving

---

[36] Li Zhaojian 李兆坚, *Woguo chengzhen zhuzhai kongtiao shengming diaoqi nenghao yu ziyuan xiaohao yanjiu* 我国城镇住宅空调生命周期能耗与资源消耗研究 [Research on life-cycle energy consumption and resource consumption of China's urban air conditioning, Doctoral dissertation] (Beijing: Tsinghua University, 2009).

lights, one can save 500 RMB per ton in CO_2 reduction. From aspect technological perspective, promotion of negative-cost technology can rely 100% on market mechanisms. However, subsidy for energy-efficient light bulbs during the 11th FYP was as high as 27% of the sales price, draining large amounts of government funding.

An overheating in subsidies for energy-saving products can have two negative impacts: one is that subsidies will encourage higher consumption. Energy-saving products only save energy when they replace non-energy saving products of the same kind. Adding unnecessary demand, even for energy-saving products, it is still a waste of resources. Secondly, unlike lighting, in China automobiles are not a necessity of everyday life, and car owners mostly belong to China's middle class. Heavy subsidies for vehicles are thus subsidizing a part of the population that is better-off to begin with, which could lead to widening income gaps. When energy-saving products have gained a sufficient market share, the government should stop handing out subsidies to these products, the funding should then be allocated to sectors where lowering costs is not possible through factor price.

4.3 Policy Features of 11th FYP

4.3.1 *Policy Measures*

During the 11th FYP, administrative and incentive-based policies assumed a leading role which illustrates the double role taken by the government: on one hand, when the market lacks the necessary economic incentives to create spontaneous demand for energy conservation, the government must step in via administrative means, and rapidly and thoroughly realize energy conservation tasks using a top-down approach. On the other hand, the government enforced direct and extensive investments in the energy conservation sector, with the aim to provide enterprises with financial incentives, promote entrepreneurial skills and improve their competitiveness in the energy conservation sector. Economic, financial, fiscal, and informatics measures have also reached their initial application phase. For instance, the energy resource tax was already undergoing reform, but the pricing mechanism for natural resources, which should reflect the degree of resource scarcity, has not yet been established.

Complementary effect of policy measures: For instance, the "Elimination of Outdated Capacity Campaign" was mainly led by constrained command-control policy measures, but it was also supplemented by incentive-based government subsidy measures, differential energy prices (market-based), increase export taxes on high-energy consuming products (tax measure), restrictions on "high energy consumption, high pollution" bank loans (credit measure), and other policy measures.

Although various policy measures were implemented, the degree to which they were applicable varied. For instance, although several policy measures were implemented simultaneously during the "Elimination of Outdated Capacity Campaign," the emphasis was on administrative measures and restrictions, while incentive-based measures lacked guidance and scope. Although in the short run, administrative measures can prevent the development of outdated capacities, market-based measures and tax measures cannot yet optimize the allocation of resources, making the creation of a permanent efficiency mechanism for eliminating outdated capacity difficult.

4.3.2 *Policy Formulation*

During the 11th FYP, the State Council and other departments intensively announced a series of policies in the low-carbon field. For example, with regards to structural adjustment, the State Council between 2006 and 2010 issued 17 policy documents specifically aimed at structural adjustment, the NDRC and other departments issued an additional 30 documents. In 2010 alone, the State Council promulgated seven documents on structural adjustment, and the NDRC and other department formulated another three policy documents.

Moreover, democratic and scientific elements have improved in the process of policy decision making. For instance, the NDRC planned to launch a gradational electricity tariff to guide awareness for energy conservation. In order to formulate the pricing mechanism for the gradational electricity tariff, the NDRC issued *Guidelines Regarding Implementing Residential Stepwise Electricity Tariff (Seeking for Public Comments)*, inviting comments from the public. This resembles the democracy element within policy decision making processes.

There are also noticeable scientific elements within policy decision making. Expert evaluation, technology consultation, pilot programs, and other measures were implemented. For example, before the implementation of the "1,000 Enterprises Energy Conservation Campaign," Shandong province's Economic and Trade Commission initiated a voluntary pilot program in cooperation with Jinan Iron & Steel Co. and Laiwu Iron & Steel Co in 2003. The success of this program provided reference material for the subsequent "1,000 Enterprises Energy Conservation Campaign."

Nonetheless, the consistency of policy decision making still requires improvement, problems such as arbitrariness, repetitiveness, and inconsistency in policy decision making remain to be addressed. For instance, in order to prevent the rapid development of high energy consumption, China revoked 553 export tax refunds for energy-intensive or high-pollution products on July 1, 2007. But in response to the financial crisis, China on April 1, 2009 once again promoted export tax refunds for these products. The financial crisis can be seen as an opportunity to prevent the development of high energy consumption, had the export tax refund policies remained unchanged during the period, which would have resulted in a positive contribution towards structural adjustment. However, the re-issuing of export tax refunds offset the early results.

4.3.3 *Policy Implementation*

The significant results of China's conservation-related policies can in part be attributed to the advantages of a socialist political framework with Chinese characteristics and a unitary political system. More specifically, this refers to the set of regulations and decomposed energy conservation targets, scientific energy accounting, monitoring and evaluation methods, enhanced frequency and intensity of energy conservation evaluations, rewards and penalties, as well as publicly announced evaluation results.

However, the pace, dynamics, and long-term effectiveness of policy implementation need to be improved. For instance, controlling the pace and intensity of eliminating outdated capacities is important, elimination that is conducted too rapidly or too slowly is itself a waste of energy. Statistics show that starting from 2006, 30%–40% of steel, cement, and other energy-intensive manufacturing industries faced market demand for new production capacity in the industrial sector. In contrast, only one third of new production capacity is for making up the left out market shares of eliminated outdated capacity.[37] To give another example, during the last season of the last year of the 11th FYP, Hebei, Zhejiang, and other provinces encountered power outages. These incidents show the lack of long-term planning mechanisms with regards to energy conservation and carbon reduction during the process of policy implementation. Problems such as inconsistent policy pacing remain.

Overall, increase in domestic demand and international pressure has moved China's low-carbon development forward. Low-carbon development in China still faces challenges such as energy shortages, energy

---

37 China Economics Times, "Taotai luohou channeng ying heli bawo lidu he jiezou 淘汰落后产能应合理把握力度和节奏 [Keep intensity and pace when eliminating outdated production capacities]," http://www.chinanews.com/ny/2010/12-27/2748655.shtmlp2010-12-27.

security, environmental protection, transforming the path of economic development, and global climate change. To tackle these challenges, China has committed itself to turn low-carbon development into a national strategy. The 11th FYP has established the initial framework for a low-carbon policy; this provides the foundation for energy conservation and carbon reduction in the long run. These policy measures will be carried forward into the 12th FYP with further development and innovation to support the transformation of China's economic development path.

Appendix 4.1 Calculation of Effects of Energy Conservation Actions and Programs

No.	Action	Data source	Energy savings (104 tce)	Coefficient (tCO_2/tce)	Carbon reduction (104 tCO_2)
1	Top 1000 enterprises	NDRC, 2011a	15000	2.55	38841
2	Ten key energy saving projects	NDRC, 2011b	34000	—	90483
	Building	Calculated	6601	—	18503
	Industry	Calculated	27399	—	71980
	Contribution of Top 1000 enterprises	Calculated	12878	2.55	33353
	Contribution of other enterprises	Calculated	14521	2.66	38627
3	Elimination of outdated production capacity	NDRC, 2010	11000	2.66	29260
4	Energy performance contracting	NDRC, 2011d	1300		3485
	Building	Calculated	250	2.77	692
	Industry	Calculated	1050	2.66	2793
5	EE design codes of new buildings	Calculated	2861		7753
	Building envelop		1717	2.71	4653
	Central heating		1144	2.71	3100
6	Retrofitting existing buildings	Calculated	195		528

Table (*cont.*)

No.	Action	Data source	Energy savings (104 tce)	Coefficient (tCO_2/tce)	Carbon reduction (104 tCO_2)
	Building envelop		117	2.71	317
	Central heating		78	2.71	211
7	Integrating renewable energy into buildings	Calculated	1201		3323
	PV heaters		598	2.71	1621
	Biogas		555	2.71	1504
	Others		48	2.71	130
8	EE lighting	Calculated	994	2.71	2750
	Receiving subsidies		257	2.71	711
9	Energy efficient standards of appliance	Calculated	1350	2.71	3735
	Air conditioner		25	2.71	69
10	Energy efficient vehicle	NDRC, 2011c	30 (gas)	2.32	70
11	Fuel economy	Calculated	379		837

CHAPTER FIVE

INSTITUTIONAL INNOVATION OF LOW-CARBON DEVELOPMENT

Abstract: The Energy Conservation Target Accountability System, energy performance contracting, and renewable energy development mechanism are the three most representative examples of institutional innovation concerning low-carbon development in China during the 11th FYP. The energy conservation target accountability system holds lower-level governments accountable for meeting energy savings targets set by higher-level governments, which strengthens the pressure of energy saving to trickle down the administrative hierarchy. It ensures effective implementation of national policies, and facilitates policy innovation with local governments.

All levels of governments use energy performance contracting (EPC) as an innovative market-based mechanism to promote low-carbon development, complemented by fiscal subsidies. This new mechanism takes advantage of the creation of new markets, effectively attenuating financing difficulties of medium and small enterprises (SMEs) and facilitating the utilization and application of energy-saving technologies and the development of the energy-saving industry.

Development and application of renewable energy technologies brings huge public benefits, yet is inhibited by high costs, which is a typical example of market failure. Renewable Energy Law is an innovative measure that the government has taken to provide institutional support for the provision of public goods. This law establishes five regimes for renewable energy development, and clearly regulates the responsibilities and obligations of relevant legal entities including the governments, enterprises and the public, which to certain extent solves problems of grid-connection, financing, and costs of renewable power and facilitates the development of renewable energy industry.

Institutional innovation of low-carbon development in the 11th FYP demonstrates China's institutional advantages in innovating management scheme of the provision of public goods and public administration. China's institutional innovation in the face of climate change provides a notable example to the world.

5.1 Pressure to Save Energy and Reduce Carbon Emissions Calls for Institutional Innovation

5.1.1 *Energy-Saving Becomes an Increasingly Important National Strategy, but Local Governments Lack Internal Incentives to Implement Energy Conservation*

Confronted with issues of energy shortages, energy security, environmental protection, and climate change, China is paying increasing attention to energy conservation and has enacted a series of laws to ensure its implementation. Structured implementation of energy-saving measures in China originated in the 1980s when the question of energy shortage began to play an increasingly relevant role. In 1980, the State Council released "the decision to strengthen energy conservation in China" which for the first time in history regulated energy conservation in the form of a formal authoritative legal document. The document also set the guiding principle for energy development: "pay equal attention to development and conservation, and prioritize conservation in the near term," which validated the strategic status of conservation in energy development. In 1986, the State Council issued a "Provisional Regulation on Energy Conservation," clearly specifying how the State Council and governments of all levels should assign and coordinate tasks related to energy conservation. The regulation proposed a quota management on energy consumption per unit of product and associated energy savings targets with planned allocation of materials through performance evaluation, rewards for meeting energy conservation targets, and penalties for exceeding targets.[1] During the next decade, the "Provisional Regulation on Energy Conservation" established the foundation for energy conservation in China. In 1997, the "Energy Conservation Law of the People's Republic of China" was passed. The law verified the guiding principle "pay equal attention to energy conservation and development, and prioritize energy conservation," and established that "energy conservation is a long-term strategic principle of national economic development." Even though energy conservation is playing a bigger and bigger role for national economic development, it is difficult to provide internal incentives for local governments to implement energy conservation, since energy conservation to a large extent is a public good. In 2007, the revision of the 1997 "Energy Conservation Law" set up a series of energy conservation management schemes, including the energy conservation target accountability system, which to some extent solves the problem of low incentives and insufficient pressure for local governments to comply with energy conservation policies.

5.1.2 *Despite the Increasing Number of Energy Conservation Policies, Market Mechanism for Energy Conservation is not yet Established*

Since the 1980s, the Chinese government has enacted a series of energy conservation policies in order to achieve its goal of reducing energy

[1] Wu Guohua 吴国华, *zhongguo jienengjianpai zhanlue yanjiu* 中国节能减排战略研究 [A strategic study on energy conservation and emission reduction in China] (Beijing: Jingji kexue chubanshe, 2009): 182.

consumption per unit of GDP.[2] These policy measures include: supporting energy efficiency investment, executing a quota management system for energy consumption and energy efficiency, increasing R&D in energy efficiency, establishing energy conversation service centers,[3] controlling energy-intensive technologies and industries, enhancing energy conservation management,[4] raising energy prices, and implementing categorized feed-in tariffs for high energy consumption.[5] Although China has implemented a fair number of energy conservation policies and measures, most existing energy conservation policies are administrative in nature, i.e., "command and control" measures, while market-based measures are missing. The Energy Conservation Law of 1997 was created from the perspective of government management and supervision, which to a certain extent set the legal foundation for government-mandated energy conservation. Under the current institutional framework, local governments play a crucial role as the administrators of energy conservation. They can allocate energy conservation tasks to enterprises according to administrative orders from higher-level governments, as well as supervise existing energy conservation policies based on their individual economic development status. Administrative command-based energy conservation management schemes have notable advantages as well as disadvantages: they emphasize effectiveness but neglect efficiency, and disregard the fundamental role of the market in energy conservation, which contributes to the insufficient availability of energy conservation services. Due to deficiency in funding and the lack of financing sources, a large number of small and medium enterprises (SMEs) are confronted with difficulties when it comes to energy conservation. Energy performance contracting (EPC) is an important mechanism by which China creates an energy conservation market. It is a new measure that is conducive to attenuating financing difficulties of SMEs in implementing energy conservation.

---

2 J.E. Sinton, M.D. Levine, Q. Wang, "Energy Efficiency in China: Accomplishments and Challenges", *Energy Policy* 26, no. 11 (1998): 813–829.

3 Ibid.; also: Zhao Xiaoli, Hong Dongyue 赵晓丽, 洪东悦, "Zhongguo jienengzhengce yanbian yu zhanwang 中国节能政策演变与展望 [Evolution and outlook on China's energy conservation policy]," *Ruankexue* 4 (2010) 29–33.

4 Chen Jiabin 陈甲斌, "Woguo xianxing jieneng zhengce shuping 我国现行节能政策述评 [Commentary on current energy conservation policies in China]," *Zhongguo Nengyuan* 3 (2003): 28–30; Chen Heping 陈和平, "Zhongguo Zhengfu de jieneng zhengce 中国政府的节能政策 [Energy conservation policy of the Chinese government]," *Jieneng Jishu* 5 (1995): 2–3.

5 Zhao Xiaoli, Hong Dongyue 赵晓丽, 洪东悦, "Zhongguo jienengzhengce yanbian yu zhanwang 中国节能政策演变与展望 [Evolution and outlook on China's energy conservation policy]," *Ruankexue* 4 (2010) 29–33.

5.1.3 *Existing Policy System in China Cannot Easily Adapt to the Fast Development of Renewable Energy Industry*

The development of a global industry for renewable energy has accelerated since the 1990s. Renewable technologies are constantly updated, and the cost of renewable power generation continues to drop. Unit capacity for wind turbines, for example, has grown from several hundred watts to several thousand watts since the 1980s. Average unit capacity of wind turbines installed globally in 2003 was 1,300 kW. Cost of wind power generation fell from 20 cents/kWh in early 1980s to 5 cents/kWh at the beginning of the 21st century. Despite the dramatic drop in costs, costs of renewable energy are still higher than conventional power. In the face of global climate change, all countries in the world regard the development of renewable energy as an important strategy and support renewable industries with policies and financial assistance. Renewable energy is no longer viewed as auxiliary energy, but alternative energy, illustrating its higher status in China's energy strategy. Compared to developed countries such as the United States, Japan, and Europe, China's policy system cannot easily adapt to the demand for fast development of renewable energy. Since the development and application of renewable technologies involves huge public benefits but is constrained by high costs, and the market potential needs to be discovered and extended, market-based mechanisms alone are unlikely to bring about a successful renewable industry. Development of renewable technologies requires mandatory administrative measures by governments and financial and policy support in all aspects in order to solve problems related to costs and grid connection in renewable energy development.

5.2 Institutional Innovation in Low-Carbon Development in the 11th FYP

During the 11th FYP, China established and refined the energy savings target accountability system in order to restrict local governments' behavior and incentivize local governments to implement national policies so that they can play a key role in energy conservation management. As a fundamental regime in energy conservation management, the target accountability system significantly contributed to the realization of the energy savings targets in the 11th FYP.

To make up for the lack of market-based mechanisms for energy conservation, the Chinese government provided EPC alongside substantial financial and policy support. Even though EPC only had a moderate influence on China's low-carbon development, as a new marked-based regime EPC is expected to play a more important role in the future implementation of energy conservation.

In the field of renewable energy development, the Renewable Energy Law, created in 2005 and put into action in 2006, clearly specified the responsibilities and obligations of relevant legal entities including the central and local governments, enterprises, and the public, and defined a series of renewable energy development regimes, including a renewable portfolio standard, compulsory grid connection, categorized feed-in tariffs, cost-sharing, as well as designated funding. A complete legal institutional system was created centering on the five fundamental regimes to support the development of renewable energy, especially renewable power.

5.2.1 *Energy Conservation Target Accountability System*

In August 2006, the State Council enacted "the decision to strengthen energy conservation in China," specifying concrete means by which China can meet its energy savings target in the 11th FYP, i.e., establishing the energy conservation target accountability system and a performance evaluation matrix. In September 2006, the State Council issued "the State Council's approval of the designations of targets of reduced energy consumption per unit of GDP to different regions in the 11th FYP," achieving the assignment of energy savings targets at the provincial level by proposing concrete targets for energy conservation in the 11th FYP. In June 2007, the State Council released the "Notice on publishing and distributing comprehensive working plans for energy conservation," making it clear that major government officials of every given region are primarily responsible for energy conservation and meeting their respective targets. In October 2007, the Standing Committee of the National People's Congress approved the revision of the Energy Conservation Law, putting the energy conservation target accountability system into law. In November 2007, the State Council released the "Notice on Accounting and Supervision and Examination Plans and Measures to Energy Conservation and Emissions Reductions from State Council." This notice specified the accounting and monitoring plans for energy conservation and the evaluation system for energy consumption per unit of GDP. The release of this notice marked the basic establishment of the energy savings target accountability system (see Figure 5.1).

The energy savings target accountability system fits well into the Chinese governmental structure. It holds lower-level governments accountable for

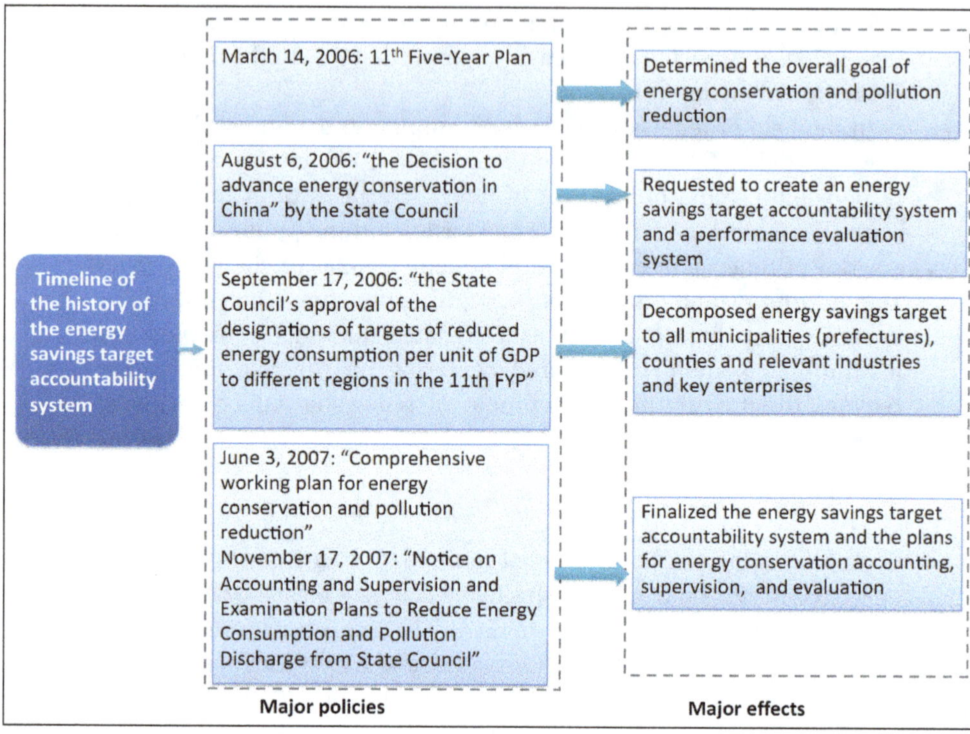

Figure 5.1. Formation process of energy savings target accountability system.

meeting energy savings targets determined by higher-level governments, which causes energy-saving policies to trickle down the administrative hierarchy, ensuring the effective implementation of national policies, and facilitating policy innovation at lower levels of government. Compared to previous energy conservation management schemes, this accountability system is groundbreaking with regards to the following aspects:

5.2.1.1 *Specifying Energy Conservation Targets of All Levels of Local Governments*

Since the official release of China's 11th FYP, the central government and lower-level governments have started to negotiate and adjust targets on energy conservation and consumption reduction assigned to different regions. According to the State Council, the NDRC, and more concretely the Department of Development and Planning within the NDRC, is responsible for decomposing and allocating energy conservation targets to provinces, municipalities and autonomous regions. The decomposition and allocation of energy savings targets are composed of several stages: First, in reference to the national goal of reducing energy intensity (defined as energy consumption per unit of GDP) by 20% and considering the actual situation of individual provinces, every province/municipality/autonomous region voluntarily proposes an energy conservation target to the NDRC. The NDRC then summarizes the promised reductions in energy intensity made by all provinces. Second, the NDRC verifies energy intensity reduction targets of higher than 20% committed to by some provinces. It then negotiates with provincial governments with energy intensity reduction targets of less than 20% or with no target at all, then adjusts these targets and finalizes them.

When energy conservation targets for provincial governments have been determined, the provincial governments then further decompose and allocate their respective targets to city-level governments, which then allocate their targets to county-level governments. The level-by-level decomposition of energy conservation targets ensures that governments of all levels have clear-cut targets, which points local governments in the right direction for their energy conservation implementation.

5.2.1.2 *Specifying the Responsibility of Major Local Government Officials*

The target accountability system makes it clear that major local government officials take the primary responsibility for energy conservation in their respective regions. As a key means to achieve energy conservation targets, the central government in June 2007 established the "National Leading Group to Address Climate Change and Energy Conservation and Emissions Reduction." This leading group was led by premier Wen Jiabao, and comprised of all the key leaders of all ministries within the central government. The key responsibility of the leading group is to "organize and manage the implementation of energy conservation and emissions reduction policies made by the State Council, uniformly arrange energy conservation and pollution reduction tasks, study and discuss key policy recommendations, and coordinate to solve key problems in energy conservation."

Only several months after the establishment of the "National Leading Group to Address Climate Change and Energy Conservation Pollutant Discharge," the provincial equivalents followed suit. The organizational structure of the provincial leading groups to address energy conservation and pollution reduction is similar to that of the national leading group, i.e., a provincial leading group is led by the highest executive officer of the province and comprised of primary leaders of the provincial-level

Development and Reform Commission, Bureau of Economics and Information Technology, Environmental Protection Bureau, Department of Finance, Department of Transportation, and Department of Construction.[6] The State Council and provincial governments set examples for municipal, county, and village governments who then created leading groups at the corresponding levels. The creation of leading groups of energy conservation and emissions reduction strengthened the power structure of energy conservation, enhanced the leadership and coordination capability of energy conservation, thus improving the efficacy and efficiency of the execution of energy conservation policies.

5.2.1.3 *Enhancing the Accounting and Monitoring Systems of Energy Conservation*

The "Implementation Plan of a Statistical Indicator System for Energy Consumption per Unit of GDP" regulates the statistical methodology in recording energy consumption. The general concept is to create and improve the energy accounting and survey system based on the accounting methods of total energy consumptions at all levels of government from the perspective of energy supply accounting and energy consumption accounting. An important precondition for the energy savings target accountability system is access to real and reliable energy consumption data provided by statistics bureaus. At present, 31 provinces and municipalities in China have created energy and resource statistics divisions to increase staffing and funding for energy accounting. This is gradually improving the originally weak energy accounting capacity in China and has formed a supportive force for the successful completion of energy accounting tasks. Meanwhile, China has continued to improve the quality of statistical data on energy consumption so that it reflects the progress made in energy conservation and consumption reduction in different regions and results achieved in key industries and enterprises. High-quality statistical data on energy consumption serves as an important basis for performance evaluation regarding energy consumption.

The "Implementation Plan of a Monitoring System for Energy Consumption per Unit of GDP" specifies rules on the monitoring methodology

[6] Qi Ye et al.齐晔等, *Zhongguo huanjing jianguantizhi yanjiu* 中国环境监管体制研究 [Researches on China's Environmental Governance] (Shanghai: Sanlian Shudian, 2008): 52.

of energy consumption and proposes that higher-level statistical bureaus verify and monitor energy conservation and consumption reduction indicators as well as the quality of data provided by lower-level statistical bureaus. The purpose of monitoring is to ensure data quality. The creation of an energy conservation accounting and monitoring system provides data support for the performance evaluation system of energy conservation, and is conducive to measuring the progress made by local governments in energy conservation.

5.2.1.4 *Linking Performance Evaluation of Governments to Energy Conservation Targets*

The "Implementation Plan of a Performance Evaluation System for Energy Consumption per Unit of GDP" provides detailed rules on subject evaluation, content and methodology, evaluation procedures, and rewards and penalty measures. Evaluation content is divided into energy conservation target performance and implementation of energy conservation measures. An appraisal and evaluation working team composed of the NDRC, the Ministry of Supervision, Ministry of Human Resources and Social Security (MOHRSS), the State-Owned Assets Supervision and Administration Commission Of The State Council (SASAC), the General Administration of Quality Supervision, Inspection and Quarantine (AQSIQ), and the National Bureau of Statistics and National Energy Office of China will assess energy conservation performances through on-site verification and random checks. After evaluation results are authorized by the State Council, the NDRC will make a public announcement of the results. Evaluation results also serve as crucial evidence for the evaluation of government officials. Under the current performance evaluation matrix, government officials face a "one ballot veto" and are held accountable for meeting energy conservation targets. Figure 5.2 is a flow chart demonstrating how the state government evaluates provincial governments. In reference to the evaluation plan issued by the state government, provincial governments also created their respective versions on the evaluation of municipal governments.

Before the introduction of the accountability system, local governments were evaluated based on their performance in GDP growth, investment attracted to the corresponding region, and investment projects. The previous GDP-oriented performance evaluation criteria encouraged some local governments to approve energy-intensive and highly polluting projects resulting in environmental deterioration and resource depletion which posed tremendous difficulties to their energy conservation targets.

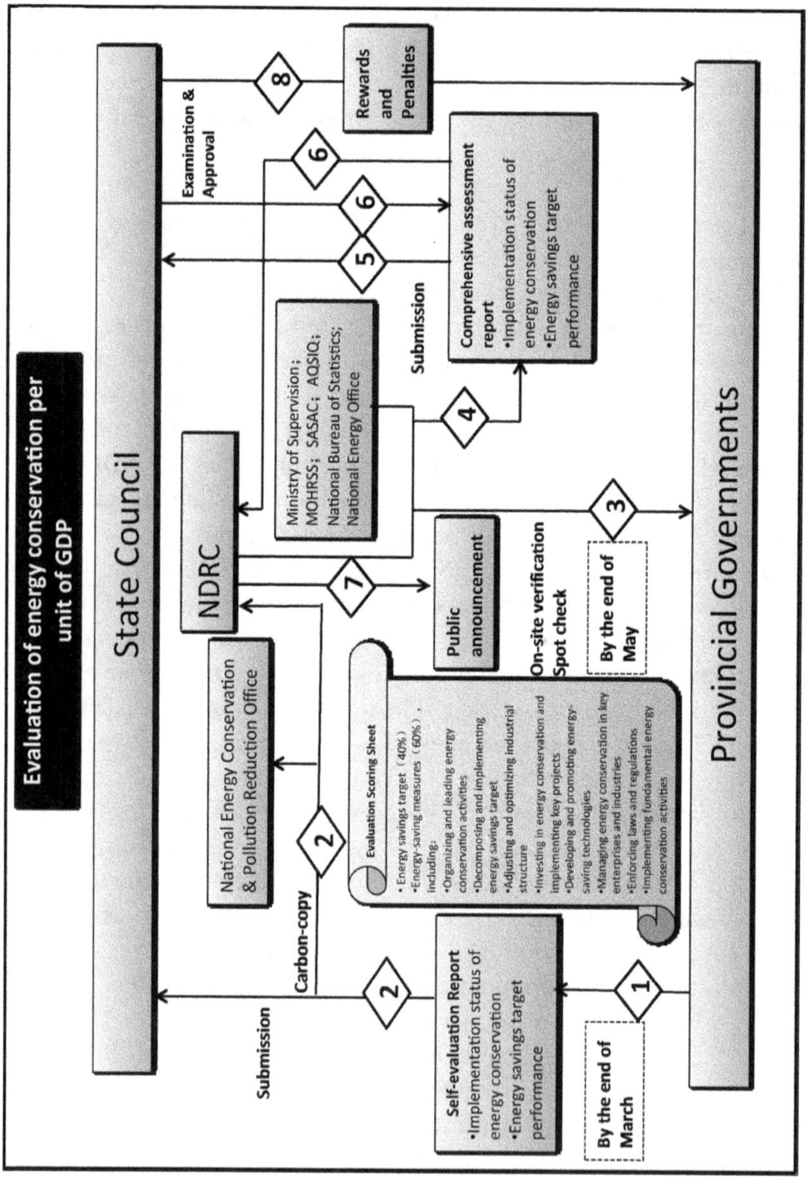

Figure 5.2. Evaluation procedure.

The target accountability system, on the other hand, views the performance of energy conservation indicators as a key indicator and brings it into the performance evaluation matrix of local government officials. The target accountability system to some extent altered the incentives of local government officials. Meanwhile, the public announcement scheme for energy conservation performance improves transparency of energy consumption information and is beneficial to public supervision of energy conservation performance.

5.2.2 *Energy Performance Contracting*

Energy performance contracting (EPC) is a new market-based mechanism commonly used in developed countries to promote energy saving services. EPC was first brought to China in the late 1990s. Under an energy performance contract, an Energy Service Company (ESCO) conducts energy audits and identifies the energy savings potential for the client, commonly referred to as the host company, then provides the financing for upgrades in energy efficiency and performs the upgrades. The ESCO shares the resulting savings with the host company, paying off the capital investment. The EPC model dramatically reduces funding and technological risks of the host company, encouraging host companies to explore their energy savings potential. Therefore, EPC is a win-win mechanism that generates not only public benefits for energy conservation and pollution reduction to the whole society, but also creates economic benefits for the ESCOs and host companies.

The 11th FYP witnessed increasing attention paid to EPC by central and local governments, which resulted in the creation of a series of policies in support of the EPC industry. The "Notice on promoting energy performance contracting and the energy service industry" (State Council Office [2010] No.25) and other supporting documents created an ideal policy and institutional environment for the promotion of EPC, marking an important milestone in the promotion of EPC in China. The Notice illustrates that the Chinese ESCO industry is in line with national industry policies which has stepped into a brand-new stage of development. Meanwhile, government divisions responsible for energy conservation are increasingly aware of the important role that EPC plays in energy conservation. Economic commissions in Shanghai, Yunnan, Gansu, and Sichuan hosted exhibitions, forums, and promotion shows which served as platforms on which ESCOs effectively exchange and cooperate with local enterprises.

Table 5.1. EPC market size in China (2006–2010).

Year	2006	2010	Growth factor
EMCA membership	89	560	5
Number of ESCOs	76	782	9
Workforce in ESCO industry (ten thousand)	1.6	17.5	10
Industry scale (100,000 RMB)	47.3	836.29	16
Investment (100,000 RMB)	13.1	287.51	22
Standard coal capacity savings (10,000 tons)	86.18	1064.85	11
CO_2 emissions reductions (10,000 tons)	215.45	2662.13	11

Source: EMCA official website, *Shiyiwu zhongguo jienengfuwuchanye fazhan baogao* '十一五' 中国节能服务产业发展报告 [Report on energy service industry in China during the 11th Five-year-plan period] (2011-01-13), http://www.emca.cn/bg/hyxw/qt/20110113122955.html [accessed: 2011-08-02].

China's ESCO industry, supported by the ESCO financing guarantee platform and energy management company association (EMCA), achieved sustainable and fast growth during the 11th FYP by taking advantage of market forces, technologies, and policies. There are currently four types of EPC business models in China: shared savings, guaranteed savings, outsourcing energy management, and financing lease. From 2006 to 2010, EMCA membership increased from 89 to 560, the number of EPC-based ESCOs in China increased from merely 76 to 782, nine times more than in 2006. Workforce in the ESCO industry increased from 16,000 to 175,000, the volume of the ESCO industry grew from 4.73 billion RMB to 83.629 billion RMB, 16 times more than in 2006; investment in EPC projects increased from 1.31 billion RMB to 28.751 billion RMB. Annual energy savings of EPC projects went up from 0.8618 Mtce to 10.6485 Mtce, equivalent to increasing CO_2 emissions reduction from 2.1545 Mt to 26.6213 Mt. Social capital investment leveraged by the ESCO industry amounted to more than 180 billion RMB during the 11th FYP (see Figure 5.1).

EPC is a new market-based mechanism used in energy conservation. It is innovative in that an ESCO financing guarantee platform not only removes the financing obstacles to energy conservation and pollution reduction that governments and energy-consuming entities encounter, but it also offers new investment opportunities.

5.2.2.1 *Create an ESCO Financing Guarantee Platform and Reduce Financing Pressure of Small and Medium Enterprises (SMEs)*

The greatest obstacle in promoting EPC is financing. Statistics suggest that 92% of ESCOs have encountered difficulties in this regard.[7] Since most ESCOs are young SMEs with low credit rating, banks normally request tangible assets such as land and equipment as collateral in order to hedge risks when ESCOs apply for bank loans. However, since most ESCO are technology-oriented firms, they normally cannot meet the request for tangible assets. Furthermore, EPC has a short history in China, and banks are thus not familiar with the working process of this new business model. When a complete loan collateral or guarantee mechanism is not yet in place, commercial banks are not willing to issue loans to ESCOs or their clients. As a result, most Chinese ESCOs either depend on support from designated funds set up by the World Bank or have to rely on their own capital to finance EPC projects. Since ESCOs need to finance the investment total in advance, and the service chain is relatively long, ESCOs typically have a high debt to asset ratio, which often inhibits further development.

(1) *Create an ESCO loan guarantee mechanism.* In order to promote EPC and improve Chinese ESCOs' access to financing, the World Bank/Global Environment Facility (GEF) partnered with the China National Investment and Guaranty Company (I&G) to create a commercial loan guarantee program under the 2nd phase of the China Energy Conservation Project. The program provided one-to-one assistance to ESCOs in need, allowing them to receive loans or equity financing through the guarantee program.[8] The World Bank established a designated guarantee fund for the ESCO financing guarantee program. The World Bank provides financing guarantees to ESCOs that perform EPC-based energy conservation projects, and is responsible for the repayment when ESCOs default on their loans. ESCOs improve their credit rating through a third party (a guaranty company), thus receiving access to financing from the bank. While improving the

7 Hu Xuyang 胡旭阳, "Jingwai EPC dianzhan gongcheng hetong zhuyao fengxian fenxi yu duice chutan 境外 EPC 电站工程合同主要风险分析与对策初探 [Risk analysis and strategy investigation for overseas EPC power station project contracts]" *Jianzhu Jingji* 8 (2008): 79–80.

8 China Social Economic Investigation & Research Center, "2010 nian Zhongguo hetongnengyuanguanli (EPC) shichangdiaocha yanjiu fenxibaogao 2010 年中国合同能源管理 (EPC) 市场调查研究分析报告 [2010 China EPC market survey study analysis report]" (2010): 45–46.

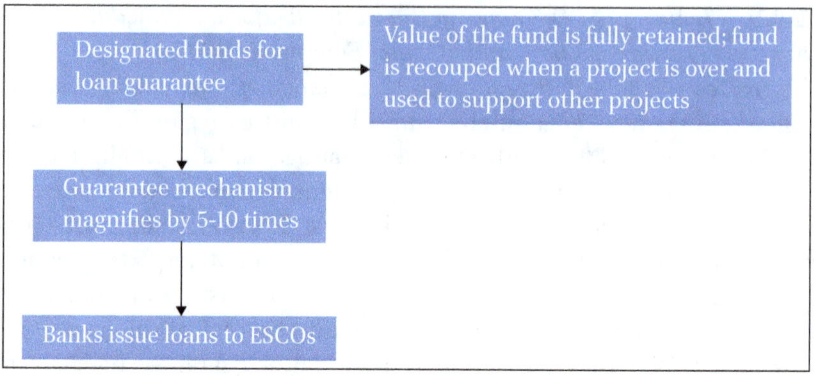

Figure 5.3. EPC loan guarantee mechanism.

credibility of the ESCO and earning commissions from the guarantee, the guaranty company also diverted the risks for the banks. Therefore, this loan guarantee model is beneficial for all three parties, ESCOs, guaranty companies, and banks (see Figure 5.3). This three-in-one credit guarantee financing model has played a major role in improving ESCOs' access to commercial loans and unlocking market-driven financing sources.

(2) *Start of a financing lease model.* Based on the implementation status of EPC projects in China, most projects belong to three types: shared savings, guaranteed savings, and energy management, among which the shared savings model—mostly found in the building sector—is the most widespread one; guaranteed savings projects have become increasingly popular, and are concentrated in the industrial sector; outsourcing energy management projects are mainly conducted in hospitals, hotels, and supermarkets of a certain scale. All three models involve only two parties, i.e. ESCOs and their clients; there are no third parties involved. On June 2, 2006, the companies Siemens, the Inner Mongolia Wulan Cement Group, and the Nan Fang International Leasing Company created a new EPC model called "financing lease" for the energy conservation project of the Inner Mongolia Wulan Cement Group. In the financing lease model, a leasing firm provides capital to the host company; an ESCO conducts the energy audit and the technological upgrades and engineering services for the host company, while the host repays the leasing firm via energy cost savings. The difference between the financing lease model and the other three models is that this new model introduces a third-party leasing firm to directly participate in the implementation of the EPC project. The leasing firm shares the financial risks and thus reduces the pressure for ESCOs.

(3) *Initiate an EPC investment and financing platform.* In order to address the financing problems in EPC, *Beijing Environment Exchange* and EMCA in June 2010 co-initiated an EPC investment and financing platform: this is the first platform in the world through which social funding enters into energy conservation. This platform makes use of market forces and innovative financial tools to promote EPC by transferring accounts receivable from energy savings to secure cash flows to fund new EPC projects. The EPC investment and financing platform is particularly useful to investors because most investors are not aware of the different degrees of profitability of various ESCOs. The platform assesses EPC projects based on a comprehensive evaluation system, which will help investors decide on the financial feasibility of EPC projects. In addition, the platform bridges the ESCO industry and the finance industry, integrates resources from both industries, and adds value to EPC projects by facilitating full information exchange between project developers and investors. The platform also provides a wide range of service options and integration service to governments and enterprises that are considering energy efficiency upgrades.

However, the greatest achievement of the creation of the loan guarantee mechanism, the invention of the financing lease model, and the initiation of the ESCO investment and financing platform is neither the total amount financed nor the performance in terms of energy conservation and emissions reduction. Instead, these innovative financing guarantee programs update the conventional business culture of ESCOs, introduce new financing concepts, familiarize them with the existing financing and credit procedures of Chinese banks, renew their perceptions about their own finance, tax, and assets management, and lead the ESCO industry towards a more regulated and market-driven development track.

5.2.2.2 *Create EMCA and Provide Technical Support to ESCOs*

Another obstacle restraining the development of EPC is technical capacity. In April 2004, a large international joint initiative called the *Energy Management Company Association* (EMCA), co-organized by the NDRC, the World Bank, and the GEF, was created to support the ESCO industry in China. The EMCA advances EPC business models by providing strong and customized technical support to emerging and potential ESCOs, helping them create and improve their operational capacity, and facilitating the creation and growth of ESCOs. The ultimate goal is to create an ESCO industry in China. The major responsibilities of the EMCA are: to promote new energy-saving technologies; adopt technologies and products that are suitable to the Chinese market; host promotion shows and exhibitions for energy-saving technologies and products; and solve technical difficulties involved in EPC.

5.2.3 *Renewable Energy Development Mechanism*

Due to China's unbalanced energy resource profile, its energy-consumption structure (which is still dominated by coal) is confronted with rising pressure from the international community to reduce emissions. Greenhouse gas emissions for every additional tce consumed in China are 50% higher than compared to the rest of the world.[9] While China's rising coal consumption stands in contrast to the urgent need for environmental protection, and its mounting energy consumption faces the reality of oil and natural gas shortages, calls for a restructured energy consumption profile, substantial energy savings, and pollution reduction in China have grown louder. In particular, the government has provided legislation for a renewable energy development mechanism to stimulate the renewable industry.

5.2.3.1 *Five Regimes for Renewable Energy Development were Created through Legislation*

On February 28, 2005, the 14th meeting of the Standing Committee of the National People's Congress passed the "Renewable Energy Law of the People's Republic of China" (hereafter referred to as "the Renewable Energy Law") which came into effect on January 1, 2006. The renewable energy law is of special significance: It is the very first law on renewable energy in China and the most authoritative law specifically written to guide the development and utilization of renewable energy. The law clearly specified the responsibilities and obligations of relevant legal entities including the government, enterprises, and the public, and defined a series of renewable energy development regimes, including a total quantity target, mandatory grid connection, classified electricity prices, expense apportioning, as well as designated funding. A complete legal institutional system was created centering on these five fundamental regimes to support the development of renewable energy, especially renewable power generation.

---

[9] Mo Shenxing 莫神星, *Jienengjianpai jizhi falvzhengce yanjiu* 节能减排机制法律政策研究 [Research on the legal policies of energy savings and emissions reduction] (Beijing: Zhongguo Shidai Jingji Chubanshe, 2008).

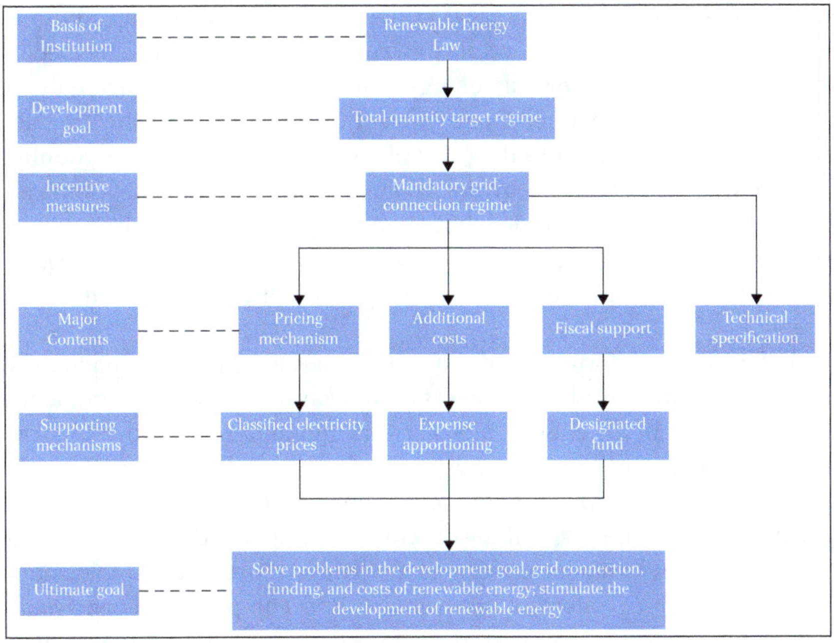

Figure 5.4. Operational model of the renewable energy development mechanism.

Among the five regimes, a total quantity target proposes the goal of renewable energy development strategy in the medium and long term, which is expected to be met in the not-too-near future, meaning ten to twenty years or even more. The five regimes provide a clear guidance for market capacity and a direction for the development of renewable energy; the creation of the state law in particular shows the determination of the government to push its renewable energy agenda. The target is set at a relatively high level; it is therefore the dominating and overarching regime among the five. The remaining four regimes are merely auxiliary means by which the total quantity target can be achieved. Provision 14 in the Renewable Energy Law states that "grid companies should sign grid-connection agreements with authorized renewable power generating enterprises or those that have filed license applications. The grid companies shall purchase all on-grid power generated from grid-connected renewable power generation projects within the coverage of the grid company and provide on-grid service for renewable power generation." Mandatory grid connection has hence to be understood as a stimulus and a necessary measure that ensures the development of renewable energy industry in the context of monopolized and franchised energy sales in China. According to Provision

20 in the Renewable Energy Law, "if the grid company incurs additional expenses in purchasing renewable power at the on-grid prices specified in Provision 19 of the renewable energy law as opposed to average on-grid prices for power generated from conventional sources, the incremental cost can be apportioned to the price of electricity. Detailed apportioning methodology is formulated by the pricing divisions in the State Council." The classified on-grid electricity prices and expense apportioning provide support to mandatory grid connection, and the designated fund regime provided fiscal support to the development and utilization of renewable energy.

As far as the performance of the five regimes is concerned, mandatory grid connection proved effective, yet also involved conspicuous problems. In the context of an energy sales network monopolized by the Chinese grid companies, mandatory purchasing of all grid-connected renewable power is realized through a grid connection agreement between power-generating companies within the coverage of a grid and the corresponding grid company. But since regulations and control mechanisms that manage the conflicting interests between the grid company and renewable power enterprises do not exist, the interests and responsibilities of both parties are not clearly defined, which means that grid companies are not subject to effective administrative oversight or supporting purchase contracts, which does not guarantee the mandatory full purchase of renewable power in the law. In order to improve the situation, the Environment and Resources Protection Committee of the National People's Congress (NPC) of the People's Republic of China in 2009 proposed the "Amendments Of Renewable Energy Law (Draft)" to the Standing Committee of the National People's Congress, suggesting a mandatory grid connection regime and a supporting purchase regime, which is the one of the biggest changes made to the original law. This purchase regime is an important means by which China cultivates its renewable energy market. It stresses the responsibility and obligation of grid companies to purchase renewable energy, which is helpful for solving the conflicts between grid companies and power plants generating renewable energy. The revision also points out that the energy department within the State Council should join electricity regulatory organizations and departments of finance to draft an annual purchase target for national renewable power generation as well as an implementation plan based on the National Renewable Energy Development And Utilization Plan, and determine and announce a minimum requirement for renewable power purchase that the grid company must meet. National electricity regulatory organizations are responsible for the implementation of the minimum purchase requirement. The minimum purchase requirement empowers the government to increase its macro–level control over the renewable energy market.

In addition, in order to resolve the improper allocation of electricity price increment targeted at renewable power, the "Amendments of Renewable Energy Law (Draft)" proposes a renewable energy development fund to guarantee coordinated use of a national support fund. The Renewable Energy Law states that the additional costs which a grid company incurs during its purchase of renewable power are apportioned to electricity prices paid by all consumers. Moreover, grid companies engage in so-called mutual "quota trading" (*pei'e jiaoyi*), in which the quota for each grid company is the difference between additional costs incurred during renewable power purchases and additional income from higher electricity prices. The purpose of this "quota trading" system is to compensate grid companies whose additional electricity income cannot cover the higher costs. However, it is difficult to implement this regulation because (1) electricity price increment charged is recorded as income of the grid companies, so the value-added tax and income tax combined account for one-third of the total additional income from the higher electricity price; (2) since renewable power price subsidies and "quota trading" transactions among grid companies are only conducted every half year, the transaction period is too long to timely and fully compensate grid companies in provinces where incremental electricity prices cannot cover the subsidy that they pay to renewable power generating companies, which imposes financial pressure on power generating companies. Therefore, the "Amendments of Renewable Energy Law (Draft)" proposes that the government will establish a Renewable Energy Development Fund. The Renewable Energy Development Fund makes use of previous funds designated for renewable energy development and the electricity price increment charged for renewable energy, but it is managed in a different way. The Renewable Energy Development Fund is a special policy used to support the development of industries which for their growth cannot totally depend on market forces. It is used to compensate grid companies for the additional costs of renewable power purchases in comparison to on-grid power from conventional energy sources, and to subsidize the electricity price for power generated from state-invested or subsidized public independent energy producers. Even though the name of this particular fund is similar to the designated fund in the past, the Renewable Energy

Development Fund has special significance, because it is directly managed by the state, which "vertically" passes on capital to lower-level organizations. The fund not only dramatically reduces administrative costs, but is also centrally distributed by the state, which ensures that companies investing in renewable energy can receive timely returns. A Renewable Energy Development Fund is an effective way that many countries use to advance sustainable and fast growth of renewable energy. The creation of China's Renewable Energy Fund is an important economic regulation and control measure. It provides stable funding sources for technological development and grid construction, while stimulating increased purchase of renewable power by grid companies.

5.2.3.2 *The Nature of Renewable Energy Development Institutions is to Promote the Growth of the Renewable Industry through Government Intervention*

During the 11th FYP, China established the renewable energy development institution through legislation. Moreover, the government encouraged the development and utilization of renewable energy through government interventions such as refining policies in support of renewable energy institution and overseeing renewable energy development plan. Overall, government interventions in renewable energy effectively reduce pressure of resources bottleneck and contribute to low-carbon development of China.

(1) *Issue relevant administrative laws and ministry-specific regulations.* To ensure effective implementation of the five fundamental regimes in renewable energy development, all relevant departments in the State Council need to discuss and formulate coherent administrative laws, administrative rules, and technical specifications, according to provisions in Renewable Energy Law. At present, the NDRC has announced the "Guided Catalog of the RE Industry," "Management Scheme for Renewable Power Generation," "Preliminary Implementation Plan for Pricing and Expenses Apportioning of Grid-connected Renewable Power," and "Medium-to-Long-term Development Plan for Renewable Energies;" State Electricity Regulatory Commissions released the "Supervision Plan for Full-Amount Purchase of Renewable Power by Grid Companies;" the Ministry of Finance formulated laws, regulations, and policies in support of the overall development of renewable energy such as the "Management Scheme of the Designated Funds for Renewable Energy." The NDRC, the Ministry of Finance, and Ministry of Construction also enacted ministry-specific

regulations or guides. For example, the NDRC and the Ministry of Finance jointly released the "Implementation Plan to Promote the Development of the Wind Power Industry;" the Ministry of Finance published "Preliminary Implementation Plan For the Management of Funds Designated for Application of Renewable Energy in Buildings;" five ministries including the Ministry of Finance jointly released the "Implementation Plan of Fiscal and Tax Policies in Support of Biomass Energy and Biochemical Engineering." In addition, various lower-level governments echoed national policies by enacting policy measures suitable for their own provinces or cities. These supporting policies, regulations, and technical specifications played an important role in promoting renewable technologies and the growth of the industry

(2) *Conduct overall planning of renewable energy development.* The planning and approval of renewable energy projects and the arrangement of designated funds are carried out by a number of ministries including the NDRC, the Ministry of Finance, and Ministry of Science and Technology. Therefore, there is no single one ministry that plans the administrative management of renewable energy in a centralized way and takes responsibility for the approval and pricing of resource planning projects. This results in a gap between the total quantity target of renewable energy and supporting measures and its implementation. Mounting problems arise in the mismatch between national renewable energy planning and local planning, and the mismatch between the planning of the development and utilization of renewable energy and the planning of other energy resources. Meanwhile, the planning of renewable power generation and the planning of electricity and grid are out of synch. For example, a wind power fad has spread across China in recent years. A lot of local governments have enacted plans for wind power projects development, categorized regions according to their respective wind resources, attracted wind power investors to bid for projects, and competed for wind power concession projects. Guangdong, Fujian, Shandong, Hebei, Gansu, and Xinjiang are all planning large-scale wind power projects. Gansu has planned two GW-scale wind power plants. Almost every single county along the coastline of Jiangsu is planning to build 100–200 MW wind power facilities, wind-energy-rich Ulanqab City in Inner Mongolia states in its 11th FYP that the total installed wind generation capacity in the city can reach 24 GW, and will prioritize the establishment of three GW-scale wind power plants in the next five years, making Ulanqab the biggest wind power base in China. As enthusiasm for energy investment grows, problems such as wasteful spending of limited resources emerge.

To cope with the problems in renewable energy development, China conducted an overall planning of renewable energy development, strengthened the comprehensive coordination between renewable energy planning and national energy development strategy; enhanced the guidance, regulation, and control role of national planning and local planning, and actively guided local governments to rationally develop renewable energy industries. Grid companies can make preparations for future development plans by improving their management capacity and quality and ensure smooth grid operation, which will prove useful in inhibiting repetitive constructions and excess production capacity in renewable energy industry. Solving the difficulties in wind power grid connection and digestion requires coordinated planning of wind power and other power sources, and coordinated planning of wind power and the grid. The fundamental solution is to blend large wind power sources into large grids. Installed capacity of wind power in Shandong and Jiangsu provinces exceeds 1GW, which is big enough to allow timely adsorption of wind power. By making interconnecting ties between the Western Inner Mongolia Grid and Jing Jin Tang Grid, wind power capacity of Western Inner Mongolia during demand peaks increased by 50%.

An institutional framework for renewable energy development was established for the first time in history through the Renewable Energy Law during the 11th FYP, and has been gradually refined and improved. In the context of energy shortage and mounting pressure from environmental deterioration, this framework provided powerful institutional support for encouraging various economic entities to participate in the development and utilization of renewable energy, growing the share of renewable energy in China's total energy consumption, optimizing China's energy structure, mitigating fossil-fuel-induced environmental pollution, and dealing with difficulties in renewable power grid connection and week enforcement of supporting policies.

5.3 Characteristics and Influence of Low-Carbon Development Institution in the 11th FYP

5.3.1 *The Establishment of the Institution had Important Impact on Low-Carbon Development*

The energy savings target accountability system was established through a series of official documents, and had important impact on the energy-saving behavior of all levels of governments and enterprises through the enforcement of mandatory targets. All levels of local governments enacted measures to ensure that energy conservation targets are met, and enterprises also pursued energy-saving equipment upgrades and increased investment in energy conservation, which together contributed to a much higher efficiency in terms of energy consumption. Specifically, the energy savings target accountability system played an important role in the following aspects during the 11th FYP:

Creating energy conservation management organizations in governments and enterprises of all levels. Currently, five levels of governments (state, province, municipal, county, and village/town) and key energy-consuming enterprises on all levels have basically established departments responsible for energy conservation, and enacted corresponding energy conservation management measures which set a firm organizational foundation for energy conservation in China.

Restraining energy consumption of all levels of governments and enterprises. Under a GDP-oriented performance evaluation matrix, all levels of local governments are keen on high GDP growth, but overlook their performance in terms of energy conservation. The creation of the energy savings target accountability system allows energy conservation to become a "one-ballet-veto" indicator in the performance evaluation matrix of officials, linking the performance in meeting energy conservation targets to the evaluation of officials, especially the highest official in a government. Therefore, the accountability system is an effective warranty for the stimulating and restraining mechanisms in energy conservation and pollution reduction. It effectively constrains energy consumption behavior of local governments, as well as providing incentives for energy conservation. From the perspective of enterprises, the accountability system is also an important restraining force because whether or not they meet energy conservation targets concerns not only their eligibility for awards, but also the approval of new projects.

Improving energy conservation by all levels of government and enterprises. Constrained by the energy savings target accountability system, all levels of local governments prioritize energy conservation in key energy-consuming enterprises, which has proven effective in raising the energy efficiency of these enterprises. During the 11th FYP, the top 1,000 energy-intensive enterprises met their energy conservation targets in advance. Key energy-consuming enterprises within the jurisdiction of local governments strengthened the allocation of energy-saving investment and energy-saving upgrades, leading to dramatically improved energy efficiency.

The implementation of the accountability system facilitated energy conservation, but also reflected some problems. The level-by-level decomposing of energy conservation targets has imposed huge pressure on governments at the grassroots level. Under the energy savings target accountability system, the central government decomposed the energy savings targets to lower-level governments through energy savings responsibility contracts, so the governments at the grassroots (i.e., county governments) bear tremendous pressure to save energy. The enforcement of the energy savings target responsibility system weakens as the energy savings target is further decomposed to lower-level governments, and support for energy saving management that the respective government can provide also becomes weaker, which calls for even more energy savings in technologies to facilitate reduction in energy consumption per unit of GDP. This poses hardships to the enforcement of energy savings targets. In the later period of the 11th FYP, under the tremendous pressure of energy conservation, some local governments had to meet their energy savings targets at the expense of the livelihood of their local residents.

5.3.2 *Government is the Leading Force behind the Formation and Development of Low-Carbon Development Institution*

Given this special transitional period China's economy finds itself in, constructing a market for energy conservation necessitates a special case of institutional innovation: combining market-based measures and administrative measures while ensuring coordination between governmental administrative orders and operational market mechanisms. During the 11th FYP, governmental administrative measures played an important role in constructing an EPC-oriented energy service market and a renewable energy development market. EPC has been known for only a short period of time in China, therefore it has a long way to go in order to improve its operation in the market. In this case, the government made use of supportive policies, fiscal measures, and taxes, and also enacted a series of policies and regulations. It took initiative in creating a market for energy conservation, which led to a lot more ESCOs, expedited market development, facilitated market competition, and made preparations for the construction of energy service market and EPC. This new policy mechanism is beneficial for alleviating financing difficulties of SMEs in energy conservation and facilitated the utilization of energy-saving technologies and the development of the energy-saving industry.

Due to the high development costs, imperfect technologies, and low energy density of renewable energy sources, and also the intermittency and unreliability of renewable power generation, renewable power is confronted with difficulties in grid connection, which has become the bottleneck in the development of renewable energy. However, in face of the pressure of energy conservation and pollution reduction, the government has created a market for renewable energy through administrative measures such as legislation, overall planning, and the refinement of laws, regulations, and policies, all of which effectively boosted the development of renewable energy. The central government gave clear-cut signals to local governments and enterprises through administrative orders and guided local governments to proactively and consciously develop the renewable energy industry. Meanwhile, the government maintains order in the energy conservation market by strengthening the enforcement of laws, releasing information, and enhancing supervision. In summary, the government has gradually created an operational mechanism for the development of renewable energy which can accommodate the demand of a market economy.

Our study demonstrates that the government plays an indispensable role in constructing an energy conservation market and a renewable energy market through administrative measures. Meanwhile, however, problems inevitably arise in local governments' implementation of energy conservation. For example, some local governments regard the development of renewable energy as an indicator for their performance evaluation matrix, therefore installed capacity of renewable power has become a "display" that local governments use to show their achievement in renewable energy development. Some departments that examine and approve power projects even require that conventional thermal power generation projects have to be accompanied by certain renewable energy projects, which has led to chaotic and overheated development of the renewable energy industry sector. China has to more intensely explore how the government can apply its economic leverage, take advantage of market characteristics, and create new energy conservation mechanisms under a market economy.

CHAPTER SIX

LOW-CARBON INVESTMENT AND FINANCING

Abstract: Over the course of the 11th FYP, China's investments in the fields of renewable energy and energy efficiency exceeded the amount invested in any previous Five-Year Plan. China has accumulatively invested 1.73 trillion RMB in new energy and renewable energy sources from 2006 to 2010, leading the world. More specifically, China invested 621.8 billion RMB (35.9%) in hydropower, 469.9 billion RMB (27.1%) in wind power, 366.8 billion RMB (21.2%) in nuclear power, 199.7 billion RMB (11.5%) in solar photovoltaic power, and 74.9 billion RMB (4.3%) in biomass. In contrast, the United States, which ranked second globally, invested 14% less than China during the same period.

China invested a total of 859.2 billion RMB to improve energy efficiency during the 11th FYP. Central and local government's fiscal funding accumulated to 126.1 billion RMB, accounting for 14.7% of the total investment. The industrial sector attracted 551.1 billion RMB, representing 64.1% of the overall investment, while the building sector amounted to 259.3 billion RMB, representing 30.2% of the total.

Fiscal funding and bank loans are China's chief low-carbon financing sources. In terms of bank loans, debt financing is the dominate channel, while the use of equity financing is less common. To support China's low-carbon development in the long run, application of diversified financing channels and innovative financing approaches need to be established.

6.1 Low-Carbon Investment

Low-carbon investment refers to funding invested to achieve the economic and social objectives involved in China's low-carbon development. It includes funding aimed at the improvement of energy efficiency, reduction of greenhouse gas emissions, and the promotion of new and renewable energy sources in pursuit of a green GDP. Relevant fields for low-carbon investment include but are not limited to: wind, solar, biomass, nuclear, hydro, and other new energy and renewable energy sources; the conservation of electricity, petroleum chemicals, building materials, iron and steel, nonferrous metal smelting products, and others; also included are the building, and transportation sector, low-carbon living schemes and energy saving management. This report divides China's low-carbon investment in two parts: new and renewable energy investment and energy efficiency investment (mainly in the industrial and building sector).

Over the course of the 11th FYP, China has increased its investment in new and renewable energy sources, and promoted energy-saving and emission reduction in the industrial, building, and transportation sectors. China has accumulatively invested 1.73 trillion RMB in new and renewable energy sources from 2006 to 2010, the largest investments made by any country worldwide. More specifically, China invested 621.8 billion RMB (35.9%) in hydropower, 469.9 billion RMB (27.1%) in wind power, 366.8 billion RMB (21.2%) in nuclear power, 199.7 billion RMB (11.5%) in solar photovoltaic power, and 74.9 billion RMB (4.3%) in biomass.[1] The United States, ranked second globally, invested 116.3 billion US dollars over the same period; equivalent to 86% of China's spending.[2]

China spent a total of 859.2 billion RMB to support schemes that would enhance energy efficiency during the 11th FYP. Fiscal funding by the central and local governments amounted to 126.1 billion RMB, accounting for 14.7% of the investment total. The industrial sector invested 551.1 billion RMB, 64.1% of total energy saving investments nationwide, while the building sector spent 259.3 billion RMB, or 30.2%.

6.1.1 *Investment in New and Renewable Energy Sources*

Table 6.1 compares key results of this report with other published reports which examine China's renewable energy investments in 2010.

Table 6.1. Global and China's renewable energy investment in 2010.

	This report	UNEP	PEW
Renewable energy investment	Wind, PV, biomass power, small hydro	Wind, solar, biomass, geothermal, small hydro and marine	Wind, solar, biofuels, energy efficiency and other renewables
Global (100 million $)	—	2110	2430
China (100 million $)	521.4	489	544
100 million ¥	3389	3179	3536

Note: 1$=6.5 ¥

Sources: UNEP, *Global trends in sustainable energy investment* (2011): 5; PEW, *Who's winning the Clean Energy Race?* (2011): 4.

[1] Data source: See 6.1.2 of this report.

[2] UNEP, 2010. *Global trends in sustainable energy investment* (2010): 6; and: UNEP, *Global trends in sustainable energy investment* (2011): 18.

Table 6.2. China's investment in wind farms from 2006 to 2010.

Year	2006	2007	2008	2009	2010	Total
New installed capacity (10 MW)	128.76	331.12	615.37	1380.32	1892.8	4348.37
Unit price of construction (10,000 RMB/kW)	1.1	1.0	0.95	0.9	0.9	
Total investment (100 million RMB)	141.64	331.12	584.6	1242.29	1703.52	4003.17

Sources: Chinese Wind Energy Association, "China Wind Power Installed Capacity Statistics in 2010" (2011); China Renewable Energy Society, *China's New Energy and Renewable Energy Yearbook 2010.*

6.1.1.1 *Wind Power Investment*

Wind Farms:
The following calculations of investments in wind farms are based on two factors: the amount of newly installed capacity per year, and the cost per unit of installation. Here, only the installed capacity is considered, while ongoing installation projects are excluded. During the 11th FYP, China's wind farm investments amounted to 400.3 billion RMB (Table 6.2).

Wind Turbine Manufacturing:
The main share of investments by wind turbine manufacturers focused on R&D and increasing production capacity. According to the *China New and Renewable Energy Yearbook 2010*, this report assumes that for an annual one GW production capacity, 2 billion RMB need to be invested. During the 11th FYP, investments in wind turbine production amounted to 69.6 billion RMB (Table 6.3).

Features of Wind Energy Investment:
China's total wind energy investment was 469.9 billion RMB during the 11th FYP. Wind energy investment displays two specific characteristics:

First, investment was mainly driven by domestic funding. Since wind power projects are costly and the cost recovery period is relatively long, private and international investors who seek quick profits from short-term investment remain cautious. Currently, large domestic energy enterprises and investment groups dominate the wind investment sector. The top-five wind investors in 2009 were the Guodian Corporation, Datang Corporation, Huaneng Group, Huadian Corporation, and Guangdong Nuclear

Table 6.3. China's investment in wind turbine production from 2006 to 2010.

Year	2006	2007	2008	2009	2010	Total
New capacity (10 MW)	60	120	600	1400	1300	3480
Unit price of construction (100 million RMB/1,000 MW)	20	20	20	20	20	
Total investment (100 million RMB)	12	24	120	280	260	696

Sources and notes:

1. Estimation based on relevant data from *China's New Energy and Renewable Energy Yearbook 2009* and *China's New Energy and Renewable Energy Yearbook 2010*.
2. China Renewable Energy Society, 2009. China Renewable Energy Society, 2010.
3. Economy&Nation Weekly, "Fengdian quanhai: Touzi huanyanzhong de xinyilun channeng guosheng? 风电圈海: 投资欢宴中的新一轮产能过剩? [Wind Power Enclosure of the Seas: A New Round of Overcapacity in the Investment Festivity?]" (2010-4-26).
4. Production line construction is mainly investment in equipment manufacturing; The report assumes that the price of manufacturing equipment is stable during the "Eleventh Five-Year" period.

Power Holding Corporation, holding market shares of 18.8%, 12.6%, 11.9%, 8.3%, and 6.2%, respectively.[3]

Two factors enable state-owned enterprises to maintain their monopoly role in the field of wind investment. First, smaller enterprises are unable to compete with the abundant funds of their larger competitors. Second, the Chinese government mandates that wind power must constitute a certain percentage within the annual newly installed power capacity of state-owned enterprises. Therefore, state-owned enterprises are forced to first invest in wind power in order seize new shares in other energy markets.

Second, investment in domestic wind-energy equipment manufacturers overtook foreign investment during the 11th FYP. The market share of domestic manufacturers has been soaring in the past five years. In particular, Chinese manufacturers acquired 55.9% in China's newly added market share in 2007, surpassing foreign manufacturers for the first time. In terms of cumulative market shares, domestic equipment reached 44.8% at the end of 2007.

[3] China Renewable Energy Society, *China's New Energy and Renewable Energy Yearbook 2010*.

Newly installed capacity of domestic and joint-venture manufacturers totaled 4.72 million kW in 2008, representing 76% of the market. The Goldwind Corporation, Dongfang Turbine Corporation, and Sinovel Corporation all produced more than one million kW in 2008. Cumulative installed capacity of domestic and joint-venture manufacturers achieved 7.51 million kW in 2008, accounting for 62% of the market and surpassing foreign manufacturers for the first time. Through continuous investment in wind power equipment manufacturing, China has progressively established an advantage in this field, which is at display not only in the domestic, but also more and more so on the international market.

6.1.1.2 *Photovoltaic Power Investment*

Investment in solar power consists of two parts: Investments in solar photovoltaic systems (on-grid and off-grid), and investments in photovoltaic production. The main data used in this report is derived from *China New and Renewable Energy Yearbook 2009* and *China New and Renewable Energy Yearbook 2010*, some amount of data is based on the assumption that products' production capacity between adjacent years remained unchanged.

On-Grid and Off-Grid Solar Photovoltaic Systems:

During the 11th FYP, total investment in solar photovoltaic (on-grid and off-grid systems) totaled 19.9 billion RMB (Table 6.4, Table 6.5, and Table 6.6).

PV Production:

Solar photovoltaic production includes polycrystalline silicon production, wafer production, silicon solar photovoltaic cell and module production,

Table 6.4. China's investment in on-grid PV systems from 2006 to 2010.

Year	2006	2007	2008	2009	2010	Total
New installed capacity (10 MW)	0.1	0.22	2.13	14.22	55[a]	71.67
Unit investment (10,000 RMB/kW)	5.5	5	4.5	2.5	2	
Total investment (100 million RMB)	0.55	1.1	9.59	35.55	110	156.79

Sources and description: China Renewable Energy Society, *China's New Energy and Renewable Energy Yearbook 2010*; Wang Wei, "Taiyangneng guangre fadian kaiqi shangye zhengcheng 太阳能光热发电开启商业化征程 [Commercialization of Solar Thermal and Power]," *State Grid News* (2011-6-16).

Table 6.5. China's investment in off-grid PV system from 2006 to 2010.

Year	2006	2007	2008	2009	2010	Total
New installed capacity (10 MW)	0.9	1.78	1.9	1.8	3[a]	9.38
Unit investment (10,000 RMB/kW)	6	5.5	5	4	3.5	
Total investment (100 million RMB)	5.4	9.79	9.5	7.2	10.5	42.39

Sources: China Renewable Energy Society, *China's new energy and renewable energy yearbook 2010*; Wang Wei, "Taiyangneng guangre fadian kaiqi shangye zhengcheng 太阳能光热发电开启商业化征程 [Commercialization of Solar Thermal and Power]," *State Grid News* (2011-6-16).

Table 6.6. China's investment in PV system from 2006 to 2010 (Unit: 100 million RMB).

Year	2006	2007	2008	2009	2010	Total
Total investment of grid-connected PV system	0.55	1.1	9.59	35.55	110	156.79
Total investment of off-grid PV system	5.4	9.79	9.5	7.2	10.5	42.39
Total	5.95	10.89	19.09	42.75	120.5	199.18

Table 6.7. China's investment in PV production from 2006 to 2010 (Unit: 100 million RMB).

Year	2006	2007	2008	2009	2010	Total
Polysilicon production	0.8	30.4	117.6	200	135.6	484.4
Wafer processing	42.86	83.44	75	180	415.5	796.8
Crystalline silicon solar cells and modules production	10.01	19.5	30	82.5	150	292.01
Thin-film cell manufacturing	1.6	6.52	6.3	24.37	185.5	224.29
Total	**55.27**	**139.86**	**228.9**	**486.87**	**886.6**	**1797.5**

and thin film solar cell production. This report assumes the cost per unit of production capacity remained stable in between 2006 and 2010; due to the fact that investments in production line construction were largely price-stable (most PV manufacturing equipment is being imported from abroad). During the 11th FYP, investment in solar photovoltaic production amounted to 179.8 billion RMB (Table 6.7). More specifically, investment in polycrystalline silicon production was 48.4 billion RMB (26.9%), wafer production 79.7 billion RMB (44.3%), silicon solar photovoltaic cell and module production 29.2 billion RMB (16.2%), and thin film solar cell production 22.4 billion RMB (12.5%) (Table 6.8).

Table 6.8. China's investment in polycrystalline silicon production, wafer processing, crystalline silicon solar cells and modules and thin-film cell manufacturing from 2006 to 2010.

Category		2006	2007	2008	2009	2010	Total
Polycrystalline silicon production	New capacity (ton)	100	3800	14700	25000	16950	60550
	Unit investment (100 million RMB/1,000 ton)	8	8	8	8	8	
	Total investment (100 million RMB)	0.8	30.4	117.6	200	135.6	484.4
Wafer processing	New capacity (ton)	5715[a]	11126[a]	10000	24000	55400[a]	106241
	Unit investment (100 million RMB/1,000 ton)	7.5	7.5	7.5	7.5	7.5	
	Total investment (100 million RMB)	42.86	83.44	75	180	415.5	796.8
Crystalline silicon solar cells and modules	New capacity (MW)	668[b]	1300[b]	2000	5500	10000	19468
	Unit investment (10,000 RMB/MW)	150	150	150	150	150	
	Total investment (100 million RMB)	10.01	19.5	30	82.5	150	292.01
Thin-film cell manufacturing	New capacity (MW)	11.43[c]	46.6[c]	45	174.1	1325	1602.13
	Unit investment (10,000 RMB/MW)	0.14	0.14	0.14	0.14	0.14	
	Total investment (100 million RMB)	1.6	6.52	6.3	24.37	185.5	224.29

Sources and description:
China Renewable Energy Society, *China's New Energy and Renewable Energy Yearbook 2009*; China Renewable Energy Society, *China's New Energy and Renewable Energy Yearbook 2010*;

a. Estimation based on wafer processing capacity from 2007 to 2009 and wafer processing actual output from 2005 to 2008.
b. Estimation based on crystalline silicon solar cells and modules capacity in 2007 and crystalline silicon solar cells and modules actual output from 2005–2007.
c. Estimation based on Thin-film cell manufacturing capacity in 2007 and crystalline silicon solar cells and modules actual output from 2005–2007.

Impacts of the Global Financial Crisis on China's Solar Photovoltaic Investment:

Due to the global financial crisis, the European solar photovoltaic market stagnated for a short period of time. This stagnancy of foreign markets tremendously influenced China's solar photovoltaic industry. Since oversea orders decreased sharply, the sale of solar photovoltaic products plummeted. The overstock of products led to a huge price fall in the Chinese domestic market. Solar photovoltaic cells sold at 40–44 RMB/Watt-peak (Wp) at the beginning of 2008, yet their price was only 12–16 RMB/Wp by the end of that year.[4] Moreover, a series of investment plans were stranded due to shrinking stock prices of solar photovoltaic companies. However, as the economic stimulus package went into effect and the government enhanced its support of the industry, the domestic market—particularly the construction of PV power generators and PV equipment manufacturers—has blossomed since the second quarter of 2009. Looking back, the global financial crisis has to be considered more of an opportunity than a disaster for the Chinese solar photovoltaic industry. The resulting decline in prices enabled China to deploy solar photovoltaic products to a greater extent than previously expected.

6.1.1.3 *Bioenergy Investment*

Investments in bioenergy include biomass, biofuel, and biogas. During the 11th FYP, investment in bioenergy totaled 74.9 billion RMB. In particular, investment in biomass energy, biofuel, and biogas was 41.7 billion RMB, 12 billion RMB, and 21.2 billion RMB, respectively (Table 6.9).

Table 6.9. China's investment in bioenergy from 2006 to 2010 (Unit: 100 million RMB).

	Biomass	Biodiesel and fuel ethanol	Biogas	Total
Investment **(billion RMB)**	41.7	12	21.2	74.9

[4] China Renewable Energy Society, *China's New Energy and Renewable Energy Yearbook 2009.*

Table 6.10. China's investment in large and medium-sized hydropower from 2006 to 2010.

Year	2006	2007	2008	2009	2010	Total
New installed capacity (10 MW)	0	1870	2011	0	4123	8004
Unit price of construction (10,000 RMB/kW)	0.7	0.7	0.7	0.7	0.7	
Total investment (100 million RMB)	0	1309	1407.7	0	2886.1	5602.8

Sources and description: China Electricity Council, *National Electric Power Industry Statistics Express* (2010).

Table 6.11. China's small hydropower capacity and price from 2006 to 2010.

Year	2006	2007	2008	2009	2010	Total
New installed capacity (10 MW)	1150	90	2	420	388	2050
Unit price of construction (10,000 RMB/kW)[1]	0.3	0.3	0.3	0.3	0.3	
Total investment (100 million RMB)	345	27	0.6	126	116.4	615

Sources and description: Ibid.

6.1.1.4 *Hydropower Investment*

From 2006 to 2010, China's investment in hydropower projects amounted to 621.8 billion RMB. More specifically, large and medium-sized hydropower projects attracted 560.3 billion RMB, and small-scale hydropower projects received 61.5 billion RMB (Table 6.11). It should be noted that the cost per unit of newly installed capacity and the costs for raw materials and reallocation correlate, and that despite a rising trend the fluctuation has been stable. This report therefore considers the average value.

6.1.1.5 *Nuclear Power Investment*

Since 2009, China has accelerated its construction of nuclear power facilities. The Sanmen, Zhejia, Haiyang, and Taishan nuclear power projects began construction successively. By the end of 2009, China had approved

28 units of installation out of 10 nuclear power projects, of which 20 units with 30.57 million kW are currently under construction. China has the largest scale of nuclear power projects under construction in the world. Assuming construction costs of 12,000 RMB/kW,[5] China invested a total of 366.84 billion RMB in nuclear power during the 11th FYP.

6.1.2 *Energy Efficiency Investment*

During the 11th FYP, China spent 859.2 billion RMB on schemes meant to improve energy efficiency. In particular, fiscal funding from the central government amounted to 85.1 billion RMB, while fiscal funding by local governments (on the provincial and prefectural level) amounted to 41 billion RMB, and private investors invested 733.1 billion RMB (Table 6.12). A study published by the Energy Research Institute (ERI) of the NDRC shows that China allocated a total of 846.6 billion RMB to improvements in energy efficiency over the course of the 11th FYP.[6]

It should be noted that the scope of this report and the ERI study differ in some regards. In this report, investment in energy efficiency within the industrial sector includes industrial projects within the ten major energy conservation projects, the Top One-Thousand Enterprises Program, and the phase-out obsolete capacity campaign. Investments to improve energy efficiency within the building sector includes the new construction of energy-saving buildings, retrofitting of residential buildings in northern China, application of renewable energy technologies within buildings, and energy-saving monitoring systems in public buildings. Investment in the transport sector only considers subsidies for the promotion of energy-saving vehicles, while the promotion of energy savings and alternative-energy vehicles and pilot programs for alternative-energy passenger vehicle subsidy are excluded.

The ERI's report on investments to improve energy efficiency in the industrial sector includes technology overhauls and the phasing-out of obsolete capacities. Investments to improve energy efficiency in the building sector includes retrofitting of residential buildings in northern China and other regions, application of renewable energy technologies

---

[5] China Investment Consulting Website, *Zhongguo hedian chanye 'Shiyiwu' huigu ji 'Shi-erwu' guihua touzi fenxi ji yuce baogao* 中国核电产业"十一五"回顾及"十二五"规划投资分析及预测报告 [Investment Analysis and Forecast of 11th Five Year Review and 12th Five Year Plan of China Nuclear Power Industry].

[6] Dai Yande, Xiong Huawen, Jiaojian, *China Energy Efficiency Financing and Investment 2010 Report* (2011): 1.

Table 6.12. China's investment in energy saving during the 11th FYP period (Unit: 100 million RMB).

The report					Energy Research Institute National Development and Reform Commission				
	Central government	Local government	Society Funds	Total		Central government	Local government	Society Funds	Total
Key energy saving projects	315	158	1860	2333	Technical transformation	305	171	5680	6156
Top-1000 enterprises	0	0	2874	2874					
Phase-out of obsolete capacity	202	102	0	304	Phase-out of obsolete capacity	219	112	0	331
Building energy conservation	152	134	2307	2593	Residential building retrofitting in northern areas	64	30	300	394
					Renewable energy building demonstration	70	35	350	455
					Public energy saving monitoring system construction	18	5	20	43
					Existing residential building retrofitting in other regions	0	12	123	135

Table 6.12 (*cont.*)

The report					Energy Research Institute National Development and Reform Commission				
	Central government	Local government	Society Funds	Total		Central government	Local government	Society Funds	Total
Energy saving lights subsidy	16.2	0	0	16.2				27	424
Energy saving air-conditioner subsidy	115.4	0	0	115.4	Capacity building	318	79		
EMC	20	16	290	326					
Fuel-efficient vehicles subsidy	30.4	0	0	30.4	Transportation	22	36	470	528
Total	851	410	7331	**8592**	Total	1016	480	6969	**8466**

Note: See Appendix 4.2 and 4.3 in Chapter 4 for calculation.

Table 6.13. China's energy saving investment in sectors during the 11th FYP period (Unit: 100 million RMB).

	Central government	Local government	Societal Funds	Total
Industry	517	260	4734	5511
Building	152	134	2307	2593
Transportation	30.4	0	0	30.4
Others	151.6	16	290	457.6

in buildings, and energy-saving monitoring systems in public buildings. Investment for capacity-building includes subsidies for energy-saving monitoring capacity-building and rewards for EMC. Investment in the transport sector includes the demonstration and promotion of energy-saving and alternative-energy vehicles and pilot programs for alternative-energy passenger vehicle subsidies.

From a sectoral perspective, investment in the industry sector, the building sector, the transport sector, and others (ESCOs, energy efficient lights, energy efficient air conditioner) amounted to 551.1 billion RMB, 259.3 billion RMB, 3 billion RMB, and 45.8 billion RMB, respectively (Table 6.13).

6.1.3 *Features of Low-Carbon Investment*

6.1.3.1 *Fast Growth Rate of Low-Carbon Investment*

Investments from the government as well as from the private sector in realms such as new and renewable energy sources as well as energy efficiency increased rapidly during the 11th FYP period. For example, investment in wind energy increased with an average annual rate of 82.4% from 2006 to 2010. More specifically, investments in wind energy almost doubled every year from 2006 to 2009, the 2010 figure being 12.8 times that of 2006 (Figure 6.1). Investment in solar photovoltaics annually grew by 101.4% and investment in 2010 was 16.5 times that of 2006 (Figure 6.2). Similar patterns could be observed in the field of energy efficiency. The Top 1000 Enterprises invested with an annual average growth rate of 65.5% during the 11th FYP, a 7.5 fold growth from 2006 to 2010 (Figure 6.3). The rapid growth of investment indicates the acceleration of China's low-carbon development in the 11th FYP period; low-carbon economy has become an essential economic development direction of China.

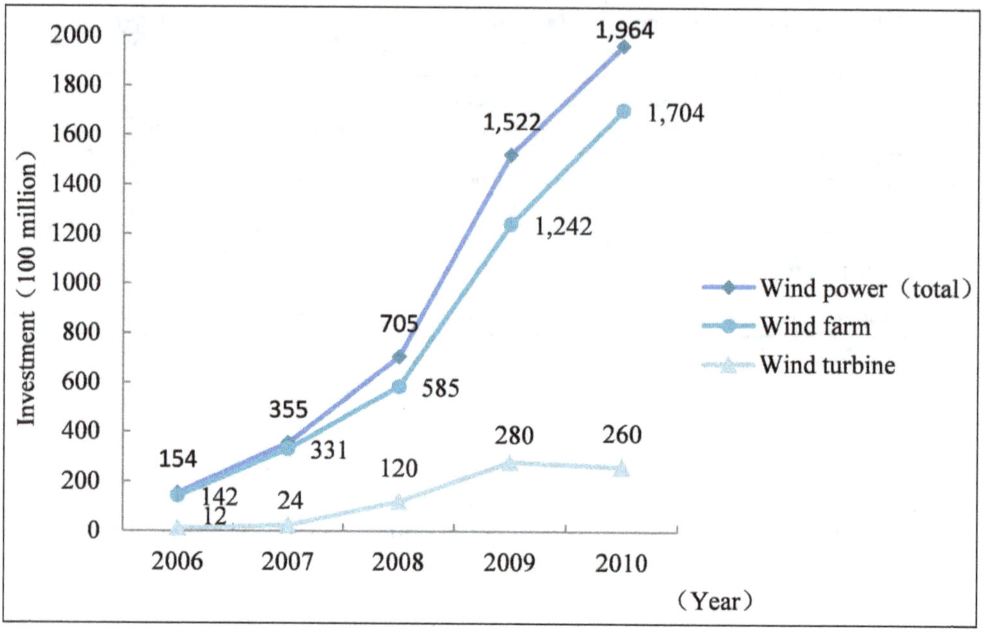

Figure 6.1. China's investment in wind power during the 11th FYP period.

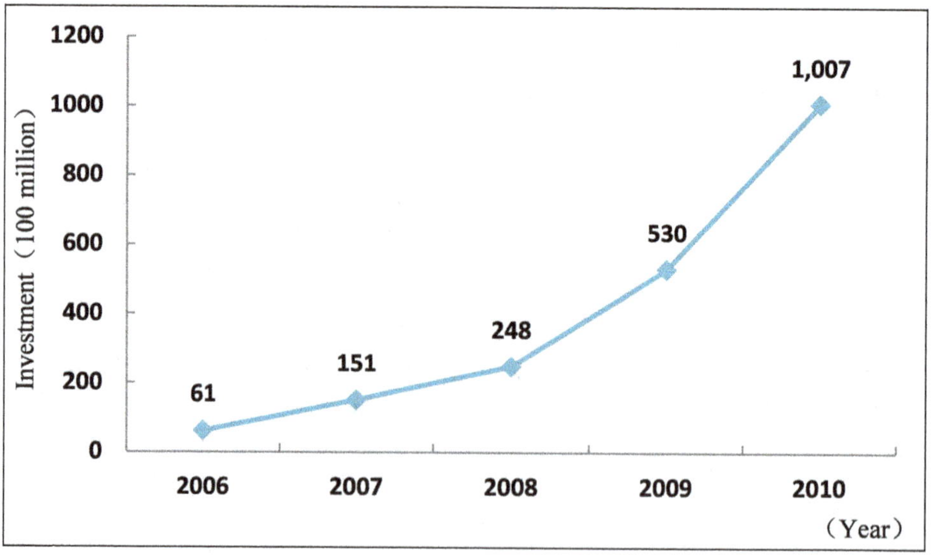

Figure 6.2. China's investment in PV during the 11th FYP period.

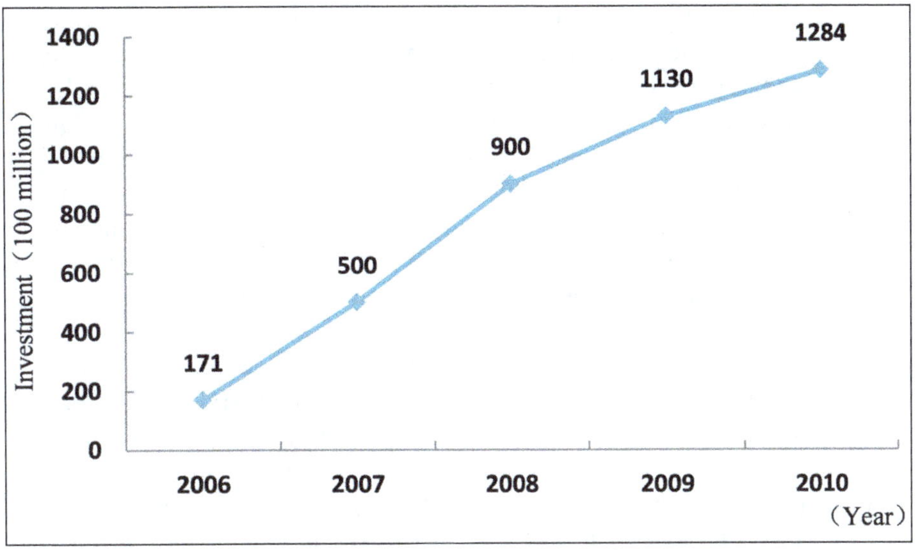

Figure 6.3. Investment of "Top-1000 enterprises energy saving action" during the 11th FYP period.

6.1.3.2 *The 4 Trillion RMB Economic Stimulus Package had Profound Impact on China's Low-Carbon Investment*

To address the global economic crisis, the Chinese government implemented a four trillion RMB economic stimulus package in 2008. The package mainly focused on infrastructure projects, thereby stimulating the cement, construction, iron and steel, and other energy-intensive industries. This made the challenge of carbon emission reductions even more difficult in the short run. On the other hand, the construction of energy-saving and environmental projects have had certain stimulating effects on the new energy industry, which means that the fruits of energy-saving and carbon emission reductions may be harvested in the long run, which would have a profound impact on China's low-carbon development in the future.

About 80% of the four trillion RMB stimulus package was allocated to infrastructure and residential building construction projects.[7] This resulted not only in a recovery but a boost for energy-intensive industries and hence negatively impacted the adjustment of China's economic

[7] WWF, *China's Stimulus Package: Climate and Energy Impact* (2010): 4.

structure in the short run.[8] However, the four trillion RMB stimulus package will have a positive impact on energy saving and emission reduction after 2014, due to the expected gains in energy savings after the completion of the rail transportation network, and the non-sustainability of high capital investment in energy-intensive industries. In addition, a total of 210 billion RMB has been spent on implementation of ecological projects, energy saving, and emission reduction. According to the estimation based on the first six months data of 2009, nearly 20% of the funding was directly allocated to schemes improving energy efficiency such as the "Ten Major Energy Saving Projects."

Furthermore, although the four trillion RMB were not directly invested in new energy industries, the central government's investment in energy conservation and structure adjustment will indirectly affect the investment in the new energy sector; the state grid, the top-five energy corporations, the top-three petroleum group and other large state-owned enterprises, as well as some energy investment enterprises, have all expanded their input into new energy manufacturing industries and power generation projects. New energy enterprises are also favored by all kinds of private equity funds and venture capital funds within the capital market, large new energy generation projects have launched in succession, and active investment in wind and solar energy generation equipment is also unprecedented. Because the large amount of investment into infrastructure projects stimulated the growth of energy-intensive industries, China's industrial structure was not well adjusted during the 11th FYP. This will have a profound future impact on China's low-carbon investment, as China will divert more funding towards structural energy-saving and carbon reduction projects. It is all but certain that low-carbon investment in China will maintain its rapid growth.

6.2 Low-Carbon Financing

To develop China's low-carbon economy requires not only substantial monetary input, but also favorable basic investment conditions such as a benign low-carbon financing environment and financing capability. Improving energy efficiency and developing new energy industries are important steps to realize China's energy development strategy. In the

[8] Ibid., 6.

11th FYP, the development of energy saving and new energy industries was rapid; China leads the world in terms of new energy investment, yet considering China's current low-carbon financing rate and its long-term development toward a low-carbon economy, China will have to diversify and improve the financial channels through which it can achieve these goals. This section summarizes the main channels of low-carbon financing in China and the progress of low-carbon financing during the 11th FYP.

6.2.1 *Low-Carbon Financing Channels*

In the initial development phase of China's low-carbon economy, the main funding source of low-carbon projects was provided by state fiscal funding plus the support of international institutions. As the low-carbon economy underwent rapid development, the demand for investment exceeded the amount of budgets and fiscal funding provided by the central government and direct/indirect investment from international institutions. Thus, in addition to state funding and traditional financing approaches through banks, the long-term development of China's low-carbon economy will also rely on market-based financing channels. China will need to make use of a series of diversified financing channels and approaches so that private funding (including private equity and venture capital) is effectively invested in low-carbon industries and to encourage listing of low-carbon enterprises on stock markets and issue corporate bonds.

Low-carbon financing channels can be divided into two categories: internally and externally funded investments. Internally supported financing refers to the process in which an enterprise retains earnings and converts profits into new investment, while externally financed investment refers to the process in which an enterprise acquires funding from other economic entities and converts them into own investments. As technology advances and production scale expands, solely relying on internal capital will not be able to satisfy the capital demand of enterprises, making external financing mechanisms significantly more viable to maintain a rapid developmental pace. External financing channels include: fiscal financing, debt financing, equity financing, and "Clean Development Mechanisms" (CDM) financing. Table 6.14 summarizes the characteristics as well as the advantages and disadvantages of each channel.

6.2.2 *The Current Status of Low-Carbon Financing*

During the 11th FYP period, large investments were made in the fields of energy efficiency and new energy. Due to the limitation of available data

Table 6.14. The characteristics of China's low-carbon financing.

Financing Channel		Financing Mode	Feature	Advantage	Disadvantage
Internal Financing		Accumulation of enterprise profit or capital market offering	Most of the new energy and renewable energy companies preferred channel.	Small cost, flexible, low policy and currency risk, simple procedure and the biggest advantage is free.	Difficult to meet funding requirements if only relying on internal financing as the scale of renewable energy increases.
Fiscal Financing	Government investment and subsidies	Direct investment	Direct funding of the project through financial investment. Not in the interest of profit.	Provide primary driving force for new energy and renewable energy, and energy conservation and emission reduction.	Information transmission and processing capacities are relatively fragile compare to the market and cannot accurately evaluate business and prompted adjustment.
		Subsidies	Include project loan interest subsidy, tax incentives, increase on-gird electricity price etc.	Several national promotion policies have become the main funding source of renewable energy projects.	
Debt Financing	Financial institution loans	Domestic policy banks	Playing an important role in coordination and guidance of national low-carbon economic development.	Focus on the realization of low-carbon economy policies, low interest rate, policy leading, publicity and easy to provide long-term loans in guarantee a follow through of renewable project's funding demand.	Mainly applicable to major projects, difficult for small energy companies to apply; government-led and coordination is necessary.
		Domestic commercial banks	The most important financing channel for renewable energy enterprises in China.	Focus on low-carbon projects with sound economic profit, offer project evaluation, credit check and commercial loans. most commonly used instrument, relatively low costs, fast speed and flexible on loans.	The banks are more cautious on risky new energy small-scale projects. There is no unified and filed loan standard.
		International loans	Playing a leading role in the early stage of low-carbon economy development.	Solve long-term funding that some construction projects urgently needed; capable of introduce advanced technical support.	Interest seems low, but has additional terms such as purchase loan provider country's equipment.

Table 6.14 (*cont.*)

Financing Channel		Financing Mode	Feature	Advantage	Disadvantage
	Bonds		An important direct financing channel for low-carbon economy in the future	It is more active and offers lower costs compare to bank loans and stock market financing.	Financing scale is very limited; and China's bond market development insufficient, structurally not justified and flawed operation mechanism.
	Private	Borrowing	Important channel of renewable energy financing.	In the absence of external capital and low supply of formal financing. Private borrowing may become the choice of many private enterprises.	Limited funding, low degree of specialization and lack of venture capital experience, need to be appropriately guided and orderly access into the new energy industry.
Equity financing	Stock market	Main Board	Major direct financing market for low-carbon economy.	The most commonly used and the most effective way to achieve capital accumulation, large sum of equity funds, improve enterprise management.	Low-carbon energy companies, especially the emerging small and medium-sized enterprises in the main board market are constraint from large-sum financing.
		Growth Enterprise Market (GEM)	Another important platform for direct financing.	Relatively flexible mechanism and listing conditions, sound financing channel for new energy enterprises, especially small and medium sized.	Relatively larger risk, higher demand for information disclosure.
	VC&PE		Appropriate financing channel for new energy.	Not only provide financial support, but also significantly improve the financial situation to reduce risk.	The size of venture capital firm is generally small, and not strong enough.
CDM			China is the main market of CDM.	Through the development of CDM, certain compensation is acquired for financing, while increase the speed of technology transfer.	CDM projects are selective, with complex procedures and high risk.

Table 6.15. China's low-carbon financing during the 11th FYP period.

Investment Field \ Financing channel		Internal Financing	External Financing							
			Fiscal Financing	Debt Financing				Equity financing		CDM
				Domestic	International	Bonds	Private	Stock	VC&PE	
New energy and renewable energy	Wind Power	★	★★	★★★★	★★	★	★	★★	★★	☆☆☆
	PV	★	★★	★★★	★★	★	★	★★★	★★	☆
	Biomass	★★★	★★★	★★	★	★	★★	★	★	☆
	Large and medium-sized hydropower	★★	★★★★	★★★	★	★	★	★★	★	☆☆☆☆
	Small hydro	★★★	★★★	★★	★	★	★★	★	★	☆☆☆
	Nuclear power	★★	★★★★	★★★	★*	★★	—	—	—	—
Energy conservation		★★★	★★	★★★★	★★	★	★★	★★	★	☆☆☆☆
Forest carbon-sinks		★	★★★★	★★	★	—	★	—	—	☆
CCS		★★★	★★★★	—	—	—	—	—	—	—

Note: (1) The quantitative relationship between financing channels is not a precisely proportional and additive relationship.
(2) The qualitative proportions of all kinds of new Chinese CDM energy projects were approved by July 12, 2010.
* Mainly refers to the international cooperation projects

concerning relevant financing channels, this report has synthesized a large volume of related research papers, and qualitatively determined the significance of low-carbon finance during the 11th FYP. The number of stars in each column indicates the level of significance for the respective measure (Table 6.15). The following conclusions can be drawn from the table:

1. The main channels of low-carbon financing that are currently employed in China are fiscal financing and bank loans.
2. Governmental fiscal funding is essential, particularly for large projects in the field of large and medium-sized hydropower, nuclear power, carbon sink, and CCS, as well as for some small projects such as biomass and small hydropower that encounter relatively more difficulties in obtaining bank loans.
3. In terms of bank loans, debt financing is the dominating channel for low-carbon financing, while the use of equity financing is rare.
4. CDM financing has the potential in guiding China low-carbon financing towards a carbon trade market.

6.2.3 *Characteristics of Low-Carbon Financing*

Low-carbon financing differs substantially from traditional financing in the sense that the development of low-carbon industries has had a relatively short history and to a large degree involves novel technologies. China's low-carbon industry has yet to build a comprehensive R&D and innovation system. Its capability to develop innovative technologies independently is still immature, since key manufacturing equipment it still mostly imported, technological and operational risks are still widespread, and the commercial aspect of low-carbon technology is also novel and unique. Compared to traditional industries, the development and production costs of low-carbon technology at its current stage are too high; low-carbon industries not only require a considerable quantity of capital, but also have to endure relatively slow investment returns. For an investor, the technological performance of a low-carbon project is associated with a high degree of risk and uncertainty alongside relatively low market share expectations, not to mention that the majority of projects are small in scale and difficult to assess with regards to their technological potential, feasibility and energy-saving accounting. In addition, Chinese managers are inexperienced in the supervision of low-carbon projects and are generally unaware of the importance of credibility in business activities.

Aside from the points mentioned above, China's low-carbon financing has the following characteristics:

6.2.3.1 *Governmental Funding and Private Loans are the Main Financing Channels*

The government plays a leading role in promoting China's low-carbon development. Fiscal financing is the main financing channel; the main source of this funding comes from fiscal budgets provided by the central and local governments. In addition, the government has established funds designated for energy conservation, implemented preferential tax discount mechanisms, issued national debt and other policies to support low-carbon development.

Bank loans also are a prominent financing channel. During the 11th FYP, China's domestic policy banks (the banks that support China's medium-to-long-term development strategies and policies) played a supportive and guiding role, while commercial banks strove to innovate their financing practices and acquire some experience in this field. According to estimations, Chinese banks approved 1.96 trillion RMB in green loans during the 11th FYP, 24.9% of which was assigned to improve energy-saving and energy efficiency projects.[9] After the release of China's green loan policy in 2007,[10] state-owned commercial banks, domestic policy banks, regional banks, and joint-stock banks all attempted to support energy efficiency projects with various financing approaches (Table 6.16).

6.2.3.2 *Prominent Input from International Organizations*

During the 11th FYP, various international actors such as foreign governments, NGOs, banks, financial institutes, consultant firms, or large multinational corporations supported the funding of China's energy efficiency improvements in the form of grants, loans, and guarantees. Through funding provisions and technical support, they also cooperated and assisted in capacity building and policy research concerning energy efficiency. International organizations invested 18 billion RMB in the form of energy efficiency funds during the 11th FYP. More specifically, a total of 11.7 billion

---

[9] Dai Yande, Xiong Huawen, Jiaojian, *China Energy Efficiency Financing and Investment 2010 Report* (2011).

[10] 2007, China Banking Regulatory Commission, State Environmental Protection Administration and People's Bank of China published *Guidelines Regarding Improvement And Enhancement Of Financial Service In Energy Saving And Environmental Protection Sectors* (*Yin Fa* [2007] no. 215).

Table 6.16. Loans for energy efficiency projects by major domestic banks during the 11th FYP period (Unit: 100 million RMB).

Banks	Loans
Industrial and Commercial Bank of China	1987.17
China Development Bank	1566.98
Bank of China	282
Industrial Bank	269.24
Agricultural Bank of China	150
China Minsheng Banking	142.65
Bank of Communications	120
China Construction Bank	100
Bank of Beijing	67.54
China Merchants Bank	55.5
Agricultural Development Bank of China	45.9
Huaxia Bank	30.27
Shanghai Pudong Development Bank	29
China CITIC Bank	20
Others	18
Total	4884.25

Sources: Dai Yande, Xiong Huawen, Jiaojian, *China Energy Efficiency Financing and Investment 2010 Report* (2011): 63.

RMB was assigned to improve energy efficiency and energy saving, 2.7 billion RMB of which were used for capacity building such as research, assistant, and guidance, while 9 billion RMB were provided as loans and loan guarantees to support market development. Domestic energy efficiency projects also received an additional 6.3 billion RMB from the Clean Development Mechanism.[11]

6.2.3.3 *Large State-Owned Enterprises have Easier Access to be Financing than Small and Medium-Sized Private Enterprises*

Compared to small and medium private enterprises, large state-owned enterprises are in a better position to obtain financing for low-carbon

[11] Dai Yande, Xiong Huawen, Jiaojian, *China Energy Efficiency Financing and Investment 2010 Report* (2011): 66.

development. Currently, 70% of enterprises in China's energy-saving service sector are small and/or medium-sized, and the difficulty to secure funding has always been the bottleneck that restraints their growth.

Large state-owned enterprises not only receive direct support from the central and local governments, but also inherently have sound financing conditions, allowing them to invest in low-carbon projects and making bank loans easier to obtain. In addition, China's stock market favors state-owned enterprises and makes it easier for them to be listed on the stock exchange.

By contrast, due to their size and limited internal funding, small and medium enterprises barely meet the requirements for bank loans, which means that often a third-party loan guarantee is needed in order to obtain financing. Often, such guarantees are difficult to obtain as most enterprises cannot provide proof of sufficient fixed assets. Although the ChiNext stock listing provides a direct platform for small and medium-sized enterprises to attract investment, the current mechanism is far from mature and has to be improved upon. The 12th FYP will shift its energy-saving and carbon reduction targets from energy-intensive industries to the dispersed field of small and medium-sized enterprises. However, the financial obstacles confronting small and medium-sized enterprises remain a shackle for the further development of China's low-carbon economy.

6.2.3.4 *The Need to Establish a More Diversified Financing Mechanism*

Insufficient funding has been a major restraining factor in China's low-carbon development. The Chinese government advocates a diversification of financing mechanisms, yet low-carbon-related projects barely have had any funding source other than government subsidies. In other words, when enterprises in the energy sector conduct R&D related to low-carbon technology, the required investment mainly comes in the form of bank loans. The significance of debt financing, stock markets, and bonds has not yet been recognized. China should build relevant incentive mechanisms and safeguard measures as quickly as possible, improve low-carbon credits, stock markets, and bond markets, and at the same time duly and adequately attract private investment and develop its carbon trading market.

According to the joint projection of the US Energy Foundation and the NDRC, it is estimated that China would need to invest 18 trillion RMB in its energy sector between 2005 and 2020, in which energy efficiency improvements, new energy, and environmental protection account for 40% of energy demand, about 7.2 trillion RMB.[12] To meet such a scale of low-carbon investment, more diversified financing mechanisms are required.

[12] Zhang Liang, "Financing Model of Energy Conservation and New Energy Industry in China," *Development Research* 7 (2009): 38.

CHAPTER SEVEN

LOCAL GOVERNMENT AND LOW-CARBON DEVELOPMENT

Abstract: The energy conservation target of 11th FYP was realized through decomposition of the 20% binding indicator from the central down to local the governments. In the meantime, local governments achieved enormous progress in areas such as establishment of low-carbon development awareness and energy conservation administration, creation of transfer mechanisms for energy conservation pressure, formulation of low-carbon oriented resource allocation, innovation in low-carbon development, and so forth. On the other hand, because of the pressure for economic growth and the effects of the economic stimulus package in the later period of the 11th FYP, local governments continued to support a number of energy-intensive projects. Local governments are faced with an ongoing dilemma between economic expansionism and low-carbon development, and its activities are representative of this dual character.

Local government is a state institution located within each administrative division at various levels; it is responsible for local administrative management.

In the course of 30 years of economic reform, local governments have achieved substantial GDP growth, fiscal revenue growth, and results in urban construction, all of which have significantly contributed to China's national economic growth. This growth-led developmental approach has driven local governments into a phase of accelerated economic activities and urban management. During the 11th FYP, China has formulated a series of policies regarding low-carbon development, represented by the 20% overall energy conservation (binding) target. This set of policies, to some extent, has suppressed the impulse towards economic expansion by local governments, and has simultaneously led to some changes in local governments' behavior. According to the report on the 11th FYP's energy conservation target published by the NDRC and the National Bureau of Statistics (NBU) in June 2011, all provinces have completed their binding targets (apart from Xinjiang province for which a separate evaluation was applied), thus local governments have played a critical role in China's low-carbon development and energy conservation.

During the 11th FYP, the behavioral adjustment of local governments with regards to low-carbon development reflects an important shift in awareness with regards to low-carbon development, as well as China's energy conservation administration, creation of a transfer mechanism for energy conservation pressure, formulation of low-carbon-oriented resource allocation, innovative approaches by local governments and the behavior of local governments.

7.1 Local Government's Low-Carbon Development Behavior during the 11th FYP

7.1.1 *The Establishment of Low-Carbon Development Awareness*

The 11th FYP witnessed a growing awareness by local governments regarding low-carbon issues. In particular, local governments' perception of energy conservation has undergone radical changes. Energy conservation and lowering the energy consumption target was included as a binding indicator in national planning, a measure that has dramatically altered the old incentive system for local officials, making energy conservation a key objective for local governments.

During the 11th FYP, the central government established an energy conservation target accountability system with a top-down approach, supplemented by a corresponding evaluation system. The relevant central departments also conduct annual progress evaluations of local energy conservation in accordance with the target decomposition. They then publicly announce the evaluation results and dispense rewards and penalties codified in the accountability system. The central government's "Implementation Plan of Energy Consumption per Unit GDP Evaluation System" states that "provincial governments that are evaluated as having completed or outperformed [their target], will be commended at national energy-saving events. On the other hand, for provincial governments whose efforts have been evaluated as incomplete, local officials will not be eligible for annual rewards or honorary titles, and the central government will temporarily suspend examination and approval procedures for all new local energy-intensive projects." Individual provinces and municipalities have similar evaluation plans as the national one, although regulations in some places are more strict and straightforward. For instance, in May 2008, the Shijiazhuang city government office issued a "Notice Regarding Issue and Print of Implementation Planning of Shijiazhuang's Key County (City) And Key Enterprises Energy Conservation and Emissions Reduction

Target Evaluation" saying that "if the energy conservation and emissions reduction target should not be realized within three years, the main director of a key county (city) shall be ordered to resign, corporate representatives of state-owned enterprises shall immediately resign, and private enterprises shall be ordered to suspend production for rectification." This meant that local governments were facing immense pressure to complete their target; it has also led to the establishment of basic energy conservation awareness.

Furthermore, energy conservation indicators were also included in several performance evaluation systems for local party and government leaders; local leaders must now pay close attention to energy conservation in order to receive a promotion. A research group at Tsinghua University's CPI Center conducted a survey in Wuxi City and learned that completion of energy conservation targets is included as an indicator in several sets of cadre evaluation systems such as the scientific development concept assessment; the China sustainable development experimental district construction index; the emerging industrial sector evaluation; and the party leaders and main governors work performance comprehensive evaluation system. The indicators attach certain importance to each of the respective evaluation systems above; this led to an even greater attention to the issue of energy conservation from the side of local government officials.

The commitment of local governments to practice energy conservation is also explicitly stated in speeches and presentations given by local government officials. The provincial governors of Gansu, Shandong, and other provinces stated that if they could not be able to meet their respective conservation target, they would resign immediately. Others, like Zhejiang's provincial governor, stated that: "A GDP growth rate of 8% or 10% is just a projected indicator, but a 20% in reduction of energy consumption per unit of GDP is a binding indicator, it must be accomplished." Zhejiang's governor has specifically written letters to the eleven mayors within Zhejiang province, demanding an assurance to complete their energy conservation tasks.[1]

[1] Southern Weekly, Jieneng jianpai daojishi—'Shangbannian bu guan ni, xiabannian guansi ni' 节能减排倒计时—'上半年不管你, 下半年管死你' [Energy conservation and emissions reduction countdown] (2010-09-22), http://www.infzm.com/content/50374 [accessed: 2011-08-13].

7.1.2 *The Establishment of Local Government Energy Conservation Administration*

7.1.2.1 *The Establishment of Local Energy Conservation Leading Group*

During the course of 11th FYP, local administrations tasked with energy conservation were gradually established and developed, forming a management framework that exists at the central, provincial, city and county levels.

Before the formulation of the 11th FYP, institutions related to energy conservation were only limited to organizations such as the NDRC and the State Economics and Trade Commission. In June 2007, the State Council published the "Notice Regarding Print And Issue Of Comprehensive Plan Of Energy Conservation And Emissions Reduction," it emphasized that "the top priority at the moment is the establishment of a sound energy conservation and emissions reduction work accountability and evaluation system at every administration level, and to form a dynamic implementation structure with the efforts at various levels. Local governments of each level are responsible for the overall task of energy conservation and emissions reductions within their own administrative districts, the major government officials are at the frontline of responsibility." In the meantime, building on the National Climate Change Coping and Coordinating Group, the central government established the National Climate Change, Energy Conservation and Emissions Reduction Leading Group. The group serves as China's advisory and coordinating organ for coping with climate change, energy conservation, and emissions reduction. Soon after, the provincial Climate Change, Energy Conservation and Emissions Reduction Leading Groups were established in succession.

The structure of the provincial energy conservation and emissions reduction leading groups mostly resembles the structure of the national leading group. While the chief administrator serves as the group director, members include institutions such as the NDRC and the Environmental Protection Bureau. Hainan province may serve as an illustration: In June 2007, Hainan province established its climate change, energy conservation, and emissions reduction leading group; the provincial governor served as the group director; the vice governor served as vice director; leaders of the provincial general offices and other related departments served as members of the leading group. The Hainan provincial government clearly stated the obligations of the leading group in the "Notice Regarding Print and Issue of Comprehensive Work Plan of Hainan Province Energy Conservation and Emissions Reduction." The main responsibility of the leading group is "to organize, implement and fulfill the key national strategic policies and countermeasures of coping climate change, energy conservation and emissions reduction; to coordinate efforts to address climate change, energy conservation, and emissions reduction for Hainan province; to engage in international cooperation and negotiate countermeasures; to coordinate and solve key issues in coping with climate change, energy conservation and emissions reduction." Moreover, to notify and request all city and county government bodies within Hainan province to establish leading groups to address climate change, energy conservation and emissions reduction; this will further enhance the organizational and the leadership capabilities to deal with these issues.

Under the direction of provincial governments, prefectural and city level governments have established corresponding levels of leading groups accordingly. For instance, Fujian province's Putian city established its leading group in November 2007. Huzhou City in Zhejiang province established its leading group one month later. Other leading groups at various levels have been established from there on.

7.1.2.2 *The Establishment of Local Energy Conservation Monitoring and Supervision Administrations*

During the 11th FYP, some provinces and cities have established energy conservation monitoring and supervision administrations, and opened up the issue of energy conservation to local law enforcement.

According to the "Energy Conservation Law of the People's Republic of China" (revised in 2007), within the field of energy conservation "every local government at county level or above should enhance its guidance, deployment, coordination, monitoring and supervision, inspection and promotion of energy conservation tasks. The department of energy conservation management and the related departments should, within their obligations, strengthen the monitoring, supervision, and inspection of energy conservation laws, regulations and implementation progress of energy conservation standards, as well as investigate and prosecute energy usage violations according to law." Currently, 30 provinces (autonomous regions and municipalities with the exception of Tibet) have established energy conservation monitoring and supervision administrations. Their organizational structure, personnel management, and other features differ from province to province; Table 7.1 shows the settings of each energy conservation monitoring and supervision administration in detail:

Table 7.1. Provincial government and sub-provincial cities' energy conservation monitoring and supervision administration settings.

Provinces/city/ autonomous regions	Name of energy conservation center	Sub-provincial cities	Name of energy conservation center
Beijing	Beijing Energy Conservation Monitor and Supervision Unit Beijing Energy Conservation and Environmental Protection Center	Harbin	Harbin Energy Conservation Monitor and Supervision Center
Tianjin	Tianjin Energy Conservation Monitor and Supervision Center	Changchun	Changchun Energy Conservation Monitor and Supervision Subunit
Hebei	Hebei Energy Conservation Monitor and Supervision Center	Hangzhou	Hangzhou Energy Conservation Monitor and Supervision Center
Shanxi	Shanxi Energy Conservation Monitor and Supervision Unit	Dalian	Dalian Energy Conservation Monitor and Supervision Subunit
Inner Mongolia	Inner Mongolia Autonomous Region Energy Conservation Monitor and Supervision Center	Shenyang	Shenyang Energy Conservation Monitor and Supervision Center
Liaoning	Liaoning Energy Conservation Monitor and Supervision Center	Chengdu	Chengdu Energy Conservation Monitor and Supervision Center
Jilin	Jilin Energy Conservation Monitor and Supervision Center	Qingdao	Qingdao Energy Conservation Monitor and Supervision Center
Heilongjiang	Heilongjiang Energy Conservation Monitor and Supervision Center	Xian	Xian Energy Conservation Monitor and Supervision Center

Table 7.1 (*cont.*)

Provinces/city/ autonomous regions	Name of energy conservation center	Sub-provincial cities	Name of energy conservation center
Shanghai	Shanghai Energy Conservation Monitor and Supervision center	Ningbo	Ningbo Energy Utilization Monitor and Supervision Center
Jiangsu	Jiangsu Energy Conservation Monitor and Supervision Center	Wuhan	Wuhan Energy Conservation Monitor and Supervision Center
Zhejiang	Zhejiang Energy Conservation Monitor and Supervision Unit	Jinan	Jinan Energy Conservation Monitor and Supervision Subunit
Anhui	Anhui Energy Conservation Monitor and Supervision Center	Nanjing	Nanjing Energy Conservation Monitor and Supervision Center
Fujian	Fujian Energy Conservation Monitor and Supervision Center	Xiamen	Xiamen Energy Conservation Monitor and Supervision Center
Jiangxi	Jiangxi Energy Conservation Monitor and Supervision Center		
Shandong	Shandong Energy Conservation Monitor and Supervision Unit		
Henan	Henan Energy Conservation Monitor and Supervision Center		
Hubei	Hubei Energy Conservation Monitor and Supervision Center		

Table 7.1 (*cont.*)

Provinces/city/ autonomous regions	Name of energy conservation center	Sub-provincial cities	Name of energy conservation center
Hunan	Hunan Energy Conservation Monitor and Supervision Center		
Guangdong	Guangdong Energy Conservation Monitor and Supervision Center		
Guangxi	Guangxi Energy Conservation Technology Service Center		
Hainan	Hainan Energy Conservation Monitor and Supervision Unit		
Chongqing	Chongqing Energy Utilization Monitor and Supervision Center		
Sichuan	Sichuan Energy Conservation Monitor and Supervision Center		
Guizhou	Guizhou Energy Conservation Supervision Unit Guizhou Energy Conservation Monitor Center		
Yunnan	Yunnan Energy Utilization Monitor Center		
Shaanxi	Shaanxi Energy Conservation Monitor and Supervision Center		

Table 7.1 (*cont.*)

Provinces/city/ autonomous regions	Name of energy conservation center	Sub-provincial cities	Name of energy conservation center
Gansu	Gansu Energy Conservation Monitor and Supervision Center		
Qinghai	Qinghai Energy Conservation Monitor and Supervision Office		
Ningxia	Ningxia Energy Conservation Monitor and Supervision Center		
Xinjiang	Xinjiang Energy Conservation Monitor and Supervision Center		

Source: Xiamen Energy Conservation Monitor and Supervision Center website: http://xmecc.xmsme.gov.cn/2010-8/20108181l1825.htm [accessed: 2011-09-20].

With legal regulations such as the "Energy Conservation Law" and its enforcement by local administrations, China has constructed a primary management framework for monitoring and supervising energy conservation in the industrial, transportation, building, energy-saving products, and public institution sectors.

7.1.3 *The Creation of Energy Conservation Pressure Transferring Mechanism*

The institutional arrangement adopted by the central government during the 11th FYP to ensure the realization of a reduction in energy consumption per unit of GDP by approximately 20% was mainly the energy target accountability system. The target accountability system is a binding administrative policy measure that is frequently applied to all levels of governments to complete a specific task. The energy conservation target accountability system includes tasks such as affirmation and decomposition of energy conservation target; establishing, monitoring, accounting, review and evaluation of binding target agreements. In September 2006, the State Council published the "Approval Regarding 11th FYP Regional Energy Consumption per Unit of GDP Reduction Indicator Plan," which formally set the target for each provincial government's energy conservation and emissions reduction indicator, and requested individual provincial government to decompose the energy conservation target to its cities, counties, related industries and major enterprises. The local governments have established corresponding pressure transfer mechanism for energy conservation.

As requested by the central government, individual provincial governments have assigned energy conservation targets to the city and prefecture-levels of government. For instance, on the basis of a comprehensive evaluation on industry structures of 14 cities and their energy consumption levels, Liaoning Province decomposed the energy consumption target to each city by an uneven reduction rate. Hubei Province, on the other hand, determined the energy conservation target for each city, prefecture, and forest district after a comprehensive balance taking account of the ratio between light and heavy industries, ratio of total value added of the large-scale industry, energy consumption share of the whole province, and energy processing and conversion input of gross energy consumption ratio. Hubei Province assigned targets to each city after a comprehensive evaluation of each city's data from the previous year. Local governments have also included major energy-consuming enterprises at various stages in the process of the implantation of the energy conservation accountability system.

This led to a substantial increase in the number of enterprises that have signed the energy conservation target accountability agreement after the central government launched the "1,000 Enterprises Energy Conservation Campaign." Table 7.2 summarizes the enterprises included in each provincial government's energy conservation accountability system.

Table 7.2. Enterprises included in each provincial government's energy conservation accountability system.

Provinces / cities / autonomous regions	Enterprise target
Beijing	Enterprises that that have signed energy conservation accountability agreement with city government in accordance with national and Beijing's related regulation and policy (include enterprises within the 1,000 major energy consumption enterprises list monitored by central government)

Table 7.2 (*cont.*)

Provinces / cities / autonomous regions	Enterprise target
Tianjin	Enterprises within the 1,000 major energy consumption enterprises list monitored by central government
Hebei	112 national major energy consumption enterprises and 109 provincial major energy consumption enterprises (so called the double-hundred enterprises)
Shanxi	200 "double-hundred" key energy consumption enterprises include 86 enterprises within the 1,000 major energy consumption enterprises list
Inner Mongolia	50 major energy consumption enterprises
Heilongjiang	318 provincial major energy consumption industrial enterprises
Shanghai	Industrial holdings (group) company; each district and county economic commission (i.e. Shanghai chemical industry park); 11 enterprises enlisted in the national "1,000 enterprises" list; enterprises that consume more than 10,000 tce/a
Jiangsu	126 major energy consumption enterprises enlisted in the "100 enterprises energy conservation campaign"
Zhejiang	17 major energy consumption enterprises enlisted in the national "1,000 enterprises" list and 28 enterprises with 180,000 tce/a above and signed 11th FYP energy conservation target accountability agreement
Anhui	153 provincial major energy consumption enterprises, in that, 33 of them are enlisted in the national "1,000 enterprises" list and 120 enterprises enlisted in the provincial 100 major energy consumption enterprises
Fujian	Provincial major energy consumption enterprises
Jiangxi	100 major energy consumption enterprises (include 19 enterprises involved in the national 1000 enterprises energy conservation campaign)
Shandong	103 national major energy consumption enterprises
Henan	Enterprises with general energy consumption of 50 000 tce/a or above
Hubei	100 major energy consumption enterprises（include 31 enterprises enlisted in the national "1000 enterprises" list）

Table 7.2 (*cont.*)

Provinces / cities / autonomous regions	Enterprise target
Hunan	100 major energy consumption enterprises
Guangdong	Major energy consumption enterprises determined by the provincial "Double-thousand" energy conservation campaign
Guangxi	Major energy consumption enterprises with energy consumption above 10,000 tce/a
Hainan	12 major energy consumption enterprises that signed energy conservation target accountability agreement with provincial government
Chongqing	100 major energy consumption enterprises, of which 13 are national major energy consumption enterprises and 87 are city level major energy consumption enterprises
Sichuan	39 major energy consumption enterprises enlisted in the national "1,000 enterprises energy conservation campaign" and 200 major energy consumption enterprises enlisted in the provincial "100 enterprises energy conservation campaign"
Guizhou	Provincial major energy consumption monitored enterprises (include 18 enterprises enlisted in the national 1000 major energy consumption enterprises)
Yunnan	25 enterprises enlisted in the national 1000 enterprises energy conservation campaign, 11 groups (corporates) that are assessed by provincial government, "double-hundred" energy conservation campaign enterprises and other major energy consumption enterprises
Shaanxi	200 major energy consumption enterprises determined by the provincial government.
Gansu	Major energy consumption enterprises with energy consumption above 10,000 tce/a
Qinghai	1,000 major energy consumption enterprises
Ningxia	Major energy consumption enterprises with energy consumption above 50,000tce/a

Source: Ma Li, Li Huimin, Qi Ye 马丽, 李惠民, 齐晔, "Jieneng de mubiao zerenzhi yu ziyuan xieyi: Liangzhong moshi bijiao 节能的目标责任制与自愿协议: 两种模式比较 [Comparative Analysis of Energy Conservation Target Responsibility System and Voluntary Energy Conservation Agreements]," *Zhongguo renkou, ziyuan yu huanjing* 21, no. 6 (2011): 95–101.

Moreover, provincial governments have decomposed target accountability to each relevant administrative department. The "Advice Regarding Strengthen Energy Conservation Work" published by the Hubei provincial government in November 2006 states: "The development and reform commission, economic commission, science and technology bureau, and the information industry bureau are responsible for adjusting the industrial structure; the development and reform commission, economic commission, state capital office, quality supervision bureau, and statistics bureau are responsible for strengthening energy conservation for the industrial sector; the bureau for government offices administration, provincial department of finance, and government procurement center are responsible for energy conservation of government institutions." Hebei, Yunnan, and some other provinces have adopted similar rules.

After decomposing the energy conservation target, local governments have formulated their complementary energy conservation target accountability evaluation system. Shandong was the first provincial government to formulate the energy conservation target accountability evaluation system, as Shandong province published the "Energy Conservation Target Accountability Evaluation System" (Lu Zheng Ban Fa [2006] No. 117) in December 2006. Shanxi province in September 2007 promulgated "Shanxi Province Energy Target Accountability Evaluation, Rewards and Penalty Methods" which provide rules on evaluation methods for city-level governments and provincial major enterprises. Some provincial governments implemented specific policy measures aimed to improve energy conservation target accountability; for instance, apart from the "Implementation Plan for Energy Consumption per Unit of GDP Evaluation System," Yunnan province has also put forward an evaluation system for administrative accountability in the "Advice on Implementation of Yunnan Province Energy Conservation and Emissions Reduction Administrative Accountability" published in April 2008. The advice stipulates that should districts, city-level governments, and associated provincial departments miss their annual energy conservation and emissions reduction target, the chief officials will not be as awarded the "competent performance" or above grade in their annual public servant assessment; according to the evaluation results, the established procedures rule out any promotion of leading officials, and they are forbidden to promote or appoint cadres at the same year. As provincial governments establish accountability evaluation systems for their conservation targets, the majority of prefectures and city-level or county-level governments have established corresponding systems within their administration zones.

In the wake of the accountability system, the number of provincial governments that missed their annual energy conservation target has gradually diminished. According to the energy conservation target completion report published by the NDRC and the NBS in June 2011, all provinces (with the exception of Xinjiang province for which different rules apply) fulfilled their 11th FYP energy conservation target. 28 provinces overfulfilled their target (see Figure 1.14).

7.1.4 *The Formulation of Low-Carbon Oriented Resource Allocation*

7.1.4.1 *Establishment of Energy Conservation Funding*

To enhance the guidance and promotion of energy conservation, the revised "People's Republic of China Energy Conservation Law" of 2007 added a new article, addressing "incentive measures." This article determines energy conservation incentive measures such as fiscal funding, taxation, government procurement, pricing, and bank credits. Article 60 rules that central financial institutions and provincial financial institutions provide funding in support of energy conservation technology research, technology, and product demonstration and promotion, implementation of major energy conservation projects, energy conservation propaganda training, information services, and recognition of awards. To carry out this article in the course of 11th FYP, the central government has strengthened the financial support of the 10 major energy conservation projects. Provinces, autonomous regions, and municipalities have also established energy conservation funds, increased energy conservation investment, and asked enterprises to conduct R&D and improvements regarding energy conservation technology. In the meantime, each province, autonomous region, and municipality has established management measures for energy conservation funding since 2007, which determines sources for funding, range of application and approach, as well as supervision and management of standardized energy conservation funding.

Provincial energy conservation funding is generally provided by provincial financial budgets with other applicable budgets as a supplement. Apart from the two funding sources mentioned above, some provincial governments have explicitly stipulated the access to energy conservation funding. For example, Shanxi province has set the provincial coal sustainable development funding as supplementary funding source; Hainan, Yunnan, and Shanghai's energy conservation funding includes the revenue collected from the implementation of differentiated electricity pricing regimes; Guangxi and Yunnan province assigned a share of central government grants towards energy conservation funding, Since 2006, a majority of provinces and municipalities have included energy conservation funding in their annual budgets, and progressively increased its volume. Table 7.3 shows various provincial energy conservation funding by provinces (autonomous regions and municipalities) throughout the 11th FYP.

Table 7.3. The amount of energy conservation funding of provincial governments (2006–2011).

Province/ city/ autonomous region	Name of funding	Amount of funding in millions of RMB					
		2006	2007	2008	2009	2010	2011
Beijing	Energy conservation and emissions reduction fund			180.25	180.04	350	
Shanghai	Energy conservation and emissions reduction fund			1070	1200	1700	
Jiangxi	Energy conservation fund			30		52*	15.6
Shandong	Water and energy conservation fund	15	300		150		190
Guangdong	Energy conservation fund; Energy conservation, emissions reduction and renewable energy fund				300		
Hainan	Energy conservation fund (2010)			4	10	38	
Liaoning	Energy Conservation technology improvement, technology upgrade and electricity demand side management			50	86		
Qinghai	Energy conservation fund				35		
Inner Mongolia	Energy conservation fund				65		

Table 7.3 (*cont.*)

Province/ city/ autonomous region	Name of funding	Amount of funding in millions of RMB					
		2006	2007	2008	2009	2010	2011
Sichuan	Energy conservation fund & Industrial sector energy conservation fund + Energy conservation and emissions reduction technology R&D fund (2010)			30		103 (80+ 23)	
Chongqing	Energy conservation fund					35.2	
Yunnan	Energy conservation fund			114			
Jiangsu	Recycle economy and Energy conservation fund		100	300	300		
Hebei	Energy conservation fund		38				
Shanxi	Energy conservation fund		20				
Jilin	Energy conservation fund		10				
Hunan	Energy conservation fund			30			
Shaanxi	Energy conservation fund					Exceed 100	
Anhui	Energy conservation fund		10	32			
Fujian	Energy conservation fund			40			

Source: Tsinghua University Climate Policy Research Center concluded on the basis of data from the Ministry of Finance and NDRC website.
* Total of 120 million in 11th FYP

The energy conservation fund established by local governments mainly lends financial support in the form of direct subsidies, interest-free loans, and substitute subsidies with rewards. The list of supported fields related to energy conservation include: (1) Implementation of major energy conservation projects; (2) R&D as well as demonstration and promotion of energy-conservation technology and products; (3) Recycling and renewable energy utilization; (4) Energy contract management; (5) Transportation and building energy conservation; (6) Energy conservation propaganda and training; (7) Establishment of energy conservation monitoring, supervision, and management systems; (8) Energy conservation recognition awards. The first and the second on this list are the most important ones. The majority of provincial governments have established ground rules for subsidy standards and regarding financial support for projects that aim at the improvement of energy conservation technology. Guangdong, Shanghai, and other provinces reward energy conservation projects that have adopted energy contract management; Shanxi province provides financial subsidies to public energy-conservation schemes such as the promotion of energy-saving transportation vehicles and agricultural machinery; Hebei, Shaanxi, Hubei, and Guangxi are among those who link the amount of financial subsidy with the amount of conserved energy and reduced emissions, and they adopted "a substitute subsidy with rewards" approach to allocate funding; Jiangsu province, on the other hand, hand out rewards depending on the evaluation results of its energy-conservation target.

Local governments have also established management measures for procedures such as fund requisition and verification. Once a project has met the conditions and requests funding, the energy conservation department alongside the financial department is responsible for project acknowledgment, funding approval, transfer, as well as supervision and management. The formulation and implementation of these fund management measures is to be understood as a policy guarantee that greatly contributed to the standardization and streamlining of conservation-related fund applications.

Statistics show that during the 11th FYP, central financial institutes allocated 62.7 billion RMB from the Energy Conservation and Emissions Reduction Fund to energy conservation expenditures such as technology adoption and building retrofitting.[2] The energy conservation fund

[2] Jia Yue, Chang Hong 贾玥, 常红, *'Shiyiwu' Zhongyang caizheng anpai 627 yi yuan jieneng jianpai zhuanxiang zijin* '十一五' 中央财政安排627亿元节能减排专项资金

auditing report of 20 provinces conducted by the National Audit Office in 2010, indicates that from 2007 to 2009, a total of approximately 19 billion RMB invested by 20 provinces was assigned to the construction of 10 major energy conservation projects. 2,002 energy conservation technology improvement projects of major enterprises have benefited from it, and reduced energy consumption by 52.85 Mtce.[3]

Energy conservation rewards encouraged the adoption of energy conservation technology by enterprises with annual energy consumption below 10,000 tce, promoted local energy conservation technology R&D, as well as local energy conservation demonstration projects. Furthermore, as management approaches diversified, local governments' energy-conservation funding was also serving more diversified functions, such as supplement energy contract management, promotion of societal energy conservation, and improvement of energy conservation target accountability evaluation systems.

7.1.4.2 *Enhance Resource Allocation of Non-Fossil Energy Industry*

National policy measures generously supported non-fossil energy as a strategic emerging industry during the 11th FYP, and triggered a phase of rapid development. In support of the development of the non-fossil energy industry, local governments apart from implementing national policies launched a series of localized industry support policies to strengthen resource allocation for the non-fossil energy industry.

Support for non-fossil energy enterprises by local governments mainly comes in the form of land use approvals, credit financing, energy prices, fiscal subsidy, and R&D platforms as Table 7.4 illustrates.

[Central finance arranged 62.7 billion for energy saving and emissions reduction fund in 11th FYP] (2010-12-01). http://npc.people.com.cn/GB/13533081.html [accessed: 2011-09-12].

[3] National Audit Office of the People's Republic of China, *20 ge sheng youguan qiye jieneng jianpai qingkuang shenji diaocha jieguo* 20 个省有关企业节能减排情况审计调查结果 [The auditing results for the situation of energy saving and emissions reduction of relevant enterprises in 20 provinces] (announced 2011-05-13), http://www.gov.cn/gzdt/2011-05/13/content_1863388.htm [2011-09-13].

Table 7.4. Detailed measures of local governments to enhance non-fossil energy industry's resource allocation.

Enterprise development phase	Major difficulties	Local government supportive measures	Examples
Investment and construction phase	Land use	Prioritize approval and discount policy	Hebei province Baoding city implemented land use discount policy to promote photovoltaic industry construction.
		Establish related industry park	Jiangsu province Kunshan city Science and Technology Industry Parks, solar power and wind power are already pivotal industries, China's top 100 Science and Technology Industry Parks in 2011.
	Financing	Establish coordination mechanism with finance institute	Yunnan province Dali district government actively coordinate bank loans to support Maanshan wind farm, stroke a 200 million RMB loan with Construction Bank of China.
		Various access to funding, i.e. public listing, national project, international fund	Hebei province Chengde county explore Luanhe river hydroelectricity, signed contract with UK carbon emissions company, through CDM project financed 1.95 million Euro.
Production phase	R&D	Support establishment of R&D platform and laboratory	Jiangsu province Suzhou industrial park actively provides obstacle-free service, landing biggest domestic new energy research facility-Suzhou Xie Xin Industrial Application Research Institute.
	Personnel	Bring in top research talents	Jiangsu province Zhenjiang city published *New Energy Industry Human Resource Induction And Training Plan* in 2009, the plan estimates to induct 31,000 assorted staff and train 34,000 staff for photovoltaic industry, wind and nuclear energy equipment manufacture, bio-mass industry by 2012.
		Training necessary technical workers	
		Bring in expert consultants	

Table 7.4 (*cont.*)

Enterprise development phase	Major difficulties	Local government supportive measures	Examples
Sales phase	Electricity Grid Connection	Implement national mandatory grid connection and guaranteed acquire policies	Implement 2010 revised *Renewable Energy Law's* regulation on full price guarantee.
	Power generation cost	Electricity price subsidy	Establish photovoltaic benchmark price, for instance in 2011, Zhejiang province increased to 1.43 RMB/kWh, Suzhou province increased to 1.4 RMB/kWh
		Tax benefits, fiscal subsidy	Anhui province exempt income tax of new energy enterprises.

Column 7.1 illustrates the example of the construction of Gansu province's "wind power corridor." It explains how the local government creates policies to support the development of non-fossil energy industries.

Column 7.1. Construction of 'Wind Power Corridor' in Gansu Province.

Gansu province issued several policies to support wind power industry's development during the 11th FYP. In 2006, the Gansu government engaged in the construction of Hexi "wind power corridor," a strategic vision to reconstruct the western "three gorges of land." During the 11th FYP, Gansu province was enlisted in one of the key regions of national wind power construction projects; it became one of the biggest domestic wind power construction provinces. Gansu province's supportive measures for wind power industry resource allocation can be categorized as below:

(1) Land Use Discounts:
The Gansu provincial government published the *Circular of Gansu Province New Energy Equipment Manufacture Industry Development Guidance and Advice Notification* (Gan zheng Ban Fa [2010] No.18) which stipulates faster project verification and approval for various types of R&D projects regarding new energy equipment manufacturing, and additionally lend priority in terms of land use for projects that are included in the provincial major project plan and annual implementation plan.

(2) Financial Policy:
For the wind power industry and wind power generation equipment manufacturers, the Gansu government demands the creation of a communication and coordination mechanism with financial institutes to encourage and guide an appropriate amount of loan approval with the aim to increase credit support for the development of strategic emerging industries. In Jiuquan city for example, the city government strictly

Column 7.1 (*cont.*)

requires enterprises which have acquired the right to construct wind power facilities to partially acquire financing at a local bank. The local branch of the Industrial and Commercial Bank of China, Agricultural Bank of China, Bank of China, and China Construction Bank now all treat wind power project as a key market, each bank has established exclusive credit service teams, actively contacting enterprises to provide loan services. China Huadian Group built its 200,000 kW wind power project at Jiuquan Guazhou, a total investment of 2.1 billion RMB, 5 banks provided financial backing, and the China Development Bank and Agricultural Bank of China have reached a preliminary loan agreement of 800 million RMB.[4] Other than bank loans, Gansu province also encourages and supports small and medium-sized enterprises (SME), takes full advantage of small and medium-sized stocks at the ChiNext stock market, and advises competent enterprises to go public at both domestic and foreign stock markets.

(3) Fiscal Policy:
Gansu province established a strategic emerging industry fund and new energy equipment manufacture industry development fund to provide subsidies to eligible enterprises, and realize tax policies in support of advance technology industry development and other aspects.

(4) Price Policy:
For new energy power generations, Gansu province has always been subsidizing electricity prices. Wind power is the top priority among other subsidy targets. From 2007 onwards, the standard on-grid tariff of Gansu province's newly launched wind power enterprises (0.2508 RMB/kWh) is matched with the tariff of desulfurization thermal power. In 2007, the provincial price subsidy for new energy was 27.37 million RMB, 27 million RMB of which subsidized wind power enterprises.

(5) Human Resource Induction and Training Policy:
Gansu province demands an expansion of its innovation and entrepreneurship talent fund plan, to set up a sound incentive-based mechanism for technical personnel and managers, gather talent through new energy industry development and major projects. In recent years, Jiuquan city in Guizhou adopted a cooperation between university and enterprises, offering specific job training which included wind energy maintenance and wind energy operation. These operations have provided associated personnel training for China Power Investment Corporation, China Guangdong Nuclear Power Group and other wind power enterprises. Until end of 2010, 500 managerial and technical personnel received training; in terms of talent induction, the county government has adopted a flexible approach, hiring more than 50 experts and scholars to serve as senior consultant of wind power construction and county economic development.

Source: http://gs.people.com.cn/GB/183341/13560424.html.

[4] Lin Qinsheng 蔺秦生, "Xindai zhichi Gansu fengdian chanye fazhan qingkuang diaocha 信贷支持甘肃风电产业发展情况调查 [The survey of credit support to Gansu's wind power industry development]," *Gansu Jinrong* 6 (2009): 13–17.

7.1.5 *Local Government's Innovation in Low-Carbon Development*

During the 11th FYP period, some local governments have displayed active involvement in low-carbon development by taking innovative policy approaches.

7.1.5.1 *Spontaneous Construction of Low-Carbon Cities*

A "Low-carbon City" is defined as "a city whose economic engine is driven by a low-carbon industry, citizens embrace concepts and behaviors associated with a low-carbon lifestyle, and the government adopts the 'low-carbon society' as the blueprint of its city." During the 11th FYP, several local governments showed strong interest in building low-carbon cities and have launched first developmental steps towards this goal.

Shanghai and Baoding city are the first local governments that have proceeded with the construction of low-carbon cities: On January 28, 2008, the World Wildlife Fund (WWF) signed a cooperation agreement, the Low-Carbon City Initiative in China (LCCI) with Shanghai and Baoding. This meant that Shanghai and Baoding are the first cities to engage in any sort of low-carbon city pilot program. The highlight in Shanghai's low-carbon city construction is to develop energy-saving buildings, to improve building energy efficiency through investigation, accounting and analyze of energy consumption; increase the operational ability for energy saving through training of property management staff; and ultimately reduce carbon emissions produced by residential living through policy research on ecological housing. Baoding city, on the other hand, emphasizes the development of new energy and renewable energy industries; comprehensive utilization of new energy and energy conservation and emissions reduction measures; leading other cities in formulating a city development concept that states as its key target the construction of a low-carbon city; Baoding has embraced slogans such as "Electricity valley of China" and "The solar power city" as a basis from which to explore a low-carbon model of city development.[5]

Soon after Shanghai and Baoding embarked on their low-carbon city projects, many local governments initiated programs associated with low-carbon city development. For instance, Chengdu city in November 2009 published the "Chengdu Low-Carbon City Development Plan;" in January

[5] Liu Zhilin, Dai Yixin, Dong Changgui, Qi Ye 刘志林, 戴亦欣, 董长贵, 齐晔, "Ditan chengshi linian yu guoji jingyan 低碳城市理念与国际经验 [Low-Carbon City: Concepts, International Practice and Implications for China]," *Chengshi fazhan yanjiu* 06 (2009): 1–7.

2010, the "Xiamen Low-Carbon City Overall Planning Outline" followed; March 2010 saw the Wuxi City "Wuxi Low-Carbon Development Strategic Plan."

In its current state, low-carbon city development in China involves mainly six areas: industrial energy conservation; industry transformation and low-carbon industry park; building energy conservation; low-carbon transportation; new energy utilization; and promotion of energy saving in social consumption, Table 7.5 summarizes the six areas.[6]

Table 7.5. Six Key Areas of Local Government's Low-Carbon City Development.

Areas	Specific implementation
Industrial energy conservation	Realization of 20% energy saving indicator
Industry transformation and low-carbon industry park	Industry transformation and economic structure adjustment
Building energy conservation	Existing building energy conservation retrofitting.
	Public administration energy conservation
	Renewable energy integrated building
Low-carbon transportation	Prioritize public transportation
	New energy powered automobiles
	Promote bicycles
New energy utilization	Landfill gas power generation
	Solar power photovoltaic power generation
Promotion of energy saving in social consumption	Low-carbon propaganda and general public education
	Ten Cities Ten Thousand Units LED project
	Domestic energy saving lights distribution
	Promotion of home appliance energy saving standard

Source: Climate Organization, *Zhongguo de qingjie geming* 3 中国的清洁革命, III 2010. China's clean revolution, Vol. 3] (Beijing: Qihou zuzhi chubanshe, 2010).

6 Climate Organization, *Zhongguo de qingjie geming* 3 中国的清洁革命, III [2010. China's clean revolution, Vol. 3] (Beijing: Qihou zuzhi chubanshe, 2010).

With each city actively developing its own vision of the low-carbon city, the climate department of the NDRC on July 19, 2010 issued the "Notice Regarding Implementing Low-carbon Province and Low-carbon City Piloting Program" (Fa Gai Qi Hou [2010] No. 1587). This represented the first official opinion taken by the central government concerning the activities undertaken by local governments regarding the low-carbon city concept. This notice listed Guangdong, Liaoning, Hubei, Yunnan, and Shaanxi province, as well as Tianjin, Chongqing, Shenzhen, Xiamen, Hangzhou, Nanchang, Guiyang, and Baoding City (also known as "the 5 provinces and 8 cities") as China's first batch in the low-carbon city pilot program, it requested all participating provinces and cities to include measures addressing climate change in their 12th FYP; to study and formulate a low-carbon development plan; to present explicit goals, key tasks and detailed measures to control GHG emissions, reduce carbon emissions intensity and to actively explore a green, low-carbon developmental model.

With the arrangement of the central government 5 provinces and 8 cities begun to draw up local low-carbon city plan; the table below summarizes some of the key aspects in this context.

Table 7.6. Key Focus of Low-Carbon City Planning of Some Pilot Areas.

Area	Key Focus Of Low-Carbon City Piloting Program
Tianjin	Collaborate with Singapore and Japan, construct ecological city and low-carbon demonstration zone, explore low-carbon concept by demonstration within the low-carbon zone
Guangdong	Improve the core competence of "made in Guangdong" under international low-carbon industry framework, promote the establishment of carbon emissions trading system
Hubei	Promote city (county), industry park/zone, enterprise and community piloting demonstration
Chongqing	Adjust industrial structure, improve energy efficiency, optimize energy structure and promote research and application of low-carbon technology
Hangzhou	Engage in "six in one" integrated low-carbon city development: low-carbon industry, low-carbon building, metro transportation, low-carbon life, low-carbon environment and low-carbon society
Xiamen	Highlight communication and cooperation with Taiwan; system and mechanism innovation; promote low-carbon industrial development, low-carbon city construction and low-carbon life

Table 7.6 (*cont.*)

Area	Key Focus Of Low-Carbon City Piloting Program
Shenzhen	Continue advance technology development, modern finance, modern logistics and culture, establish low-carbon development system marketization
Nanchang	Prioritize the "big four" industries: solar photovoltaic development, green lighting, cultural tourism; focus on service outsource, new energy powered automobiles, modern logistics, aviation manufacture, new energy equipment, biology and new medicine, new materials and 6 other major industries; construct new industrial framework with low emissions.
Baoding	On the basis of rapid development of new energy and energy equipment manufacture, focus on building "China electricity valley" and "solar power city"

With the central government actively backing the pilot programs, local governments achieved relatively swift progress in their work towards low-carbon cities. Particularly the implementation plans of those participants listed in the pilot program already have gone through several stages of official approval and thus are equipped with a clear overall development vision that can provide useful guidance for other low-carbon city development projects.

7.1.5.2 *Promotion of Voluntary Energy Conservation Agreement*

During the 11th FYP, apart from decomposing energy conservation target accountability to enterprises within their jurisdiction, some local governments have actively facilitated a new non-mandatory energy conservation mechanism to enterprises—the voluntary energy conservation agreement. Voluntary agreements are a non-mandatory agreement between economic agents and governments. The agreement sets estimated goal negotiated between government and economic agents; it usually also includes corresponding supporting policies.

Before the 11th FYP, Shandong province formulated the "Shandong Province Voluntary Energy Conservation Agreement Pilot Program Plan," conducting a pilot program on voluntary energy conservation agreements within the province. The Jinan Iron and Steel Group Corporation and the Laiwu Steel Group LTD became the first two domestic enterprises to voluntarily participate in this energy conservation pilot program. During the 11th FYP, local governments promoted and developed voluntary energy conservation agreements; Table 7.7 provides a brief summary of voluntary energy conservation agreement projects that were implemented during the 11th FYP.

Table 7.7. Promotion and Development of Voluntary Energy Conservation Agreement during the 11th FYP.

Areas	Time	Events
Shandong	July 2006 to June 2007	Launch of "expanding promotion of voluntary energy conservation agreement pilot program project." Organizes 38 enterprises from Yantai, Weifang, Taian, Zaozhuang, Dongying, Heze, Binzhou and seven other cities to launch the voluntary energy conservation agreement pilot program in 2007.
	2008	Third batch of voluntary energy conservation agreement pilot program, a total of 89 enterprises signed the voluntary contract.
	2009	108 enterprises participated in voluntary energy conservation agreements, a total of 250 enterprises have participated in the pilot program to date.
Jiangsu	2006	10 enterprises signed voluntary energy conservation agreement with the local government in Nanjing.
	June 2006	10 enterprises signed voluntary energy conservation agreements with the local government in Yangzhou.
	March 2007	14 energy consumption enterprises joined the EU's "Asia environment support project"—Chinese city environment management voluntary agreement method project (phase two).
	2008	9 enterprises signed voluntary energy conservation agreements with county government in Yangzhou Baoying county.
Chongqing	October 31, 2008	Yuzhong district and its water utility, Marriott hotel, Intercontinental hotel, and other major energy consumption enterprises signed voluntary energy conservation agreements.
Tianjin	2008	Economic Commission signed voluntary energy conservation agreements with Tianjin Tiantie Metallurgical Group and Tianjin pipe corporation.

Table 7.7 (*cont.*)

Areas	Time	Events
Xinjiang Uygur Autonomous Region	April 2009	Xinjiang Oilfield Company, Kramayi Petrochemical Company, Dushanzi Petrochemical Company joined 'China city environment management voluntary agreement method' project and signed voluntary energy conservation agreement.
Guangdong	January 11, 2010	Economic and Trade Commission signed voluntary energy conservation agreement with Guangzhou Power Supply Bureau.
Jiangxi	May 19, 2010	Jiujiang City government signed voluntary energy conservation agreement with Gongqing open zone administration committee.
	May 19, 2010	Gongqing open zone administration committee signed voluntary energy conservation agreement with 2 enterprises within administration.
	June 18, 2010	16 industrial enterprises signed voluntary energy conservation agreement in Jingdezhen city.
	2010	Yingtan City Industry and Information Technology Committee mobilized 10 enterprises to sign the voluntary energy conservation agreement.

Source: Ma Li, Li Huimin, Qi Ye 马丽, 李惠民, 齐晔, "Jieneng de mubiao zerenzhi yu ziyuan xieyi: Liangzhong moshi bijiao 节能的目标责任制与自愿协议: 两种模式比较 [Comparative Analysis of Energy Conservation Target Responsibility System and Voluntary Energy Conservation Agreements]," *Zhongguo renkou, ziyuan yu huanjing* 21, no. 6 (2011): 95–101.

7.1.6 *Local Government's Game in Low-Carbon Development*

Under the current decentralized political system and the political incentive system, local governments are burdened with even greater responsibility for the development of their local economy. Hence, many local governments are reluctant to execute central government policy with regards to low-carbon development, while some opt to not act at all; numerous measures and investment decisions taken by local governments even run against the guidelines provided by the center.[7] Therefore, some dynamics

[7] Sun Liancai 孙连才, "Zhongguoshi jieneng jianpai 中国式节能减排 [Energy conservation and emissions reduction with Chinese characteristics]," *Xin Caijing* 01 (2011).

between the central and local governments have developed with regards to low-carbon development.

7.1.6.1 *Energy Conservation Target Decomposition—A Game between Central and Local Governments*

Since the official publication of the national "11th FYP Outline" in March 2006, central and local governments have engaged in negotiations to adjust targets for energy conservation and consumption reduction. The State Council has assigned the NDRC to be responsible for decomposing the energy conservation indicator from central to each province, autonomous region and municipal governments. The decomposition process can be divided into several stages: (1) With regards to the overall 20% energy consumption reduction target, the NDRC considers the actual situation of each province, collects the voluntary energy-conservation target commitments from each province, autonomous region and municipality, then summarizes and concludes each provinces' reduction commitment with regards to energy consumption per unit of GDP; (2) For those provinces that have already declared more than 20% in reductions for the 11th FYP, the NDRC will confirm its energy consumption reduction target; (3) For those provinces that have declared less than 20% in reductions, the NDRC will consult with local associated departments to make necessary adjustments, and determine the energy consumption reduction target for the province in question.[8] On September 17, 2006, the State Council published the "Approval Regarding Provinces' Energy Consumption per Unit of GDP Reduction Target Plan during the 11th FYP" (Guo Han (2006) No. 94) which confirmed each province's energy conservation and energy consumption reduction target. Table 7.8 compares each province's voluntary energy conservation promise with the ones from the State Council's final approval for the 11th FYP.

Table 7.8 shows that the majority of provincial governments have selected a relatively low energy conservation target, only four provinces pursued targets higher than the national target, fifteen provinces proposed targets equal to the national target, twelve provinces proposed

[8] Zhou Dadi et al. 周大地等, *Difang jieneng 20% de tujing he cuoshi yanjiu* 地方节能20%的途径和措施研究 [The Research of pathways and measures to 20% local energy saving] (Beijing: Kexue chubanshe, 2008); Song Yanqin, Gu Dedan 宋雅琴, 古德丹, 'Shiyiwu guihua' kaiju jieneng、jianpai zhibiao 'shiling' de zhidu fenxi '十一五规划'开局节能、减排指标'失灵'的制度分析 [An Institutional Analysis on Failure to Achieve Energy Conservation Target and Pollutant Emissions Control Target at the Beginning Stage of the 11th Five-Year Plan]," *Zhongguo ruankexue* 09 (2007): 25–32.

Table 7.8. Provinces' voluntary energy conservation promise and that of the State Council's final approval for the 11th FYP.

Province Autonomous Region and Municipality	Energy consumption per unit of GDP reduced (%)		Province Autonomous Region and Municipality	Energy consumption per unit of GDP reduced (%)		Province Autonomous Region and Municipality	Energy consumption per unit of GDP reduced (%)	
	Promised	Approved		Promised	Approved		Promised	Approved
Hainan	–7.5	12	Hebei	20	20	Shandong	22	22
Guangxi	10	15	Liaoning	20	20	Shanxi	25	25
Qinghai	10	17	Heilongjiang	20	20	Inner Mongolia	25	25
Yunnan	12	17	Shanghai	20	20	Jilin	30	30
Guangdong	13	16	Jiangsu	20	20	Total: 4 provinces		
Tibet	13	12	Jiangxi	20	20			
Beijing	15	20	Henan	20	20			
Tianjin	15	20	Hubei	20	20			
Zhejiang	15	20	Hunan	20	20			
Ningxia	15	20	Chongqing	20	20			
Shaanxi	19	20	Sichuan	20	20			
Fujian	—	16	Gansu	20	20			
Total: 12 provinces			Xinjiang	20	20			
			Anhui	20	20			
			Guizhou	20	20			
			Total: 15 provinces					

targets lower than the national target. This allows for an excellent comparison with each province's GDP plan: During the 11th FYP, a majority of provinces and municipalities reported a GDP growth rate higher than the national target, which caused the NDRC to issue documents demanding a GDP cool-down. Subsequently, with regards to energy conservation targets, the task of the central government became to coordinate with those provincial governments that had set up lower energy conservation targets to assure the fulfillment of the national target. The same problem occurred when each provincial government proceeded with the decomposition of its respective energy target to the city-level governments. Therefore, for many provinces, one important principle in decomposing energy conservation target is that "total target decomposed to each city and prefecture should not be less than the provincial energy conservation target."

7.1.6.2 *Local Negative Execution and Violation Behavior of Energy Conservation Policy*

To maintain economic growth, some local governments are unwilling to implement stringent energy conservation policies, and consequently their policy enforcement in this area is substantially lower than expected. Some instances of rule violation by local governments in the context of energy conservation were as follows:

First, a significant number of enterprises with outdated technology and equipment, inefficient resource utilization, and/or high energy consumption are protected and supported by local interest.[9] As indicated in the State Electricity Regulatory Commission's (SERC) 2009 "Annual Electricity Price Execution and Electricity Billing Supervision Report," some provinces enacted discount policies for electricity rates which have supported the reckless development of energy-intensive enterprises; a sluggish update of differential electricity rate lists means that some restricted or eliminated types of enterprises with high energy consumption were no affected by differential pricing according to regulation.

[9] Zhuang Guiyang 庄贵阳, "Zhongguo fazhan ditan jingji de kunnan yu zhang'ai fenxi 中国发展低碳经济的困难与障碍分析 [Analysis of difficulties and obstacles of China's low-carbon economic development]," *Jiangxi shehui kexue* 7 (2009): 20–26.

Particularly in China's mid-western regions, a number of localities have not yet transformed their traditional development approach that emphasizes investment-driven high energy consumption. Local governments provided energy-intensive enterprises with protection and support. For example, Guizhou province's Qianxinan prefecture promulgated a "Notice Regarding Implementation of Supportive Measures for Ferroalloy, Calcium Carbide, PVC and Other High Energy Consumption Enterprises" at the time of the global financial crisis in 2008. The prefectural government requested its department of power supply to support energy-intensive enterprises with regards to their energy purchases; the financial department was requested to provide credit support, while the tax office was requested to provide tax discounts, all with the intention to reduce the production costs for energy-intensive enterprises and help them to cope with deteriorating markets.[10] Such measures explicitly contradict the national credit, tax, and energy policies for energy-intensive industries.

Second, despite repeated orders from central government to strictly control the unchecked expansion of such industries, some local governments in violation of national law and regulations still initiate or expand copper smelting, ferroalloy, cement, steel, and other projects that are highly energy-intensive. Column 7–2 illustrates an example of Jilin province's Yanbian prefecture initiating an energy-intensive cement enterprise in violation of the law.

Third, because of unsatisfactory target completion results for the first four years of the 11th FYP, some local government adopted last-minute output and power restriction measures towards steel, cement, and other enterprises during the last year. Some governments even implemented residential power rationing to meet their target. The restrictions on output and energy use forced some enterprises to pursue independent energy-supply schemes, which reduced energy conversion efficiency and increased GHG emissions at substantial economic cost; rationing of public services and residential power has caused substantial social cost. Table 7.9 summarizes extreme energy conservation measures taken by some local governments at the end of the 11th FYP.

---

[10] See: Qian Xinan government office: http://law.baidu.com/pages/chinalawinfo/1704/51/a7a2e6f6bf5846f9a66318e608d88dd9_0.html 2011-07-20.

Column 7.2. Illegal initiation of high-energy consumption project in Jilin Yanbian.

Cement is a highly polluting, high energy consumption and overcapacity industry, and has always been one of the key industries under macro control and adjustment. According to the national regulation, Yatai Group Tumen Cement Co. LTD's new project, a production line with 5,000 tons daily output of new dry process cement clinker was not allowed to initiate. In spite of the fact that it has been instructed to halt three times by the Yanbian prefecture secretary, this cement project was eventually illegally initiated with the support of the Tumen county government where it resides:

The Yatai Tumen Corporation project was approved by Jilin province's Industry and Information Department in June 2009, although the project never initiated. Hereafter, China has made adjustment with regards to the authorization of cement projects due to issues that emerged from repeated constructions: It suspended construction of all cement projects that have not initiated before September 30, 2009; the NDRC would validate and approve any indispensable cement projects. There were other issues with the Yatai Tumen project such as the occupation of basic farmland, "excessive occupation," and avoiding examination by the Ministry of Environmental Protection and approval in terms of environmental impact assessment. Consequently, this project did not acquire the re-approval from the NDRC.

Because of the existing issues with this project, Yanbian county's secretary expressed explicit objections to the construction of this project and initiated three instructions demand ceasing. Yet, the Tumen county government insisted on going through with the project, the Tumen county water department approved water usage files for this project. In April 2010, two months after Yanbian secretary's first instruction, the project's construction was in full process. On August 19, 2010, a total of more than 100 people including Tumen secretary, magistrate, members of the "four major administration teams," and heads of each departments, office, and commissions came to Yatai Tumen Corporation for inspection. The director of the corporation alleged: "overcome challenges, strive for early production and early results." In October 2010, after Yanbian secretary's third suspending instruction, Tumen county and the Yatai Tumen Corporation not only did not halt the construction, but accelerated the construction pace.

Source: http://jjckb.xinhuanet.com/2010-11/24/content_271846.htm.

Table 7.9. Extreme energy conservation measures of some local governments adopted at the end of 11th FYP.

Local governments	Extreme energy conservation measures adopted at the end of 11th FYP
Hebei, Wuan	Shut down of 18 steel mills' partial production line
Hebei, Tangshan	Output control of steel, coking enterprises
Hebei, Chengde	Restrict electricity usage of some districts, steel mills output control, temporarily suspend construction of a batch of high energy consumption industries
Zhejiang, Wenzhou	Enterprises power rationing
Zhejiang, Huzhou	Unconventional power restriction for enterprises and public service administration
Guangxi, Yulin	Enterprises, public service administration and residential power restriction
Anhui, Quanjiao	Power restriction and power cut within jurisdiction, implement rolling power cut to residential living power usage

7.2 Promotion and Limitation Factors of Local Government's Low-Carbon Development

7.2.1 *Promotion Factors of Local Government's Low-Carbon Development*

Two main factors promote low-carbon development at the local level; one is the binding energy conservation policies imposed by the central government, the other is the potential benefits low-carbon development entails for local governments.

During the 11th FYP, the central government adopted incentive-based and binding policy measures to promote local governments' energy conservation and reduction work. The central government set a 20% reduction in energy consumption per unit of GDP as the binding factor within the national plan, to be implemented by the target accountability system. The central government assigned energy conservation targets to local governments and enterprises as well as target accountabilities. Numerous documents specified the regulation concerning the accountability and presented a detailed catalogue of penalties should the energy conservation targets not be fulfilled. This, in turn, has caused local governments to carry out energy conservation tasks.

There are several benefits for local governments to adopt low-carbon development. First of all, low-carbon development has positive impacts on adjusting local industrial structures and transforming economic growth, which benefits the sustainable development of local economies. Secondly, low-carbon development will provide local governments with more financial support. With respect to international funding, more than 50 financial institutions had joined the "climate change investment network" worldwide by the end of 2007, with a total investment of 13 trillion dollars. Competition for international funding has become a strong objective for many cities. Within China, the Central Economic Working Conference declared that in 2010, China should "adjust structure, prevent inflation," expand the development scope for new energy, energy conservation and environmental protection; engaging in low-carbon development may receive certain funding from the central government, thusly providing to some extent another incentive for local governments.[11]

Lastly, low-carbon development entails political incentives for local governments. For instance, the concept of the low-carbon city has the potential to become a motto for engaging in city planning and promotion.[12] Certain cities were not well-known by the general public before publicly committing themselves to develop low-carbon economies or to carry out a low-carbon city pilot program. The application of this new motto helps cities to explore new ways of low-carbon transformation, as well as to promote their community and attract investments. This in turn benefits local governments by bringing political achievements and competitive advantages, which means promotion for local officials. All of these factors have boosted local governments' low-carbon development measures.

7.2.2 *Limitation Factors of Local Government's Low-Carbon Development*

Progress in low-carbon development is limited by the demand of local economies. The economic demand at the local level is, to a large degree, affected by the current financial system. In 1994, China implemented a tax-sharing system, and established central and local tax revenue management systems which divide revenue and expenditures between the central and local governments. The tax sharing system gave central financial

---

[11] Zhou Hongchun 周宏春, "Ditan jingji zhi zheng 低碳经济之争 [The debate on low-carbon economics]," *Juece* 1 (2010): 60–61.

[12] Ibid.

institutes dominant power over local governments. The share of local fiscal revenue to total fiscal revenue decreased substantially.[13] In response to increasing financial pressures, local governments had to seek out new sources of fiscal revenue to finance their growing expenditures, and one of the main approaches to increase fiscal revenue was the vigorous expansion of the manufacturing sector, which is mostly driven by high-energy-consumption industries.[14] Some places enthusiastically invested in steel, electrolytic aluminum, electricity, and other high-energy-consumption industries, mainly because the amount of investment involved in these projects is considerably large, which brings about substantial benefits in terms of local GDP growth and tax revenue.

In the wake of the global financial crisis, local governments bear the responsibility for economic recovery while maintaining their economic base; the economic stimulus plan has spurred investment in basic infrastructure projects and energy-intensive industries in the short run, which also brought pressure on local governments' in terms of low-carbon development.

The second limitation can be found in the different development phases different provinces find themselves in. This has impeded low-carbon development in underdeveloped regions.

China spans over a vast territory, and thus economic development levels between localities are not identical. Since the economic reform, the Chinese economy has developed rapidly, and economic growth was accompanied by profound discrepancies in the spatial distribution of economic activity. This kind of asymmetry has resulted in imbalanced levels of urbanization and economic activities as well as widening regional income gaps.[15] The economic development gap between different regions

---

13 Li Feizhou 周飞舟, "Fenshuizhi shinian: Zhidu jiqi yingxiang 分税制十年:制度及其影响 [A Decade of Tax-Sharing: The System and its Evolution]," *Zhongguo shehui kexue* 06 (2006): 100–115.

14 Tao Ran, Lu Xin, Su Fubing, Wan Hui 陶然, 陆曦, 苏福兵, 汪晖, "Diqu jingzheng geju yanbian xia de Zhongguo zhuangui. Caizheng jili he fazhan moshi fansi- dui gaige 30 nian gao zengzhang de zhengzhi jingjixue zai kaocha he laizi 'tudi caizheng' shijiao de zhengju 地区竞争格局演变下的中国转轨: 财政激励和发展模式反思---对改革30 年高增长的政治经济学再考察和来自"土地财政"视角的证据 [China's Transition and Development Model Under Evolving Regional Competition Patterns]," *Jingji yanjiu* 7 (2009): 21–33; Tao Ran, Wang Hui 陶然, 汪晖. 2010. "Zhongguo shangwei wan zhi zhuanxing zhong de tudi zhidu gaige: tiaozhan yu chulu 中国尚未完之转型中的土地制度改革: 挑战与出路 [China's Unfinished Land System Reform: Challenges and Solutions]," *Guoji jingji pinglun* 2 (2010): 93–123.

15 Feng Junxin 冯俊新, *Jingji fazhan yu kongjian buju. Chengshihua, jingji juji he diqu chaju* 经济发展与空间布局: 城市化、经济聚集和地区差 [Economic Development

not only persists, but trends indicate that it is widening.[16] This is particularly true for the overall income gap between coastal and inland provinces. In 1978, the per-capita GDP in the wealthiest province, Liaoning, compared to poorest province, Guizhou, was at 3.89. In 2005, the difference between the wealthiest (Zhejiang) and the poorest (Guizhou) province has increased to 5.48.[17]

From the perspective of underdeveloped regions, it is only logical that the development of their local economies has top priority. Financial aid and preferential policies by the central government also provided an enormous push for underdeveloped regions' economic developmental behavior. Moreover, industries transferred from developed regions to underdeveloped regions are by and large characterized by low technological standards, high energy consumption, high pollution, and high labor costs. Therefore, underdeveloped regions and developed regions do not start off from the same point in their pursuit of low-carbon development.

The third limitation is the process of industrialization. During the 11th FYP, the rapid development of China's economy has resulted in a further boost to its industrialization level. The majority of energy-intensive industries are pivotal industries for the national economy. Whether it is the development of local economies or the improvement of living standards, they all rely on the fundamental support of these industries. With relatively intense employment and taxation pressure, local governments still face difficulties to optimize their industrial structure and accelerate the structural adjustment in the short run.[18]

---

and Its Spatial Consequences: Urbanization, Geographic Concentration and Regional Income Disparity. Doctoral dissertation] (Beijing: Tsinghua University, 2009).

[16] Lin Yifu, Cai Fang, Li Zhou 林毅夫, 蔡昉, 李周, "Zhongguo jingji zhuanxing shiqi de diqu chaju fenxi 中国经济转型时期的地区差距分析 [Economic analysis of China's regional disparity during the transition period]," *Jingji yanjiu* 06 (1998): 3–10; Wang Shaoguang, Hu Angang 王绍光, 胡鞍钢, 1999. "Zhongguo: BU pingheng fazhan de zhengzhi jingjixue 中国:不平衡发展的政治经济学 [China: unbalanced development of political economics. Beijing]" (Beijing: Zhongguo jihua chubanshe, 1999); Cai Fang, Du Yang 蔡昉, 都阳, "Zhongguo diqu jingji zengzhang de qutong yu chayi. Dui Xibu fakai zhanlüe de qishi 中国地区经济增长的趋同与差异–对西部开发战略的启示 [Convergence and divergence of Regional Economic Growth in China]," *Jingji yanjiu* 10 (2000): 30–37.

[17] Feng Junxin 冯俊新, *Jingji fazhan yu kongjian buju. Chengshihua, jingji juji he diqu chaju* 经济发展与空间布局: 城市化、经济聚集和地区差 [Economic Development and Its Spatial Consequences: Urbanization, Geographic Concentration and Regional Income Disparity. Doctoral dissertation] (Beijing: Tsinghua University, 2009).

[18] Zhuang Guiyang 庄贵阳, "Zhongguo fazhan ditan jingji de kunnan yu zhang'ai fenxi 中国发展低碳经济的困难与障碍分析 [Analysis of difficulties and obstacles of China's low-carbon economic development]," *Jiangxi shehui kexue* 7 (2009): 20–26.

The forth limitation is the process of urbanization. One of the indicators for a country's developmental state is its urbanization level. For quite some time, the guiding thought has been that China needed to implement rapid urbanization, as too big a share of its population was living in the countryside and the development of urbanization was lagging behind. All provinces consider a high urbanization rate an indicator for their political performance, and competition in this regard is fierce.[19] The Chinese urban population has increased from 50 million in 1949 to 562 million in 2005; its urbanization ratio reached 43%. According to Zhou Yixing, China's urban population ratio will have reached approximately 57% in 2020.[20] To a large extent, cities are consumption-intensive. When large numbers of rural residents move to the city, the energy consumption behavior changes, energy usage per person will dramatically increase and hence an increase in urban population will cause an increase in energy usage in transportation and housing. This will inevitably cause urban energy consumption levels to skyrocket.[21] In the process of urbanization, infrastructure is a crucial factor to improve urban functionalities.[22] As cities expand, local governments also pay more and more attention to infrastructure construction. Furthermore, during the urbanization process, problems occur, such as local governments engaging in extensive planning and reckless expansion; several local governments have entirely relocated governmental administration buildings to a new town, vanity projects and "GDP worship" have caused a large number of functional buildings to be demolished.[23]

The fifth limitation is the current political incentive system. The central government adopted conservation-related target performance as one of

---

19 Lu Dadao, Yao Shimou 陆大道, 姚士谋, "Zhongguo chengzhenhua jincheng de kexue sibian 中国城镇化进程的科学思辨 [A Scientific Thought about Urbanization Progress in China]," *Renwen dili* 04 (2007): 1–5.

20 Zhou Yixing 周一星, "Chengzhenhua kuaisu bushi yuekai kuaihao 城镇化速度不是越快越好 [Urbanization speed: faster is not better]," *Kexue juece* 8 (2005): 30–33.

21 Lei Hongpeng, Zhuang Guiyang, Zhang Cu 雷红鹏, 庄贵阳, 张楚, *Bamai Zhongguo ditan chengshi fazhan—Celüe yu fangfa* 把脉中国低碳城市发展—策略与方法 [Diagnosing Chinese low-carbon city development—Strategies and methods] (Beijing: Zhongguo huanjing kexue chubanshe, 2011).

22 Liu Yanan 刘亚娜, *Woguo chen gshihua jincheng zhong zhengfu xingwei cunzai de wenti ji duice yanjiu* 我国城市化进程中政府行为存在的问题及对策研究 [The Existing Problems and Counterplans of Government Behavior in the Course of Urbanization. Master dissertation] (Changsha: Hunan University, 2007).

23 Wang Wei, Huang Xiaowen, Lü Han 王炜, 黄晓文, 吕晗, "Zhong guo jianzhu pingjun shoumin jin 30 nian zhaiqian mai di liyi qudong shi qi duanming 中国建筑平均寿命仅30年 拆迁卖地利益驱动使其短命," *Renmin Ribao* (2010-11-1).

the criteria for assessing local government leaders during the 11th FYP, yet the current social order in China is still to a large extent dependent on economic growth. Hence, for a long period of time, the incentives for local governments to pursue economic growth was not weakened and economic performance as the key evaluation criterion within the political incentive system will continue to exert significant impact on local governments' behavior. Several research articles indicate that the promotion of government officials and the economic performance of their respective region closely correlate.[24] Thus, the centre's emphasis on economic growth and increased fiscal revenues has largely determined the growth-oriented developmental approach taken by local government officials.

7.3 Brief Remarks on Local Government's Low-Carbon Development Behavior during the 11th FYP

In the wake of its economic and social development, China's industrialization and urbanization pace has sustained a relative fast growth over a long period of time. During the 11th FYP, local governments have achieved substantial increases in fiscal revenue, GDP growth, and city construction performance. Low-carbon development policies, such as reduction measures for energy conservation and energy consumption during the 11th FYP, have slowed down the expansionary impulse of local governments to some extent. Local governments faced a dilemma between economic expansionism and low-carbon development that is still ongoing.

During the 11th FYP, local government agency was an indispensable force that delivered substantial contributions to China's energy conservation task and low-carbon development. Other than Xinjiang, all provincial governments have fulfilled their assigned 11th FYP energy conservation target. Local government's low-carbon development behavior mainly reflected in the establishment of low-carbon development awareness and energy conservation administrations, the creation of energy conservation pressure transferring mechanism, the formulation of low-carbon oriented

---

[24] Li Hongbin, Zhou Li-An, "Political turnover and economic performance: the incentive role of personnel control in China," *Journal of Public Economics* 89 (2005): 1743–1762; Zhou Li'an 周黎安, *Zhuanxing zhong de defang zhengfu: Guanyuan jili yu zhili* 转型中的地方政府: 官员激励与治理 [Local government in transition: Official incentive and Governance] (Shanghai: Gezhi Chubanshe, 2008).

resource allocation and local government's innovation in low-carbon development.

Meanwhile, several local governments were bargaining with higher-level administrations regarding the decomposition of energy conservation targets; some local governments showed a lack of interest in implementing energy conservation policies, and have instead lent active support to high-energy-consumption projects even in violation of national regulation. Some local governments had to rely on extreme measures such as power cuts at the end of 11th FYP to feign the appearance of target completion. Thus, the behavior of local governments displays a dual character.

Binding energy conservation policies by the central government as well as benefits gained from low-carbon development have facilitated the compliance of local governments. These take a largely utilitarian attitude towards low-carbon development, and focus on fields that benefit their local economies.

Nevertheless, local governments currently face serious challenges in low-carbon development; the increasing pressure on fiscal revenue and demand for economic development, the impact of the financial crisis, the effect of the economic stimulus plan, the process of countrywide industrialization and urbanization, and a political incentive system that weights economic performance as a key evaluation criterion all impede local governments' low-carbon development. Furthermore, because of China's vast territory, the different developmental stages across provinces have generated an intense demand for economic development in the underdeveloped regions. Therefore, underdeveloped regions face greater challenge in terms of low-carbon development.

"The 12th FYP Outline of Economy and Social Development of People's Republic of China" was released in March 2011, 16% reduction in energy consumption per unit of GDP and 17% reduction in CO_2 emissions per unit of GDP are listed as its main targets. This presents even more pressure, as well as new opportunities and challenges to local governments. The next phase is to further spurn local governments' commitment to low-carbon development, which would require a better incentive-based mechanism and an improved interest internalization mechanism.

CHAPTER EIGHT

ENTERPRISES' RESPONSIVENESS TO LOW-CARBON DEVELOPMENT

Abstract: During the 11th FYP period, energy-intensive enterprises improved their energy efficiency in response to low-carbon development policies. Accountability for energy-saving targets, the guidance provided by industrial policies, the reform of the energy management system and an increase in energy prices all pushed businesses to implement efficient energy-saving measures.

During this period, industrial enterprises realized 287 Mtce in energy conservation by substantially reducing their energy consumption per unit of production. The "Top 1000 Enterprises" saved 150 Mtce, which accounted for 52.3% of the industrial sector. The actions undertaken by enterprises were based on their own internal system adjustment and realized through the adoption of energy efficiency technologies and elimination of outdated capacities.

The dual pressure exerted by energy conservation policies and the increase in energy prices has caused enterprises to change their original internal rules. At the same time, increased distribution of energy efficiency technologies and the support of financial incentives have helped firms to acquire knowledge and funding in order to improve their capabilities.

China's Energy saving policy was preferential for large enterprises, and less favorable for SMEs, which undermined their capability to raise their energy-efficiency level. Fortunately, the extension of energy-saving projects on key energy-consumption firms in the 12th FYP should alleviate this problem.

The main share of energy was consumed by enterprises with high energy consumption. In order to correspond to the national low-carbon strategy, Chinese enterprises improved their energy efficiency in order to conserve energy. Under the dual pressure from both said policies and the market, enterprises reduced their energy intensity by adopting energy efficiency technologies (which additionally improved their competitiveness), while at the same time achieving their energy saving targets. Conservation measures undertaken by energy-intensive enterprises contributed too much of the completion of the national energy conservation goal. This report focuses on the conservation measures taken by key enterprises as defined by the *Chinese Energy Conservation Law*, namely enterprises with an overall energy consumption of over 10,000 tce, and partly enterprises with an overall energy consumption of over 5,000 tce as specified by local governments. For a long time, energy prices have been comparatively low due to government regulation while overall technical standards were

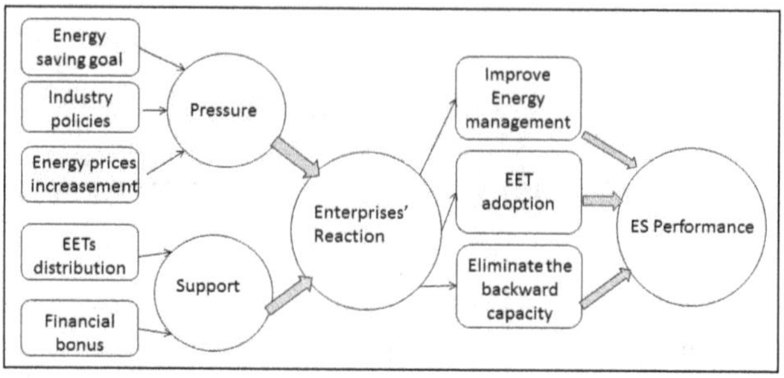

Figure 8.1. The mechanism affecting enterprises' energy consumption.

lagging behind. This caused enterprises to adopt a production mode characterized by low capital investment and high energy consumption. Due to the fact that investments in the production process in these enterprises are large while the payback period is long, enterprises lack a willingness to pro-actively update their technical equipment. In the long term, the adoption of energy efficiency technologies can reduce production costs; but most energy-intensive enterprises require continuous production, and the adoption of energy efficiency technologies may affect the scheduled production during the installation period, resulting in financial losses.

The potential risks inherent in the adoption of Energy Efficiency Technologies (EETs) combined with relatively low energy prices impeded the improvement of energy efficiency in the industrial sector. Accountability for energy-saving targets, the guidance provided by industrial policies, the reform of the energy management system, and an increase in energy prices all pushed businesses to implement efficient energy-saving measures. During the 11th FYP, in the context of changing institutional and market environments, energy-intensive enterprises responded with strategic and tactical adjustments. They achieved an improvement in energy efficiency while promoting their own competitiveness at the same time (for the mechanisms affecting enterprises' energy conservation, see Figure 8.1). The content of this chapter is as follows: the first part illustrates the measures taken by enterprises to conserve energy; the second part analyzes the mechanisms by which energy-saving policies influenced the energy conservation of enterprises; the third part points out the problems and challenges enterprises were facing during energy conservation from the perspective of energy policy.

In the 11th FYP, energy saving policies were implemented by different levels of government. The central government stressed that large enterprises

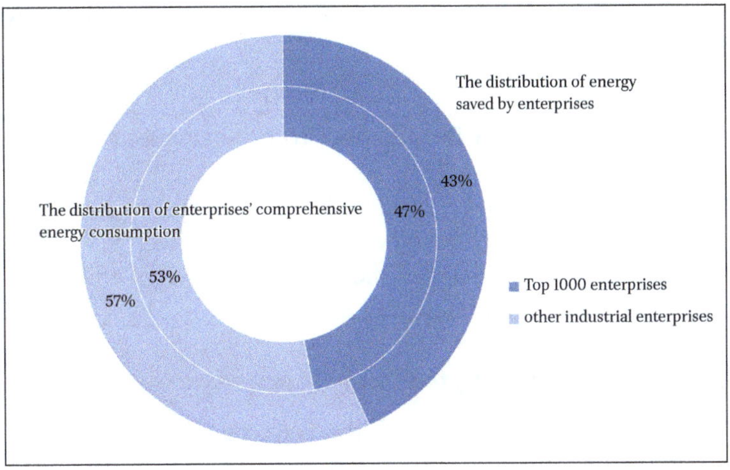

Figure 8.2. Distribution of Industrial Enterprises' comprehensive energy consumption and saved energy.

and local governments were responsible for the other enterprises. The "One-Thousand Enterprises" program was the focus of energy saving regulation for enterprises in the 11th FYP. In 2006, the NDRC confirmed the list of the Top 1,000 enterprises according to the overall energy consumption of those enterprises in 2004, and decomposed the overall target of 100 Mtce to each provincial government. Then, the provincial governments determined the target for each enterprise, and were responsible for supervision and assessment of the enterprises' energy conservation efforts. The other key energy consumption enterprises were assessed by different levels of local governments.

All industrial enterprises saved 342 Mtce by decreasing energy consumption per unit of product,[1] among which the top 1,000 enterprises saved 150 Mtce.[2] Their overall energy consumption in 2004 was 670 Mtce, which accounted for 47% of the industrial sector, and 33% of national consumption. Energy saved by the Top 1,000 enterprises in 11th FYP

[1] Energy saved through decreasing energy consumption per unit production by industrial enterprise = energy saved by 39 industrial sectors through energy intensity decrease—energy saved through product structure. The detailed calculation can be found in the appendix to Chapter 3.

[2] NDRC, *Qianjia qiye chao'e wancheng 'Shiyiwu' jieneng renwu* 千家企业超额完成'十一五'节能任务 [Top 1000 Enterprises Exceeded the Target of Energy Saving in the 11th FYP], http://www.sdpc.gov.cn/xwfb/t20110314_399363.htm [accessed: 2011-3-14].

surpassed the goal of 100 Mtce, and accounted for 43% of all energy savings in the whole industrial sector.

8.1 Schemes by Enterprises to Conserve Energy

Conservation-related measures undertaken by enterprises were based on their own internal system adjustment, and were realized via technological means, which mainly included the adoption of energy-efficiency technologies and elimination of outdated capacities.

8.1.1 *Improving the Energy Management System*

The Energy Management System is the basic institution for enterprises to manage and monitor their energy consumption. It provides data for decision-making when adopting EETs, and thus is the prerequisite for the application of financial benefits for energy saving. The Energy Management System in Chinese enterprises was established in the 1980s, and includes mechanisms for energy measurement and a statistical system for basic data collection, an energy supervision and auction system for technical support, and an energy assessment system for management safeguards (see Figure 8.3). Depending on the degree of commitment, the implementation of the Energy Management System in enterprises differed greatly. In the 11th FYP, most enterprises decomposed their energy saving target according to the requirements of the government, enhanced the priority of energy conservation in their decision making processes, and modified related rules.

8.1.1.1 *Basic Data Collection System*

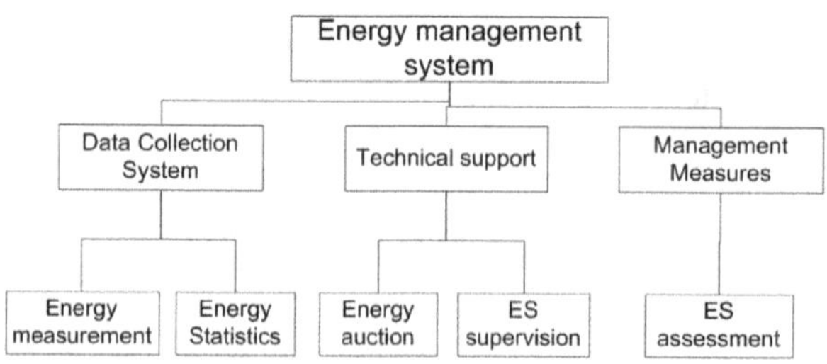

Figure 8.3. Enterprises' Energy Management System.

Improvements in basic energy data collection were reflected in a higher covering rate and an increasing precision of energy meters, alongside stricter standards in the energy statistical system. Data collected from this system would facilitate decision-making processes and effect the adoption of energy efficiency technologies. As shown in the "Announcement of Top 1,000 Enterprises Energy Consumption 2007," 95% of enterprises set up a department in charge of energy management, equipped with its own staff, in 2006, which demonstrates that the Energy Management System of most enterprises was gradually strengthened. Most Top 1,000 enterprises adopted a three-level energy measurement system; while the installation rate of the 1st and the 2nd level of this system was comparatively high, reaching 90%, and 80% respectively; the installation rate of the 3rd level was low at about 60%. Most enterprises involved produced energy statistics for the process of procurement, conveyance, transformation, utility, recycling, etc. The record and tables produced by the enterprises, including energy input/output/store records and tables, energy consumption records and tables, and tables on energy consumption per unit/product met the requirements of energy statistics. 50% of enterprises had good management systems, with complete statistical departments and qualified staff; only 20% the enterprises had weak statistical systems.[3] In December 2010, the Energy Saving Office in Shandong province invited the Shandong Investigation Center and Climate Change Center of Shandong Science and Technology Institute to conduct an investigation of the energy conservation measures of the key energy consumption enterprises. The investigation showed that most enterprises addressed energy conservation, and nearly 80% enterprises had set up energy measurement systems and energy saving assessment systems; nearly 75% enterprises had set up energy consumption statistics and report systems, and more than 70% enterprises had set up energy saving accountability and annual energy conservation regimes.[4]

---

[3] NDRC, *Qianjia qiye nengyuan liyong zhuangkuang baogao 2007* 千家企业能源利用状况公报 2007 [Announcement of Top 1000 Enterprises Energy Consumption 2007].

[4] Liu Ming 刘铭, "Shandong jieneng diaocha: Nengyuanjia shangzhang cheng qiye jieneng dongli 山东节能调: 能源价上涨成企业节能动力 [Energy Conservation Investigation: Energy Price Increase Became the Main Incentive For Enterprises to Save Energy]," *Qilu Wanbao* (2011-1-14).

8.1.1.2 *Technical Support System*

According to the requirements of the "Verification Guidance of Enterprise Energy Audits" and "Verification Guidance of Enterprise Energy Conversation Planning," 942 enterprises launched energy audits through on-site investigations, verification of related data, and testing if required. Enterprises analyzed their own energy utilization status by comparison with the average energy consumption per unit/product in large national enterprises and on the national advanced level, and then tried to identify existing problems and put forward feasible measures and suggestions to realize their potential for energy conservation.[5] Enterprises made energy conservation plans based on their overall deployment, considered the result of energy audits, and planned scaled adjustments and large equipment upgrades which helped to complete their energy savings target. Taking an oil field enterprise affiliated with Sinopec as an example, this enterprise had provided technical support for each production unit of the Energy Saving Supervision Center since 2008. The energy saving department was responsible for organizing all the production units to raise their technical standards in terms of energy saving, and made the plan for the next year. The Energy Saving Supervision Center evaluated the energy utilization during the production process (which included the water injection system, the power system, etc.) through energy audits, and provided a systematic report to assess the management system of each production unit. It also identified problems, suggested solutions to each production unit, and through supervision made assessments by studying the results of energy efficiency technologies. The data collected could then be used by the production units to compare each other's performance, and update their facilities with specific focus and goals.[6]

8.1.1.3 *Management Measures*

From a technological perspective, enterprises could decrease their energy intensity by adopting EETs. But energy conservation under actual working conditions depends on the performance of on-site operators; this requires the strengthening of internal management systems to guarantee that operators run the equipment according to established routines. Being required by the government to make energy savings assessments,

---

[5] NDRC, *Qianjia qiye nengyuan liyong zhuangkuang baogao 2007* 千家企业能源利用状况公报 2007 [Announcement of Top 1000 Enterprises Energy Consumption 2007].

[6] The data of the case is taken from the investigation material produced by the CPI.

all enterprises set up energy conversation leading groups, according to the requirements of the "Top 1000 Enterprises Energy Savings Program." Generally, the number one among the enterprises was assumed the leader of the group. The enterprises decomposed the total energy saving target into an operational index on the basis of their actual production. The indexes were used to assess the performance of each unit, from plant, workshops, production group, down to individual staff members. The first level of the energy management department generally included technical production personnel, statisticians, and measuring personnel; the second level included energy statistics and comprehensive management; most of the staff personnel in charge of the third level were part-time energy administrators.[7] Energy-saving target accountability was implemented through a top-down approach and combined the staffs' energy saving performance with their incomes; this improved the awareness of each production unit and individual, and raised their technical operation standards. The programs, such as energy conservation advocacy and training, contributed to the staffs' operational skills, and reached the goal of saving energy by decreasing unnecessary energy loss during equipment operation.

8.1.2 *Adoption of EETs Enhanced Enterprises' Performance*

Energy efficiency is a significant factor in the production rate of an enterprise. As energy prices in the market increased continuously, improvements in energy efficiency was an important measure to enhance enterprises' competitiveness. Adoption of energy efficiency technologies requires investments in new equipment alongside know-how and experience. Equipment investment includes energy conservation updates of both the manufacturing process and the auxiliary system. Such technologies were either developed independently by enterprises, or purchased as mature technologies on the free market. Energy saving know-how and experience depends on the institutional experience within the enterprise, or on the professional guidance provided by external experts who will improve operational routines and help to develop higher value-added products; this will improve energy efficiency through enhancing the production standards. For ways in which enterprises adopted Energy Efficiency Technologies, see Figure 8.4.

[7] NDRC, *Qianjia qiye nengyuan liyong zhuangkuang baogao 2007* 千家企业能源利用状况公报 2007 [Announcement of Top 1000 Enterprises Energy Consumption 2007].

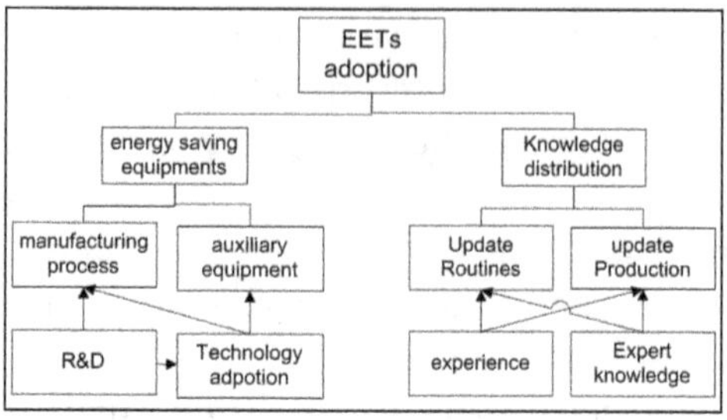

Figure 8.4. The Ways Enterprises Adopted EETs.

8.1.2.1 *Investment in Energy Efficiency Equipment*

Research and development geared towards the specific requirements of individual enterprises, conducted either independently or in cooperation with external agencies, was an important measure to reduce energy intensity. Taking one big textile enterprise in Hebei Province as an example, this enterprise found that the energy consumption of its electronic spindles was significantly too high after it had analyzed the energy consumption of each step in its production process. Then, a subsidiary company set up a research project addressing this specific problem, and cooperated with Tsinghua University to develop a much more efficient device. From 2008 on, the subsidiary company began to test the energy efficiency of its new spindle. The trial project lasted two years, and ended with the replacement of all the spindles in all textile machines of the company. Data from the test showed that the energy consumption involved in the production process was now 20–30% lower than before. As the energy consumption of the process accounts for 25% of that of the whole production, this technology yielded not only significant results in terms of energy conservation, but also economic benefits. Subsequently, the enterprise has been expanding the adoption of this technology since 2010.[8]

The expansion and adoption of mature EETs in large enterprises was another important way to realize energy conversation, which manifested itself most significantly in the iron and steel and concrete industry. Coking is a key energy consuming process within the iron and steel industry. From 2005 to 2008, 11 enterprises including Anshan Steel, Laiwu Steel, and others installed a total of 13 coke dry quenching plants which could improve the energy efficiency of 18.65 Mt production capacity. The adoption rate of coke dry quenching systems in steel enterprises increased

[8] The data of the case is taken from the investigation material produced by the CPI.

from 35% in 2005 to 83% in 2008. The utilization of by-product heat and pressure was adopted widely in cement enterprises; by the end of 2010, more than 700 of the 1300 precalcining production lines were recycling their by-product heat and pressure, and built 561 sets of by-product heat and pressure generators, with a total installed capacity of 368 million kWh which amounted to energy savings of more than 900 tce annually.[9]

Small and medium-sized enterprises would adopt EETs pro-actively due to their flexible decision making processes. In the 11th FYP, almost all the refractory enterprises in Yangquan, Shanxi Province, transformed their coal furnaces into natural gas furnaces. The working principle of coal furnaces is to utilize the heat burned by gas generated from coal burning. Due to the imperfect combustion of coal, some heat will be lost in the process; the average heat value of coal gas is 2,000–3,000 kcal/Nm^3. By contrast, the average heat value of natural gas is 8,000–9,000 kcal/Nm^3, and the steady degree of natural gas combustion is higher, without the ash generated by burning coal. The average quality and rate of the finished product soared considerately, which improved energy efficiency indirectly. These technologies would also eliminate SO2 emission generated by coal combustion, which made its adoption by enterprises entirely positive.[10]

Although auxiliary production equipment constitutes only a little share in overall energy consumption, the transformation of lighting systems was another important avenue through which manufacturing enterprises could reduce costs while at the same time conserve energy. For example, one big coal enterprise in Shanxi province spent 10.18 million RMB to overhaul its lighting system in 2010. The enterprise replaced more than 8,000 incandescent light bulbs in its workshops with energy-efficient ones, with a total power of 1,400 kW. The enterprise took this opportunity to redesign the layout of the electric circuit and switch control system, which not only saved electricity but also ameliorated the working environment. The energy conversation rate of this project exceeded 60%, and would conserve 5.26 million kWh, equal to 1,842 tce.[11]

---

[9] Shao Xiaotian 邵晓天, *Woguo 2010 nian shengchan xinxing ganfa shuini zhanbi da 80%* 我国 2010 年生产新型干法水泥占比达 80% [The Adoption Portion of Precalcining Production Process Reached 80%] (2011-03-16), http://www.yicai.com/news/2011/03/708737.html [accessed: 2011-09-23].

[10] The data of the case is taken from the investigation material produced by the CPI.

[11] Ibid.

8.1.2.2 *Dissemination of Energy Conservation Know-How and Experience*

Enterprises need to combine specific professional know-how to realize the full conservation potential of their technical equipment. Surplus production capacity in energy-intensive industries caused severe competition, which pushed enterprises to address energy conservation through increased knowledge. For energy-intensive manufacturing enterprises, the application of know-how related to energy efficiency in its daily operation and management could help save energy. The Tangshan Thermal Power Company of Heibei Datang International Co-generation Power Co., Ltd. provides a good example in this regard. The enterprise required its operators to adjust their target to be within the optimal range of the performance appraisal system. The frequent fine adjusting by the operators improved equipment safety and yielded environmental as well as economic benefits. The thermal testing unit tested leakages in the air pre-heater and at the rear of the flue at least once monthly, and would cope with it instantly if the amount of leaking air exceeded limitations. A special chemicals unit tested the amount of unburned carbon in the fly ash no less than once weekly, and if the share of unburned carbon in the fly ash increased, they would inform the operator in charge to adjust the air volume, as to ensure a reasonable particle size. The controller on duty could adjust the temperatures according to the flue gas discharging from the furnace and air surplus coefficients, and increase the temperature of the mixer of the primary air and coal powder properly, while decreasing the amount of cold air. The auxiliary control network not only saved energy but also improved the efficiency of the production process by transforming the mode of electrostatic precipitation. When incoming gas would have low-load or low-sulfur concentrations, the desulfurizing operator would take measures such as add desulfurizing synergists or shut down the circulating pump, all of which would improve the company's energy saving capacity.[12]

Since SMEs found it difficult to reduce their costs through equipment overhauls (the funds for which are harder for them to obtain due to their limited size), developing products with higher added value became the only way to survive, and it was also an important way to improve their

[12] Yu Jinwu, Zhen Dayong 于进武, 甄大勇, *Tangshan redian gongsi silunqudong zhua jieneng* 唐山热电公司"四轮驱动"抓节能 [Tangshan Thermal Plant Implemented Energy Conservation driven by four wheels] (2011-06-27), http://www.china-cdt.com/news/datangnews/17780560.html [accessed: 2011-06-30].

energy efficiency. Several cement producers in Shandong province, for example, operated on a relatively small production scale, and were aware that they would not be able to turn a profit if the continued to produce ordinary portland cement. The group adjusted its business strategy according to the scale of the plant. Assisted by professional know-how provided by external experts, the enterprises took advantage of their own strengths, and changed their product to a special sort of cement with higher added value. It took one year for the enterprise to test the manufacturing process in 2005, and it commenced test production in 2006. Every month, it spent ten days to produce special cement, and the remaining time was allocated to produce ordinary portland cement. With the maturity of the production process, annual production of the enterprise increased rapidly, with 30,000 tons in 2007, 70,000 tons in 2008, and 90,000 tons in 2009. The enterprises switched to the production of special cement completely in 2010, and annual production reached 160,000 tons. After years of experimenting and practicing, the enterprise nearly realized its transformation towards higher value-added production. The price of special cement being double the price of ordinary portland cement, the enterprise became profitable. Its energy intensity per unit of product increased slightly, but energy intensity per unit/output decreased significantly.

8.1.3 *Elimination of Obsolete Production Capacities*

For a long time, the standard of technical equipment used in China was comparative low. In order to advance China's technological level, the NDRC issued *The Guideline Catalogue for Industrial Restructuring* to control the market access of enterprises from the perspective of industrial policy. During the 11th FYP, the plan to eliminate obsolete capacities was made based on the projected records of enterprises. The NDRC informed all levels of government and respective enterprises of the list and the corresponding outdated capacity to be eliminated. The municipal government inspected the compliance with these guidelines, and took measures such as cutting off water and electricity supply for those enterprises that could not dismantle their outdated equipment in time.

The portion of obsolete capacity which was listed by the government in relation to the total production capacity determined the strategy and measures enterprises took to comply with the policy. For small enterprises, such as small thermal power plants, small paper plants, and small cement plants, obsolete production capacity constituted almost the whole estate of their enterprise, eliminating this capacity meant to exit the market.

Most of their outdated equipment had been used longer than its specified expiration date, so its fixed cost was much lower than that of advanced equipment. Although energy intensity per unit/product is much higher, the low energy price only brought limited pressure on the enterprises. The enterprises would insist on continuing production if only they could make a profit, and they would shut down and wait for chances when losing money. From another point of view, their outdated production capacities were only relatively outdated, and there have been some exceptions depending on the character of product and the region. For instance, the adoption of a blast furnace with 400–500 m^3 producing rebar used in infrastructure projects has much lower costs than that of a larger blast furnace, so it is economically reasonable. In the sparsely populated western areas of China, cement production lines with 1000t/d suit local market condition better.[13] Market demands and profitability were the main factors affecting the behavior of enterprises. The NDRC issued the *Notice of Controlling Overall Quantity, Eliminating Obsolete Capacity and Accelerating Structural Adjustment in Steel Industry* in 2006, and required enterprises to eliminate outdated capacity of less than 300 m^3. Many small and medium-sized blast furnaces resumed production, due to the strong demand of steel products.[14] This caused a rapid capacity expansion in the iron and steel industry, and excess capacity rose from 80 Mt in 2006 to 160 Mt in 2008.[15]

Faced with the responsibility to address unemployment in their districts, some local state-owned enterprise still insisted on continuing production, even when they are facing financial loss. Lacking profitability and financing channels, such enterprises had no ability to meet the requirements of industrial policy through technical updates. The control over these enterprises was held by local governments. Pressured to comply with energy saving targets, local governments usually ordered such enterprises to shut down via administrative measures. The bonus provided by the central and local governments could help the enterprises to deal with debts and lay-offs. Prior to enforcement by the government, these

[13] Xiong Huawen 熊华文, "Taotai luohou channeng ying heli bawo lidu jiezou 淘汰落后产能应合理把握力度和节奏 [The Eliminating of Outdated Capacity should grasp the strength and pace]," *Zhongguo jingji shibao* (2010-01-27).

[14] Dong Fang 东方, "Gangtie chanye zhenxing xize jijiang chutai shouxian taotai channeng wen zezhi 钢铁产业振兴细则即将出台首现淘汰产能问责制 [Detailed steel revival plan Is Going to Be Issued, First Adoption Accountability in Laggard Capacity Elimination]," *Xiandai wuliu bao* 2 (2009).

[15] Ibid.

enterprises had a chance to pro-actively shut down their outdated capacity, and often weighted the financial benefit of grants received from the government against the benefit of continuing production.

Although the energy intensity of small thermal generators and some outdated equipment used by big enterprise was higher, enterprises would not eliminate their existing equipment by themselves since they were restricted by the equipment management system and the need for continuous production. The listed enterprises need to launch a new project for replacement, and eliminate outdated capacities according to the cost and steady degree of the replacing capacity. The cotton pulp spherical digesters were outdated capacity with high energy consumption and high pollution, and half of them were required to be eliminated during 11th FYP. To guarantee continuous production, the enterprise parallely built a clean production line with low energy consumption. This project reduced costs significantly, taking advantage of the high quality of raw material and cheap labor within the region. Then, the enterprise in turn eliminated three cotton pulp spherical digesters from 2009 to 2010, which caused a reduction in energy intensity per unit/product, and helped the enterprise to complete its energy saving target. Since energy intensity in large measure is related to the production process, enterprises such as steel, thermal power, cement, alumina, and electrolytic aluminum could not realize the target of 20% in reductions of their overall energy consumption in 2005 only through technical overhauls based on their existing processes. Elimination of obsolete capacities contributed much to the completion of energy saving targets for such enterprises.

The elimination of obsolete capacities freed up market space in the process of pursuing energy efficiency. Big enterprises with advantages in terms of capital and technology expanded their production scale by adopting new processes; this accelerated the improvement of industrial energy efficiency. The cement industry for example is an industry in which the bigger the scale of production, the lower the cost. Coal and electricity are the main energy resources consumed in cement production, which accounted for about 60% of the producing cost. Compared with traditional production processes, the consumption of coal and electricity in the precalcining production process was greatly reduced, lowering producing costs for cement.[16] In the 11th FYP, obsolete capacity including dry

---

[16] Ye Jing 叶静, "Jieneng jianpai: Shuini mianlin shengsi xuanze 节能减排: 小水泥" 面临生死选择. [Energy conservation and emission reduction: Small Cement enterprises face life and death choice]," *Zhongguo jingji zhoukan* 29 (2007): 44–45.

hollow kiln, wet kiln, and shaft kiln were forced to be eliminated in great measure. The now-open market share provided good opportunity for large enterprises to expand their scale and market shares. Making use of their advantage in capital, big enterprises adopted new precalcining production processes through new construction, reconstruction, expansion, or relocation. 80% of the 1,870 Mt cement produced in 2010 were produced by precalcining production processes, this exceeded the target of 70% set for the cement industry in the 11th FYP.[17]

8.2 Mechanisms Affecting Energy-Saving Behavior of Enterprises

A firm is a group of individuals brought together by a complex set of contracts both written and unwritten, these rules and conventions govern their interactions and the environment within which the firm operates.[18] In the 11th FYP, under pressure from both energy policies and increasing energy prices, energy-intensive enterprises needed to improve their capacity by changing their internal management rules to adapt the external environmental incentives. Government agencies not only exerted pressure, but also helped to disseminate energy efficiency technologies, and provided support for capacity improvement in form of bonuses for enterprises.

8.2.1 *Pressure on Energy-Intensive Enterprises to Save Energy*

8.2.1.1 *Restriction of Energy Saving Target Accountability*

The biggest change in energy policies during the 11th FYP was the fact that energy conservation targets became binding and all enterprises above scale were assigned energy saving targets. Complying with the governmental regulation, the measures taken by enterprises to conserve energy were based on a cost-benefit analysis. Enterprises undertook measures towards energy conversation in order not only to meet the regulatory requirements, but also to occupy more favorable market and development

---

[17] Shao Xiaotian 邵晓天, *Woguo 2010 nian shengchan xinxing ganfa shuini zhanbi da 80%* 我国 2010 年生产新型干法水泥占比达 80% [The Adoption Portion of Precalcining Production Process Reached 80%] (2011-03-16), http://www.yicai.com/news/2011/03/708737.html [accessed: 2011-09-23].

[18] DeCanio, S.J., "Barriers within firms to energy-efficient investments," *Energy Policy* 21, no. 9 (1993): 906–914.

niches. The energy saving target of the Top 1000 enterprises specified by the NDRC depended on the supervision and assessment of all provincial governments, and the key energy consumption enterprises specified by the local government depended on the lower-level governments to issue their energy savings assessment. Enterprises failing to pass the assessment would be deprived of new project and land approvals, constraining their survival and development. Passing the assessment became an indispensable condition for enterprises. What is more, the result of energy savings assessments was integrated into the performance evaluation system for SOE leaders and was executed via one-vote veto. This means that if the major leader of a SOE failed to pass the energy saving assessment, not only would he or she not be considered for a promotion, but may also lose the position. Under the rigid restraints of meeting energy saving targets, enterprises had to raise capital to invest in EETs and meet their target without harming their economic bottom line.

8.2.1.2 *Industry Policy Raised the Bar for Market Access*

The government's industrial policy guided the technical update process of enterprises by legally restricting their market access. The *Guideline Catalogues for Industrial Restructuring* issued in 2005 classified enterprises into three classes, according to energy consumption, materials consumption, pollution, and their technical standards. This forced enterprises to realize technical overhauls so as to achieve a high ranking in the catalogue. The policy raised energy prices for enterprises in the "restricted" and "eliminated" categories, while maintaining normal energy prices for those enterprises in the higher category. This incentivized enterprises to accelerate their transformation towards greater energy efficiency and elimination of obsolete capacity. In the 11th FYP, the central government issued correlative industrial development policies aimed at energy-intensive industries such as cement, iron and steel, and ferroalloy, which not only set clear requirements for market access, but also set a future energy-intensity target for the industry. In the iron and steel industry for example, the *Steel Industry Policy* issued in 2005 required that the overall energy consumption in blast furnace flows should be lower than 0.7 tce, and enterprises should eliminate and stop investing in obsolete capacities, such as soil sintering, coking (containing modified coke) and other outdated technology and equipment. These polices required enterprises to adopt Energy Efficiency Technologies and resource recycling technologies, and additionally provided detailed requirements for some specific technologies involved in new projects. For instance, blast furnaces must now be equipped with by-product pressure generators and coal power injection devices. Coke ovens must be equipped with coke dry quenching plants, and matched with dust collectors and desulphurization devices. These specific requirements compelled enterprises with obsolete capacities to exit the market by means of regulation and enforced elimination, and guaranteed the low energy intensity of new projects. The *Cement Industry Development Policy* issued in October 2006 stipulated, that "the share of precalcining production at the end of 2010 should reach to 70%. The technical economic indexes of large precalcining production above 4000 t/d should reach the following targets: energy consumption per ton cement should be lower than 95 kWh, and heat consumption of clinker should be lower than 740 Kcal/kg." This policy also forced some enterprises to exit the market before the deadline by eliminating obsolete capacity. "Before the end of 2008, all regions should eliminate all kinds of obsolete capacities such as dry hollow kiln, wet kiln, etc., reduce the production capacity of shaft kiln, and eliminate all shaft kilns production if possible. All levels of local government should force cement enterprises with a scale of less than 200,000 tons and enterprises with substandard products to shut down, merge, or relocate." Industrial development policies coordinated closely with the energy saving target of the 11th FYP, and maintained a rigid standard for obsolete capacities and energy consumption per unit/product. Some enterprises with obsolete capacity were forced to exit the market; some enterprises implemented the requirements reluctantly and managed to fulfill their energy saving target; some enterprises took the opportunity offered by the "large replacing small" policy, they eliminated their obsolete capacities proactively in order to gain more production capacity for the future.

8.2.1.3 *Increasing Energy Prices Propelled Enterprises to Conserve Energy*

Cost pressure caused by increasing energy prices was a major reason for enterprises to adopt EETs. The report on key energy consumption enterprises in Shandong province showed that 75% of enterprises regarded price increases in energy as the major impetus to conserve energy, and the adoption of EETs was regarded as the most effective measure to that end.[19] The increase in energy prices and the change of policy orientation adjusted the expected energy price in the future; this improved the

---

[19] Liu Ming 刘铭, "Shandong jieneng diaocha: Nengyuanjia shangzhang cheng qiye jieneng dongli 山东节能调: 能源价上涨成企业节能动力 [Energy Conservation

expected return of energy conservation transformation. Coal and power were the major energy sources for industrial enterprises, and their prices increased to a certain extent. During the 11th FYP, the double-track pricing system for coal ended, and the trading of coal was now almost entirely left to market mechanisms. As a result, the price of coal increased swiftly: In the first half of 2010, the price of coal for a big chemical enterprise in Henan province reached 450 RMB/t, more than double the price of 2003. At the same time, the quality of coal dropped, and the reduction of actual calorific value indirectly raised the coal price. Assuming that 0.5t coal is needed to produce one ton of urea, the cost per ton of urea would increase by 0.5 RMB if the price of coal is raised by 1 RMB per ton. Due to the increase in coal prices, the enterprise was faced with additional cost pressure of at least 125 RMB/t of urea in 2010 compared with 2003. From July 1, 2008 onward, the national average price of energy increased by 0.025 Yuan/kWh, causing all energy-intensive enterprises except chemical fertilizer factories to face different degrees of cost pressure. At the same time, the NDRC directed all local governments to screen energy-intensive enterprises, and abolish all past preferential energy policies for electrolytic aluminum enterprises, chlor-alkali enterprises, and ferroalloy enterprises. By the end of 2009, 30 provinces had completed their screenings and produced comprehensive reports. The increase in energy prices caused higher pressure for enterprises with heavy energy consumption.

8.2.2 *Support and Incentives for Enterprises to Conserve Energy*

8.2.2.1 *Dissemination of Energy Efficiency Technologies Reduced Risk Expectancy*

The efficiency of energy conservation measures depended on the adoption of EETs. What technologies were available at what accompanying risks were key factors in enterprises' decision making process. The Ten Key Energy Saving Projects provided effective information for enterprises through technology dissemination and policy guidance. Enterprises actively improved their energy efficiency through equipment maintenance and adoption of EETs, according to the demands of the market and their own situation. Successful cost reduction through energy conservation measures by some trailblazing companies was an effective way to promote

Investigation: Energy Price Increase Became the Main Incentive For Enterprises to Save Energy]," Qilu Wanbao (2011-1-14).

the adoption of similar technologies throughout a particular industry, and then led to the decreasing of the average energy intensity of all enterprises in the industry. The Energy Efficiency Projects with their high return rate of investment helped enterprises to reduce production costs and improve their competitiveness.

Energy Performance Contracting projects could push the launching and advancing of energy conversation projects by mitigating risks through financing. One big corporation group conducted pilot projects which involved high-risk technologies by making use of Energy Performance Contracting projects, and evaluated the risks of the technologies afterwards. Then they expanded the adoption of these technologies in the enterprises within the group, and achieved great economic benefits as a result. The successful experience of energy conservation projects by Energy Performance Contracting helped other enterprises to accurately evaluate the risks involved in technology adoption. The wide dissemination of some energy efficiency equipment reduced acquisition costs, and subsequently improved the return rate of investment, which pushed more enterprises to launch energy conservation policies.

8.2.2.2 *Energy Saving Bonus Reduced Investment Cost*

The adoption of EETs requires a large amount of capital; therefore, the capacity of raising capital affected the development of energy conservation projects. Large enterprises command wide financial assets, and are more likely to receive bank loans, so they have both incentive and capacity to adopt EETs. Energy saving bonuses provided by the government additionally reduces investment costs. In the 11th FYP, the Ministry of Finance allocated special funds to support enterprises to implement energy conservation schemes, which included projects to upgrade coal-fired industrial boilers and furnaces, utilization of by-product heat and pressure, conservation and substitution petroleum, energy conservation of motor systems, and optimization of energy systems specified in the *Implementation Opinions of Ten Key Energy Conservation Projects of the 11th FYP*. Energy saving bonuses were rewarded to enterprises according to the actual quantity of energy saved which was calculated according to the data produced by the energy measuring and statistical system. All enterprises that applied for the bonus must provide qualified energy measurements, statistics and management systems. When the annual energy conservation quantity exceeded 10,000 tce, the enterprise could apply to the financial department and the department in charge of energy saving within the local government (usually the Development and Reformation

Committee, the Economic and Trade Committee, or the Industry and Information Technology Committee), aimed at projects launched in the current year. After verification of the energy saving quantity by the respective departments, the Ministry of Finance would reward the enterprises with 200 RMB/tce in the eastern region, and 250 RMB/tce in the central and western regions. Besides the support from the central government, some provincial governments issued additional reward measures; Hainan and Jiangxi province reduced the application threshold of the annual energy saving quantity to 2,000 tce, and provided financial support to relatively small projects.

The Ministry of Finance issued *Interim Measures of Financial Reward Funds for Eliminating Obsolete Capacity* to give fiscal subsidies to enterprises eliminating obsolete capacities in underdeveloped regions. Additionally, some local governments also actively matched funds or obtained capital on the market through capacity quota auctions to help enterprises with obsolete capacity to withdraw from the market, and to solve social problems resulting from lay-offs. Some enterprises not only obtained the reward but also expanded their production capacity by investing in big new production capacities to replace the outdated ones.

8.3 Problems and Challenges

In the 11th FYP, all levels of local governments focused on large enterprises to guarantee the fulfillment of the overall energy conservation target. Large enterprises found it easier to receive financial benefits from the government, while SMEs faced many restraints and received less support.

The industrial development policies concerning the steel and cement industry served as a model for future industrial policies in other sectors; the new requirements with regards to technical capacity and production scale compelled some outdated enterprises to withdraw from the market which opened up market space for others. Large enterprises had a clear technical and financial advantage; they expanded their market share by acquiring smaller competitors and investing in new processes to replace outdated ones. The orientation of government regulation also affected investment direction of capital on the market. Enterprises with scaled capacity always had advanced equipment and greater potential for energy conservation, which meant that the expected investment return rate was higher, so a lot of capital was invested in the large enterprises. The increasing concentration of the industry was beneficial for enterprises' structural adjustment, and to improve their energy efficiency. Despite the fact that governmental policies directed companies to merge into larger entities, many enterprises still maintained an independent position, which meant that they could not improve their energy efficiency through size advantage.

The threshold for applications for the energy saving bonus, the size requirements of the industrial policy, and the investment situation on the capital market all benefited large enterprises. The requirements in terms of quantity of energy savings within a single project could only be met by big plants. Though some provinces lowered the requirements eventually, the high costs accompanying the application process impeded SMEs to obtain the bonus. Enterprises applying for the bonus must provide a technical approval report to support their submission of the quantity of energy savings, which should be provided by the qualified third party. In one province in central China, for example, the average cost of the bonus application for enterprises was more than 70,000 RMB.[20] The energy savings quantity generated by SMEs was lower, due to poorer technical standards and smaller production scale, making the energy saving bonus for SMEs financially unviable. On the other side, most SMEs face the problem of capital shortage, and they could not receive additional financial support from banks or via other channels; this caused the suspension of some economic efficient energy conversation projects. Additionally, SMEs in underdeveloped regions still carry social responsibilities such as employment and economic development; they cannot simply disappear in a short period of time. The current energy conservation policy system lacks effective measures to incentivize SMEs to save energy more proactively. The expansion of the Key Energy Efficiency Program in the 12th FYP will be helpful to mitigate this problem.

[20] The data of the case origin from the investigation material of CPI.

CHAPTER NINE

IMPACT ON LOW-CARBON DEVELOPMENT BY PUBLIC CONSUMPTION BEHAVIOR

9.1 Change of Public Consumption Behavior Poses a Challenge to Low-Carbon Development

China is in the process of rapid urbanization, and amidst a mélange of rising residential incomes, increasing material well-being, and cultural flows, public consumption behavior is undergoing dramatic changes. Given China's huge population base, the impact of public consumption on energy use and carbon emissions cannot be ignored, and the changes in consumption behavior pose a serious challenge to China's future low-carbon development.

9.1.1 *Public Consumption Behavior Drives Energy Consumption and Carbon Emissions to Grow Rapidly*

Public consumption behavior refers to the general population's choices and actions with regards to daily life, including food, clothing, shelter, and means of transportation. Rapid economic growth and urbanization do not only cause social goods and wealth to be increasingly abundant, but also boost a variety of cultural, ideological, and intellectual interactions, which has caused China's public consumption behavior to undergo dramatic changes. For example, more and more people travel by car, possess the willingness and the capability to buy bigger houses, even villas; they pursue a Western "pastoral" lifestyle, enjoy living conditions with indoor temperature regulation, and are happy to accept energy-saving lamps, recycled paper, or other environment-friendly products. The luxury market expands rapidly. These changes generate new demands for energy consumption and result in a growth in carbon emissions.

Within the energy consumption total, leaving aside energy consumption resulting from public infrastructure construction and from production and transportation of export commodities, the remaining energy production involved in the manufacturing and transportation of domestic consumption goods functions to fulfill the need of consumption behavior in food, clothing, shelter, and means of transportation. Behavior-related

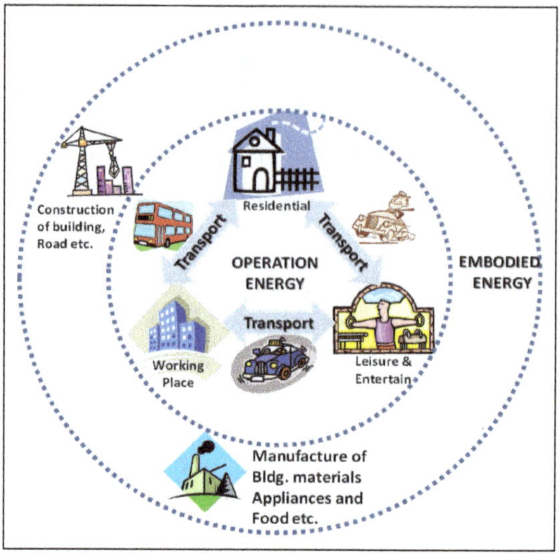

Figure 9.1. Public consumption behavior-related energy consumption.

energy consumption and carbon emissions in public consumption can be divided into direct and indirect categories that are displayed in Figure 9.1. Direct energy consumption and carbon emissions are caused by public activities, including construction and transportation. The construction part corresponds to energy consumption and carbon emissions arising from services provided for public activities in living, work, and leisure environments; the transportation part corresponds to energy consumption and carbon emissions produced by various residential travel activities. Indirect energy consumption and carbon emissions refer to energy consumption and carbon emissions arising from manufacture of all material goods required for aforesaid living, work, leisure, and travel activities, such as housing construction, production of automobiles, television sets, refrigerators, and clothing.

Public consumption behavior not only plays an important part in energy consumption and carbon emissions, but it is also the main driving force behind China's rapid growth in energy consumption and carbon emissions during the 11th FYP. Public consumption (direct and indirect) of construction, transportation, and household goods accounts for nearly half of China's total energy consumption.[1] China's direct energy consumption

[1] Xiong Huawen 熊华文, "Jiyu touru chanchu fangfa de jumin xiaofei quan liucheng nenghao fenxi 基于投入产出方法的居民消费全流程能耗分析 [Analysis of Share of Household Energy Consumption Based On Input-Output Method]," *Zhongguo nengyuan*

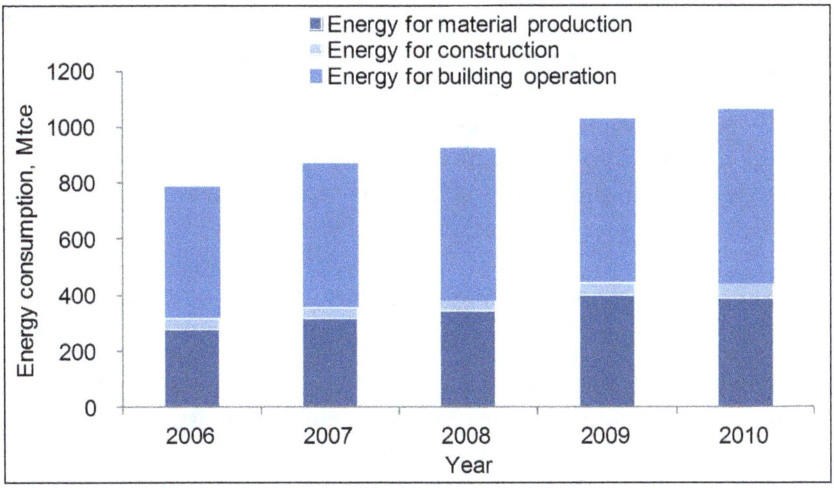

Note: The calculation method is detailed in Appendix 9.1, data for 2010 based on estimation. Data source: National Bureau of Statistics, *Summary of 2011 China Statistical Yearbook* (Beijing: China Statistics Press, 2011), Table 7.9.

Figure 9.2. Housing construction-related energy consumption in the "eleventh five-year plan" period.

in construction and transportation reached 880 Mtce in 2009, accounting for 29% of total energy consumption in that year. Construction and transportation-related carbon emissions totaled 2.15 billion tons of CO_2[2] in 2010, making up for 29.7% of total carbon emissions in that year, up one percentage point compared to 2005, and they were an important part of and played an increasing role in China's carbon emissions. During the 11th FYP, construction and transportation-related carbon emissions increased by approximately 500 Mt or by 41%, faster than the 36% growth in total emissions, and were an important factor driving the growth of energy consumption.

In terms of indirect energy consumption, taking the construction sector as an example, the average annual completed construction area surpassed 2.6 billion m² in the 11th FYP period and the total in completed construction

7 (2009): 17–20. The results are based on data from 2005, excluding energy consumed by the construction of infrastructure including housing, vehicles and roads.

[2] See the calculation method in Chapter 11 of this report. The data on transportation in 2010 are estimated based on data from 2009. In addition, transportation energy consumption hereon also includes freight transport which is not directly related to public consumption behavior, but due to data limitation it is not isolated. Thus the actual directly related carbon emissions from public consumption behavior should be about 1.8 billion tons of CO_2.

area reached 13.1 million m^2 during the same period.[3] Intense and rapid building construction caused energy consumption of building materials and construction industries to remain high. As shown in Figure 9.2, the sum of energy consumption from the production of building materials for newly-completed construction projects and energy consumption of building construction in progress totaled about 440 Mtce in 2010, accounting for 13.5% of total energy consumption that year.

9.1.2 *Direct Carbon Emissions of Public Consumption Behavior Will Become Future Mainstream*

The direct energy consumption and carbon emissions in the consumption sector are significantly different from those of the industrial production sector. In the 11th FYP period, total energy consumption of the industrial production sector grew rapidly, but its intensity continued to decline so that the curves for total energy consumption and energy consumption intensity displayed a clear "scissors" shape (Figure 1.17); however, in the consumption sector, both energy consumption as well as its intensity continued to rise (construction and transportation serve as the examples as displayed in Figure 1.18 and Figure 1.19).

During the 11th FYP period, urbanization and rise in incomes caused dramatic changes in public consumption behavior, which was reflected in rising demands for consumer products and changing consumption patterns. As a consequence, per-capita demand for services and commodities grew substantially, offsetting previous reductions in energy saving and carbon emissions which had been mainly achieved through technological progress, and causing energy consumption and carbon emissions to continue to rise. For example, fuel efficiency for private vehicles has improved significantly over the past decade; the fuel consumption of vehicles per 100 km fell by over 15%.[4] However, if public buses were to be replaced by private vehicles as means of transport, then carbon emissions intensity would increase twelve-fold, and even if the 15% improvement in fuel efficiency is taken into account, carbon emission intensity would multiply nine-fold. Similar problems also widely exist in the field of

---

[3] The data for 2010 are estimated based on the construction of newly-built energy-saving buildings and its proportion in total construction area announced by the Ministry of Housing and Urban-Rural Development in 2011.

[4] Climate Policy Initiative of Tsinghua University, *Zhongguo ditan fazhan baogao* 中国低碳发展报告 2010 [Annual Review of Low-Carbon Development in China: 2010] (Beijing: Kexue Chubanshe, 2011): 227–229.

building energy consumption. When residents choose an indoor environment with constant temperature and humidity regulation, a central air conditioning unit is required; currently, domestic advanced technologies can achieve a power consumption intensity of 20 kWh/(m2.a);[5] whereas when residents choose an indoor environment that is synchronized with outdoor temperature, a split-type air conditioning system can deliver satisfactory results, the average power consumption intensity for this type being merely 2.3 kWh/(m^2.a),[6] much lower than that of the former. However, when split-type air-conditioning systems operate in "full-time and full-space" air conditioning mode,[7] the power consumption intensity is 15 kWh/(m^2.a),[8] much higher than the average power consumption mentioned above. Therefore, in the consumption sector, changes in public consumption behavior are the root cause of increased energy consumption.

According to the experiences of developed countries, economic development and energy consumption variation can be divided into three stages, as shown in Figure 9.3. Stage 1 began with a GDP of approximately 4,000 Dollars per capita, energy consumption in the United States (1960) and Japan (1965) was mainly caused by the production sector, energy consumption of the industrial production sector and the consumption sector grew rapidly and simultaneously residents' consumption pattern began to change. Stage 2 refers to the period after the 1973 oil crisis, when energy became a non-negligible constraint to national economies, and the energy consumption levels of the two countries' production sector experienced a steady decline. However, the energy consumption level in Japan's consumption sector continued to grow rapidly, while in the case of United States, steady growth even surpassed that of the production sector, and became the main factor of national energy consumption. Stage 3 refers

---

5 Building Energy Research Center of Tsinghua University, *Zhongguo jianzhu jieneng niandu fazhan yanjiu baogao* 中国建筑节能年度发展研究报告 2009 [2009 Annual Report on China Building Energy Efficiency] (Beijing: Zhongguo jianzhu gongye chubanshe, March 2009): 98.

6 Ibid.

7 The so-called "full-time and full-space" air-conditioning mode refers to the air conditioners in all rooms operating for the whole day and the corresponding "part-time and part-space" air-conditioning mode refers to air conditioners only turned on in rooms when needed. The latter is a more common lifestyle adopted by Chinese residents, but the "full-time and full-space mode" begins to emerge, if it becomes the norm, it will cause rapid increase in energy consumption and cause severe electricity shortages in the summer.

8 Building Energy Research Center of Tsinghua University, *Zhongguo jianzhu jieneng niandu fazhan yanjiu baogao* 中国建筑节能年度发展研究报告 2009 [2009 Annual Report on China Building Energy Efficiency] (Beijing: Zhongguo jianzhu gongye chubanshe, March 2009): 98.

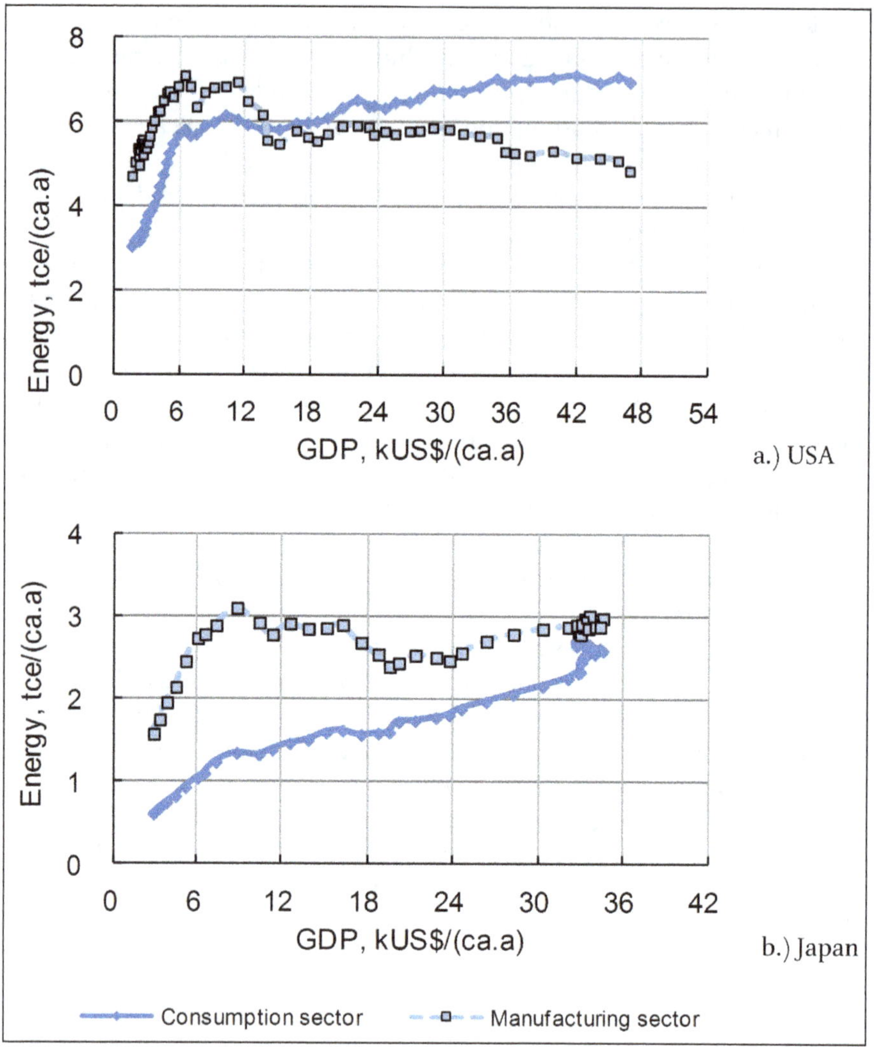

Note: Energy consumption mentioned here refers to primary energy consumption. The consumption sector includes residential housing, public buildings and public transport. See: EIA United States Energy Information Administration, *Annual Energy Review 2008* (USA: EIA Publications, 2008); MIAC Ministry of International Affairs and Communications, Statistic Bureau, "Housing and Estate Survey 2008," http://www.stat.go.jp/english/data/index.htm [accessed: 2009-08-08]; EDMC The Energy Data and Modeling Center, *Handbook of Energy & Economic Statistics in Japan* (Japan: The Energy Conservation Centre, 2009).

Figure 9.3. Historical development of GDP and energy consumption in United States and Japan.

to the 21st century, when the two countries' energy consumption within their consumption sectors stabilized at about 7 tons per capita in the United States and about 2.7 tons per capita in Japan.

China's current GDP per capita exceeds 4,000 Dollars, and similar to the development of the United States and Japan at stage 1, its energy usage structure heavily emphasizes the production sector. As incomes rise and public consumption patterns begin to change, a percentage of China's population has adopted consumption behavior resembling that of the United States. For example, a large number of buildings with glass-curtain walls has been recklessly constructed in all parts of China, in an attempt to mirror new aesthetic tastes and modern design concepts imported from overseas; in some developed cities, residential communities have adopted centralized indoor environment control systems, maintaining a constant temperature and humidity, while a traditional lifestyle was abandoned; some households have started to use energy-intensive home appliances such as dryers and dishwashers. These changes in consumption patterns have led to a rapid growth in energy consumption.

Judging from the experiences of developed countries, the energy structure of China's production sector will persist for some time (10 years in the case of Japan) while meanwhile, the energy consumption level of its consumption sector will increase simultaneously. When China will have completed its stage 1 and enters stage 2, the production sector's energy consumption will stabilize and the consumption sector's energy consumption will maintain its growth so that this sector will gradually replace the production sector and become the pivotal factor in public energy consumption.

9.1.3 *Guidance to Consumption Behavior is the Key to Solve Low-Carbon Challenge*

China is currently at stage 1 of its economic development and energy consumption growth, and the boom in infrastructure construction brought by China's rapid urbanization will persist for a relatively long period; meanwhile, China's export potential remains high, fueling the energy consumption of the industrial sector. However, as problems linked to energy shortages and climate change become more and more apparent, China faces far greater challenges in its developmental stage than the United States and Japan. Therefore, China cannot develop as freely as the United States and Japan, posing unprecedented historical challenge.

In the short run, the need for growth in the production sector can be counterbalanced with energy savings in the consumption sector. In the long run, the establishing of a low-carbon development model for China's consumption sector is the fundamental step to cope with the challenges of energy conservation and carbon emissions reduction in the future. Currently, China is in a critical phase of transforming itself from a low-income to a middle-income country; urbanization, modernization, and a variety of cultural collision and integration have led to dramatic changes in public consumption behavior. The corresponding energy-consuming sector has many possible energy consumption patterns. Hence it is an important opportunity for China to realize its low-carbon development.

To advance its economic development, to complete the process of urbanization and to meet its people's increasingly rising demand in living standards, while at the same time having to cope with issues of energy control and carbon emissions, China will need to embark on a developmental path completely different from that of other developed countries. It is a path that developed countries have never walked on. This path will be very difficult considering China's total population.

This chapter discusses the impact of public consumption behavior on energy consumption and carbon emissions from two perspectives, housing and automobiles. It also addresses ways in which to guide and induce low-carbon consumption behavior and briefly comments on the Chinese government's as well as the Chinese public's activities in this regard during the 11th FYP period.

9.2 Urban Residential Housing Space Grows Fast

With the further progressing of urbanization and modernization, Chinese residents' lifestyle has entered a period of dramatic change. Factors such as the influx of a large rural population into urban areas and rapid economic growth have resulted in huge demand for higher standards of living; extensive reform and continuous improvement of China's social security systems have strengthened residents' confidence in increasing expenditures on consumption goods; global cultural exchange has changed attitudes and caused a rethinking with regards to city planning. Residential housing has become an issue at the center of society's attention. In the 11th FYP period, Chinese urban residents' housing space grew fast; the construction scale was extremely large and was met with vigorous demand.

9.2.1 *Growth was Mainly Driven by Rigid Demands*

At the end of the 11th FYP period, living space in urban residential housing in China[9] reached 14.4 billion m^2 and average living space per capita was 21.6 m^2, a growth of more than 10% compared to 2005.[10] Living conditions for China's urban residents have significantly improved. As shown in Figure 9.4, since the beginning of the new millennium, urban residential living space has grown noticeably faster, with an average growth rate of 12.5% between 2000 and 2009, and an average annual growth of 1.5 billion to 2 billion m^2, faster than the 7% in growth from 1990 to 1999. Following an increase of 6.3 billion m^2 in the 10th FYP period, urban residential living space maintained its rapid growth in the subsequent FYP period and grew by 3.6 billion m^2 within five years, surpassing Germany's total residential living space of 3.25 billion m^2 in 2009.[11] The construction area over the same time period grew faster than the growth of urban population (average annual growth of 8.4% from 2000 to 2009) to increase per capita residential living space to grow year by year, as shown in Figure 9.4.

Strong demand drove the growth of urban residential housing. In the 11th FYP period, the annual population growth reached 100 million, including 13 to 20 million in urban population annually. To reach the average area per capita of 21.6 m^2 in 2010, annual construction of 280 to 430 million m^2 of residential housing was needed; in addition, China's urban population was about 600 million, living space per capita increased 0.5 to 0.7 m^2 annually, so 300 to 400 million m^2 residential housing needed to be built annually. Therefore, 590 to 840 million m^2 in residential housing needed to be built annually just to meet the demand produced by urbanization and improved living standards, which is close to the actual annual newly-added urban residential living space in China. Surveys show that strong demand accounted for 65% of overall housing purchase decisions in nine

[9] In accordance with international practices, buildings usually include urban residential housing and public buildings. For countries and regions such as China, with significant urban and rural gap, rural residential housing is also an inneglectable component. Urban residential buildings account for only 35% of total construction area of existing buildings in China.

[10] Data from the National Bureau of Statistics shows China's urban housing area per capita was 31.6 m^2 in 2010. The population samples did not include housing information of military personnel, students, migrant workers and other population groups, so the figure is larger than 21.6 m^2.

[11] Odysee Data, "Energy Efficiency Indicators in Europe," 2011, http://www.odyssee-indicators.org/ [accessed: 2009-08-08].

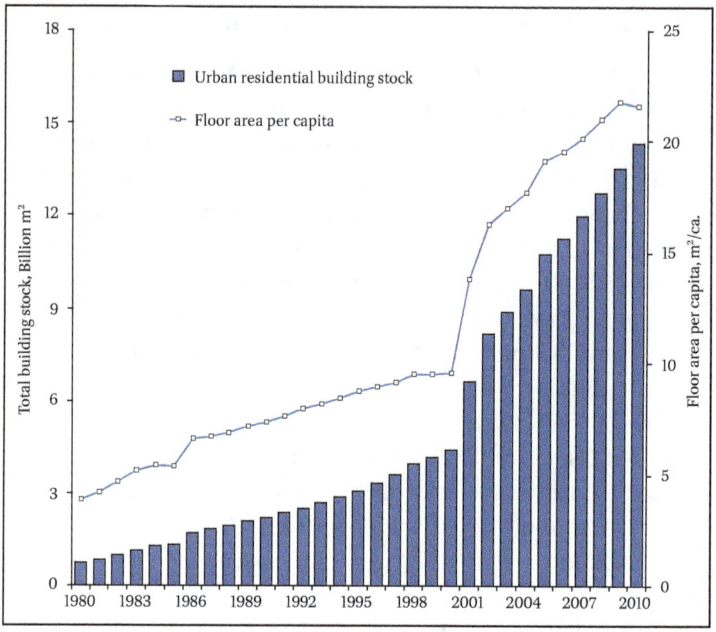

Note: In the three years from 2001 to 2003, the construction area has doubled and construction energy consumption increased 4.5 billion m^2, which may due to changes in statistical calibers.

Data Source: National Bureau of Statistics, 2011, Table 10.35.

Figure 9.4. Total and per capita urban residential housing floor-space.

cities (Figure 9.5), implying that demand features prominently as a key factor for the growth of residential housing space in China's urban areas.

9.2.2 *China Faces Choices of Different Modes*

9.2.2.1 *The Growth of Residential Housing Living Space was the Key Driving Force behind the Growth in Energy Consumption during the 11th FYP Period*

The growth of residential living space resulted in a growth of energy consumption and carbon emissions, a process that can be divided into three aspects. First, larger residential living space means residents consume more energy for heating, air conditioning, lighting, and other building services, which can be defined as operational energy consumption. Second, a certain amount of energy is needed for building construction, such as transportation of cement, mortar, and structural welding, which can be defined as constructive energy consumption. Furthermore, the production of building-related materials (such as cement, iron and steel, glass) also

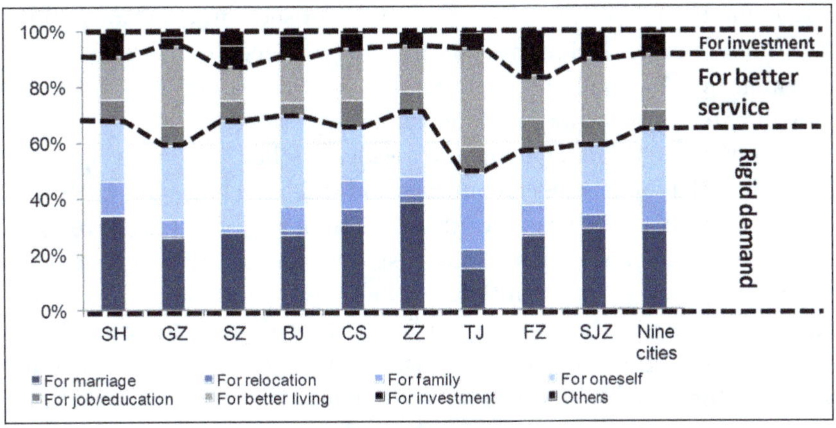

Source: China Index Academy, "2010 nian sijidu goufangzhe xiaofei diaocha yanjiu 2010 年四季度购房者消费调查研究 [House purchasing behavior investigation for the fourth quarter of 2010]," 2010-10-27, URL: http://fdc.soufun.com/report/1301.htm [accessed: 2011-07-18]: 13.

Figure 9.5. Survey of urban resident's housing purchase purposes by cities.

consumes large amounts of energy which can be called implicit energy consumption of newly-completed buildings.

In urban China, housing-related energy consumption is a vital component of total energy consumption, contributing approximately 12%. Energy consumption maintained a rapid growth rate of about 40%, from less than 280 Mtce in 2005 to 390 Mtce[12] in 2010, higher than the growth rate of total energy consumption in the 11th FYP period, contributing about 5% to the total energy consumption as an important driving force of energy consumption growth.

9.2.2.2 *China is at the Crossroad of Different Growth Paths for Urban Residential Housing Floor-Space*

In 2010, China's urban population reached 660 million, with an urbanization ratio of 49.8%, According to projections by the UNDP,[13] China's urbanization ratio will reach 60% by 2030 with an urban population of

[12] The operation energy consumption of existing buildings, implicit energy consumption of newly-completed buildings and construction energy consumption of buildings in progress accounted for 67%, 28% and 5%, respectively.

[13] UNDP, *China and a Sustainable Future: Towards a Low Carbon Economy and Society*, http://hdr.undp.org/en/reports/national/asiathepacific/china/name,19646,en.html [accessed: 2011-09-01].

900 million. To meet the housing demands of such an enormous populace, China's urban residential housing area will need to maintain a persistent growth for the foreseeable future which will continue to drive the growth of energy consumption and carbon emissions.

Furthermore, China's per capita residential housing living space is lower than that of developed countries and still has a large growth potential. However, there are significant variations among different developed countries: Per-capita residential living space is about 30 m^2 in developed countries in Asia, such as South Korea and Japan, 40 to 50 m^2 in Europe, and more than 60m^2 in the United States. China must satisfy residents' housing demands on its path to transform itself into an advanced country, With three different scales of residential housing needs provided by developed countries, which would China adopt in the future?

9.2.2.3 *The United States Model is not Suitable for China*

Scenario analysis results indicate that if the United States model was adopted, energy consumption related to urban residential housing would reach 960 million tons, as shown in Table 9.1. This is equivalent to nearly 30% of total energy consumed in 2010, just to meet the housing needs of urban residents, which is not a viable option. In addition, to meet such housing needs would require 57,000 square kilometers of land, equivalent to three times the size of Beijing. 5% of the country's arable land resources would be lost to construction, meaning that 70 million people would need to find other land for agricultural production.

Table 9.1. Scenario Analysis of different living styles in 2030.

Scenario Name		South Korea Mode	Europe Mode	North America Mode
Scenario planning	Per capita area (m^2 / per person)	30	40	60
	Total Area (100 million m^2)	270	360	540
	Energy consumption	6.9	8.3	13.1
Scenario results	Average annual growth of residential housing owned (100 million m^2 / a year)	6.3	11	20
	Residential housing-related energy consumption in 2030 (100 Mtce)	4.3	6	9.6
	Land area covered (10,000 m^2)	1.8	3.1	5.7

According to a survey covering six large Chinese cities,[14] the ideal living space size of urban residents is about 40 m^2, close to the level in Europe but still corresponding to a higher energy consumption. If the South Korean model was to be adopted, China may satisfy the expectations its residents have regarding their living standards, along with a low growth in residential housing-related energy consumption and carbon emissions, even close to the 2010 level.

Therefore, how to control and guide the appropriate growth of urban residential living space while at the same time paying attention to low-carbon issues is an enormous challenge China is facing in its process of urbanization and modernization.

9.2.3 *Opportunities to Slow Down Growth*

The fundamental driving force behind the growth of urban residential housing areas are firstly the housing demands of China's new urban population, and secondly demands for housing retrofitting and renovation by old-established urban residents. To balance this development with the necessity of low-carbon development, the South Korean model with its relatively low per-capita living area should be adopted. Because changes in private lifestyle often do not take into consideration the issues of energy saving and carbon reduction, China will need to push forward changes in public behavior towards a low-carbon style from various sides.

9.2.3.1 *The Public does not Realize the Housing Area They Needed is not as Much as They Have Imagined*

Residents' expectations towards their living standards have a decisive impact on final housing conditions. As long as people are still seeking a more spacious and comfortable living environment, the per-capita housing area will not lose its role as a factor in energy consumption growth. While such desires are only too human, it will be necessary in the interest of low-carbon requirements to curb extravagancies and excessive expectations. Research shows (Figure 9.6) that the living space residents actually do need is not as large as the one they have imagined. Those who live

---

14 Zhang Shengyuan 张声远, *Zhongguo qi chengshi geren xiaofei lingyu nenghao ji geren xingwei moshi diaocha yanjiu* 中国七城市个人消费领域能耗及个人行为模式调查研究 [Study on urban-life energy use and impact of human behaviors through investigations in seven Chinese cities, Doctoral Dissertation] (Beijing: Tsinghua University, July 2010): 75–76.

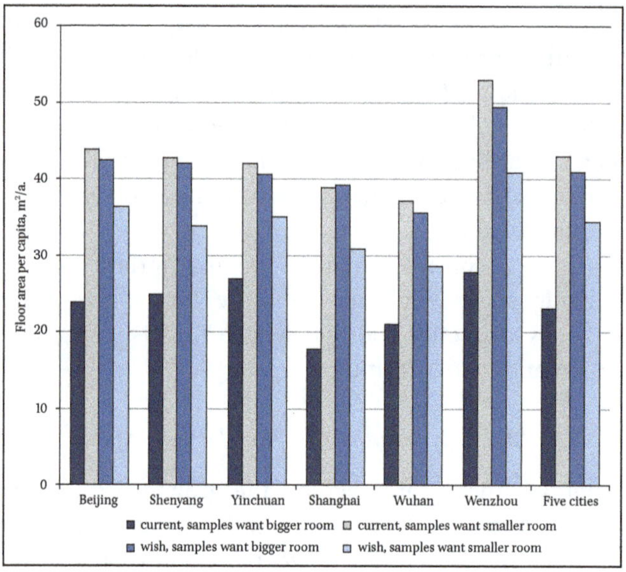

Source: China's Urban Energy Consumption Status and Energy Conservation Policy Research Group, *Chengshi xiaofei lingyu de yongneng texing yu jieneng tujing* 城市消费领域的用能特性与节能途径 [Energy Characteristics and energy-saving approach of urban consumption field] (Beijing: Zhongguo jianzhu gongye chubanshe, July 2010; and: Zhang Shengyuan 张声远, *Zhongguo qi chengshi geren xiaofei lingyu nenghao ji geren xingwei moshi diaocha yanjiu* 中国七城市个人消费领域能耗及个人行为模式调查研究 [Study on urban-life energy use and impact of human behaviors through investigations in seven Chinese cities, Doctoral Dissertation] (Beijing, Tsinghua University, July 2010).

Figure 9.6. Resident's ideal housing floor-space by cities in 2008.

with 20 m² per person name approximately 40 m² per person as their ideal, while residents living with a living space of 40 m² per person cite an ideal living space of 35 m² per person, an attitude that reflects the bias of residents when it comes to consumption. In other words, when living conditions improve towards 40 m² of living space per person, 5 m² of it may prove as superfluous. If this bias can be avoided through rational guidance, then an excessive growth and waste of construction area can be avoided.

9.2.3.2 *The Public is Price Sensitive, so a Pricing Mechanism can be a Possible Approach to Impedite Growth of Housing Floor-Space*

The income level and the size of housing are certainly correlated. As shown in Figure 9.7, higher income is accompanied with larger residential housing floor-space; the residential housing floor-space per capita for the group with the lowest income is less than half of that of the highest income group. This is because the higher the income, the higher the price

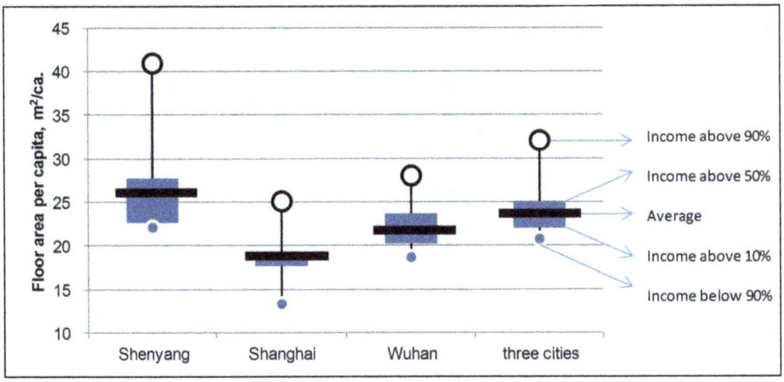

Data Source: see Figure 9.6.

Figure 9.7. Per capita housing floor-space of different income groups in several cities.

one can afford; or, at the a fixed price, larger housing floor-space become more affordable.

A macro survey (Figure 9.8) shows similar results. Because housing prices in first-tier cities (Beijing, Shanghai, Guangzhou and Shenzhen) are generally higher than prices in second-tier cities (Changsha, Zhengzhou, Tianjin, Fuzhou and Shijiazhuang), buyers in first-tier cities are more willing to buy small-sized residential housing; 66.9% buyers in first-tier cities are willing to buy small or medium-sized housing below 90 m^2, compared with only 49.2% buyers in second-tier cities. This result reflects the price impact on public's housing demands and indicates housing demand can be addressed by economic means. Although the ideal housing floor-space for Chinese residents is about 40 m^2 per person (Figure 9.6), buyers' actual desire to purchase housing dropped to 31 m^2/per person in 2010 as housing prices rose.[15] Most buyers tend to reduce the requirement on housing floor-space when facing price constraints.

However, this does not mean higher housing prices equate higher quality. Figure 9.5 shows that residents' main purpose in purchasing housing is to meet the basic requirements of living and improve living conditions.

---

[15] The survey on house buyers in nine cities including Beijing and Shanghai (Source: the survey and research of the consumption of house buyers in the fourth quarter of 2010, http://fdc.soufun.com/report/1301.htm) shows 58% buyers are willing to buy houses with construction area below 90 m^2, and based on 2.88 persons per urban household in China in 2009 (China National Bureau of Statistics, 2011, Table 10.35), the ideal housing area of most buyers is less than 31 m^2 person.

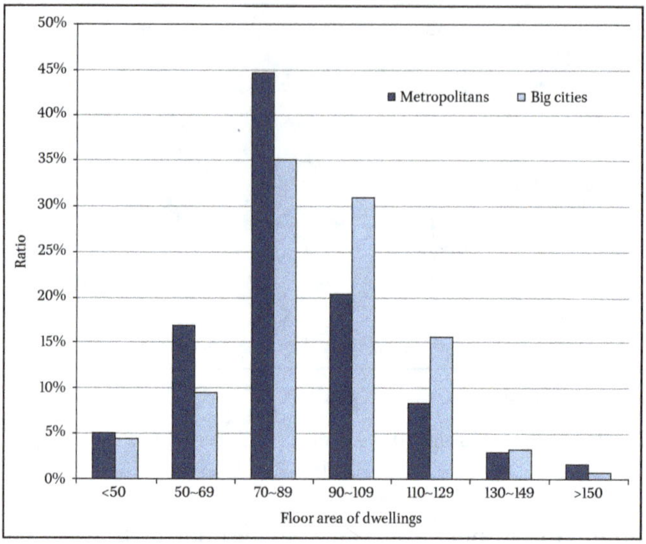

Source: China Index Academy, 2010: 15.

Figure 9.8. Survey data of housing purchase preference (percentage represents the proportion of sample population).

Higher housing prices will impede the natural improvement in residents' living standards. In addition, its characteristics, such as relatively high value, durability, and its exclusiveness, have caused real estate to become a profitable and attractive investment. As long as housing prices are expected to rise, investors gain profit returns, generating demand for real estate investment. If housing prices are too high, it may result in a housing bubble which not only will make it difficult for potential residents to bear the costs, but also create excessive growth of residential housing areas that is not beneficial for low-carbon development.

9.2.3.3 *Directly Addresses the Problem of Unnecessary Demolition and Construction in the Process of Rapid Urbanization*

Vice Minister Qiu Baoxing of the Ministry of Housing and Urban-Rural Development stated at the sixth International Green Building and Energy Conservation Conference that the average life-cycle of demolished buildings in China is merely 25 to 30 years, which is caused by factors such as changes in city planning, changes in the nature of land use, and changes in land prices. These "short-lived buildings" constantly emerged in cities at all levels during phases of excessive demolition and construction. This unnecessary demolition and construction is essentially caused by local governments' one-sided pursuit of economic development while neglecting scientific planning in the process of urbanization. Data shows that from 1996 to 2006, demolished buildings accounted for 23% of the area of completed buildings.[16] Solving this problem will make a direct contribution in slowing down the rapid growth of China's urban construction area.

9.2.4 *Housing Policy System is Established and Improved*

In recent years, China has promulgated a large number of housing policies, which can be divided into five categories according to their measures: land policy, financial policy, tax policy, housing supply policy and others. In terms of purposes, the housing policies can be categorized as regulation and control of the real estate industry, regulation of real estate markets, and housing security systems.[17] The objects of housing policies include local governments, real estate industry, and the public. In the 11th FYP period, the housing control and regulation regime for public consumption behavior mainly adopted financial policies, tax policies, and administrative measures; specific measures include property transaction taxes, differentiated credits, purchase quota policies, property taxes, and others. The main contents, purposes, and policies of each measure are shown in Table 9.2.

Since 2006, housing policies began to change to guide a health development of the real estate industry from a "pillar industry" to "promote consumption, expand domestic demand, stimulate investment growth, and maintain a sustained, rapid and health development" of the real estate industry; macroeconomic policies began to make adjustments regarding the supply structure; forced by the global financial crisis in 2008, the central government had to "rescue the market" so that housing policies changed to support the market from curbing; housing policies re-started tightening in 2009 and a series of strong measures, such as house purchase limits and property taxes, were issued in succession in 2010 to strictly curb speculative investment behavior. Generally, in the 11th FYP period, macro-control policies on housing were continuously adjusted responding to the

[16] Zhao Hui, Jiang Yi 赵辉, 江亿, *Chaichu xianxiang de wenti fenxi yu duice* 拆除现象的问题分析与对策 [Analysis and Countermeasures for the demolishing problems in Chinese cities] (Beijing: Tsinghua Daxue jianzhu jieneng yanjiu zhongxin, 2011).

[17] He Ximing 何溪明, *Fangdichan zhouqi jiqi hongguan tiaokong zhengce yanjiu* 房地产周期及其宏观调控政策研究 [Research on Real Estate Cycle and Macroeconomic Regulation, Doctoral dissertation] (Tianjin: Nankai Daxue, May 2010): 102–105.

Table 9.2. Main contents and purposes of five control measures on housing.

Policy Means	Main Content	List of Policies	Purpose
Housing transaction tax	Adjust sales tax, deed tax and personal income tax in housing transactions		Guide improvement demand and curb investment demand
Differentiated credit	Regulate down payment and interest rate of loans to second and third houses	Guo Ban Fa [2006] No. 37 Guo Shui Fa [2006] No. 74 Guo Shui Fa [2006] No. 108 Guo Shui Fa [2006] No. 144 Cai Shui [2006] No. 21 Guo Ban Fa [2008] No. 131 Guo Ban Fa [2010] No. 4 Cai Shui [2010] No. 94 Guo Shui Fa [2006] No. 37 Yin Fa [2007] No. 359 Yin Fa [2007] No. 452 Guo Shui Fa [2008] No. 131 Yin Jian Fa [2009] No. 59 Guo Shui Fa [2010] No. 4 Guo Fa [2010] No. 10	Curb investment demand
Purchase limit policy	Limit the number of houses bought by a family in cities where house prices are too high and grow too fast and the supply is tight within a certain time		Curb investment demand
Property tax	Increase taxes in the holding of houses, with a purpose to guide rational housing consumption of the public	Guo Fa [2010] No. 10 Guo Fa [2010] No. 15 Jian Fang [2010] No. 155	Guide improvement demand and curb investment demand
Others	Prevent affordable housing from coming to market and being traded Increase efforts of publicity and guide residents' rational consumption starting from the national condition	Guo Fa [2007] No. 24	Guide improvement demand and curb investment demand

macro-economic situation. In that process policies pursued a multitude of purposes, such as promoting economic development, preventing overheating of the economy and curbing investments, and establishing and improving the housing security system.[18]

9.3 Dramatic Change towards the Car

In the 11th FYP period, Chinese urban residents' means of traveling took on a "car" trend and more and more residents bought and owned cars, replacing other means of traveling with private vehicles.[19] This is the most important carbon source in urban passenger transport and a major factor in causing urban transportation energy consumption to grow rapidly in the 11th FYP period. It has had an important negative impact on low-carbon development in the transport sector.

9.3.1 *Large-Scale, Fast-Speed and Significant-Rigidity Change*

Since the Reform and Opening Era, the number of privately-owned cars has grown exponentially (Figure 9.9). By the end of the 11th FYP period, this number was close to 50 million, or 37 per 1,000 people. Meanwhile, the proportion of travel by car reached around 10%.[20] In 2010, the number of privately-owned cars was close to 200 per 1,000 people and the proportion of travel by car rapidly grew to 38% from 26.5% in 2000.

The growth in the number of motor vehicles in China, especially private cars, was huge and topped the world for two consecutive years, 2009 and 2010. In 2010, the number of newly-added private cars was 11.81 million,[21] accounting for 86% of passenger cars sold (13.7578 million)[22]

[18] Wang Yi 王艺, *Zhongguo tese de fangdichan shichang yu zhengfu guanzhi yanjiu* 中国特色的房地产市场与政府管制研究 [A Research on the Housing Market and Government Regulation with Chinese Characteristics, Doctoral dissertation] (Wuhan: Wuhan University, May 2010): 87–95.

[19] Cars mentioned in this section refer to private cars, namely, cars for private use of carrying passengers. Except as otherwise explained in this report, the cars exclude cars for civilian use, such as governments, enterprises, business and leasing but not for private use.

[20] Zhang Tieying 张铁映, *Chengshi butong jiaotong fangshi nengyuan xiaohao bijiao yanjiu* 城市不同交通方式能源消耗比较研究 [Study on Different Urban Transport Modes' Energy Consumption, Master thesis] (Beijing: Jiaotong Daxue, June 2010): 3.

[21] China National Bureau of Statistics, 2011, Table 10.17.

[22] Sina.com, "2010 nian quanqiu qiche chanliang gonggda 7761 wan liang 2010 年全球汽车产量共达 7761 万辆 [Global Auto Production Reaches 7.8 Million in 2010]," 2011-03-18, http://auto.sina.com.cn/news/2011-03-18/0736737974.shtml [accessed: 2011-09-21].

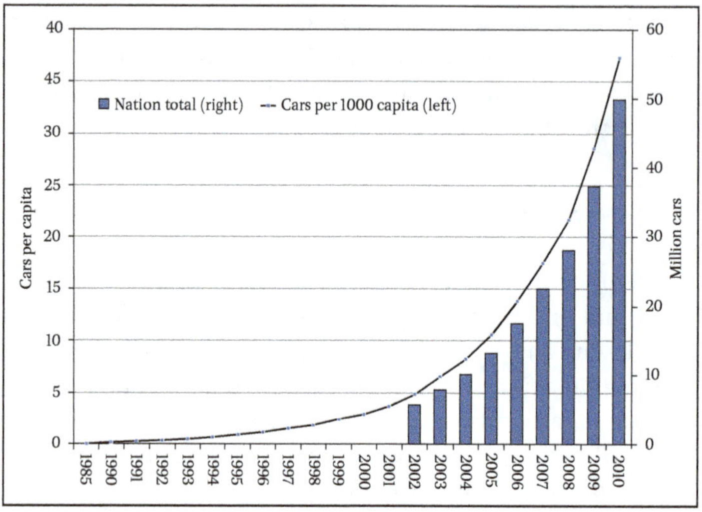

Note: Because of incomplete data, number of private cars-owning per 1,000 people from 1985 to 2001 is calculated based on the number of private passenger cars. In 2009, the number of private cars accounted for more than 98% of the number of private passenger cars.

Source: National Bureau of Statistics. 2011, Table 10.10.

Figure 9.9. Private cars-owning number in total or per 1,000 people in China in 1985–2010.

and accounting for more than 20% of global output of passenger cars (58.264 million).

In the 11th FYP period, the number of privately-owned cars in China grew much faster than residential incomes. The number of privately-owned cars grew to more than 37 per 1,000 people in 2010 from about 10 per 1,000 people in 2005, surging more than twice, much higher than a growth (82.1%) in per capita disposable income of urban residents in China during the 11th FYP period.[23] Besides, the number of privately-owned cars in China grew much faster than that in developed countries. "It took 18 years (1955 to 1973) for Japan to lift the number of privately-owned cars to 133.6 per 1,000 people from 1.7 per 1,000 people, while it took only 10 years (1993 to 2003) for Beijing to lift the number of privately-owned cars to 131.4 per 1,000 people from 2.48 per 1,000 people, nearly 50% faster than Japan."[24]

23 China National Bureau of Statistics, 2011, Table 10.2.

24 Chen Shanghe, Wang Peilin 陈尚和, 王沛霖, "Beijingshi xiaoqiche jiaotong fazhan yanjiu 北京市小汽车交通发展研究 [Study on Increase of Car Ownership and Usage in Beijing]," *Chengshi jiaotong* 6, no. 3 (2008): 53–56.

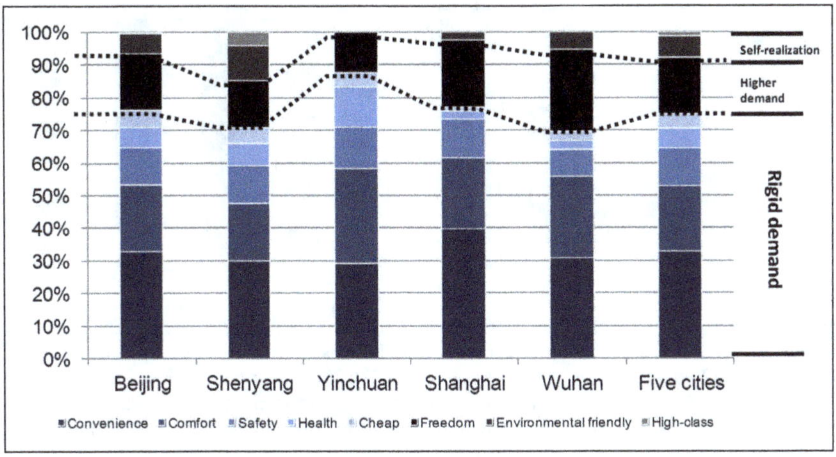

Source: Zhang Shengyuan, 2010; analysis by Climate Policy Research Center, Tsinghua University.

Figure 9.10. Mains reasons for urban residents to use cars in 2008.

China's private car consumption showed a very significant rigidity because more than 70% car consumption is driven by strong demands. As shown in Figure 9.10, "convenience," "comfort," and "freedom" are the main three reasons provided for car purchases[25] by Chinese residents; higher level of symbol-based needs,[26] such as "green" or "advanced" are not main factors driving residents to use cars and account for less than 2% of car consumption. Because "almost all travelers' ideal transport is independent, fast, and comfortable," this consumption psychology will drive residents to continue using cars as a future means of traveling.[27]

---

25 According to Maslow's Hierarchy of Needs, the "convenience" and "comfort" can correspond to "survival" type need and the "freedom" can correspond to "safe" type need that is a lower-level need and owns a strong rigidity property.

26 According to Maslow's Hierarchy of Needs, the "advanced" can correspond to "belonging" type need and the "green" can correspond to "self-realization" type need, both of which belong to the advanced level of Maslow's pyramid of needs. When the lower-level needs are greatly met, such needs are likely to become the main driving force.

27 Li Linbo, Wu Bing 李林波, 吴兵, "Jiaotong fangshi xuanzezhong xinli yinsu yingxiang fenxi 交通方式选择中心理因素影响分析 [The Analysis of Psychology Factor's Effects on Decision of Selecting Traffic Mode]," *Shadong jiaotong xueyuan xuebao* 11, no. 9 (2003): 27–31.

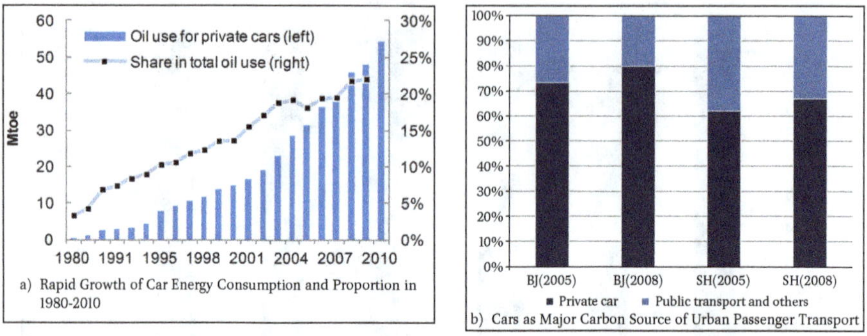

a) Rapid Growth of Car Energy Consumption and Proportion in 1980-2010

b) Cars as Major Carbon Source of Urban Passenger Transport

Sources: Zhang, M. et al., "Energy and Energy Efficiencies in the Chinese Transportation Sector, 1980–2009," *Energy* 36 no. 2 (February 2011): 770–776; Zhang Shengyuan, 2010; Lu Ximing, Zhu Yiran 陆锡明, 祝毅然, "Shanghai jiaotong nenghao xianzhuang ji fazhan qianjing 上海交通能耗现状及发展前景 [Current Situation and Development Prospect of Transportation-related Energy—Consumption in Shanghai]," *Shanghai Jieneng* 1 (2010): 4–6; Zhu Songlin 朱松丽, "Beijing, Shanghai chengshi jiaotong nenghao he wenshi qiti paifang bijiao 北京、上海城市交通能耗和温室气体排放比较 [Comparison of Transportation Energy Consumption and Greenhouse Gas Emission between Beijing and Shanghai]," *Chengshi Jiaotong* 3, no. 8 (2010). 58–63; Guan Juxiang 管菊香, *Jiyu shengtai jiaotong de chengshi jiaotong fangshi jiegou youhua yanjiu* 基于生态交通的城市交通方式结构优化研究 [Urban Traffic Mode Structure Optimization Based on Eco-transportation, Master thesis] (Wuhan: Wuhan Ligong Daxue, November 2007). Analysis by Climate Policy Research Center, Tsinghua University.

Figure 9.11. Proportion of cars and public transport in passenger transport energy consumption in surveyed cities in 2008.

9.3.2 *Cars Pose a Serious Challenge to China's Low-Carbon Development*

9.3.2.1 *Cars are the Main Factor to Drive Carbon Emissions of the Transport Sector to Grow in the 11th FYP Period*

Cars caused related energy consumption and carbon emissions to grow rapidly. As shown in Figure 9.11, car fuel consumption reached 54 Mt in 2010, more than 74% as compared to 2005. It took a rising proportion of the country's total energy consumption in the transport sector and accounted for 22% in 2009, and was an important component of the country's transport energy consumption. Cars also were the main factor that caused carbon emissions in China's transportation sector to grow. In 2009, the total energy consumption of China's transport sector reached 219 Mt, growing by more than 26.6% compared with 2005, and cars contributed 9.7 percentage points or more than 30% to the growth of China's transport energy consumption.

In cities (in the passenger transport sector), cars have become the main source of carbon emissions and their share increased significantly in the 11th FYP period. The proportion of energy consumption in passenger transport in two cities, Beijing and Shanghai, grew by more than five percentage points in three years to 80% and 67% in 2008 from 74% and 62% in 2005, which is due to the number of privately-owned cars.[28] Carbon emissions by private cars per 100 km are the highest of several major modes of transport, about 12 times those of buses.

Cars, among urban passenger transport, are not only the main source of carbon emissions, but also the main factor behind the rapid growth of energy consumption and carbon emissions. Studies have estimated the number of privately-owned cars in China increases by 1% annually and additional 80 million liters of fuels annually are required (compared to buses),[29] corresponding to carbon emissions of 340,000 tCO_2; the proportion of car travelling increases by 10% annually, Beijing urban transport carbon emissions will grow by 21.3%, and the marginal effect is largest.[30]

9.3.2.2 *Urbanization and Modernization will Continue to Promote Residents' Cars-Oriented Lifestyle*

Urbanization and modernization is the driving force behind China's economic boom. With the further advance of urbanization and modernization, a large population will flow into urban areas, providing sufficient labor capital to modern economic development and meanwhile producing a great deal of demand for consumption goods, especially transport. In terms of cars, the penetration rate usually increases with economic growth.[31] As shown in Figure 9.12, compared with the number of privately-owned cars in several countries at different stages of development, the higher a country's per capita GDP is, the larger the number of privately-owned cars per 1,000 people. As urbanization and modernization are continuing to further advance the development of the Chinese economy, it can be expected that the number of privately-owned cars will continue to grow.

---

28 Generally, the carbon emissions of a transport mode = vehicles stocked corresponding to the transport mode * average running distance of the vehicles * the carbon emissions intensity of the vehicles. The carbon emissions intensity means carbon emissions per person per km.

29 Wang Guangrong 王光荣, "Xiaoqiche yu chengshi hexie jiaotong 小汽车与城市和谐交通 [Cars and Harmonious Traffic]," Chengshi guancha 1 (2009): 165–170.

30 Zhang Tieying, 2010: 39–41; analysis by Climate Policy Research Center, Tsinghua University.

31 Qian Zhenwei 钱振为, "Lun jiaoche xiaofei yu jingji zengzhang de guanxi 论轿车消费与经济增长的关系 [Research on the relationship between car consumption and economic growth]," *Qiche gongye yanjiu* 10 (2005): 2–12.

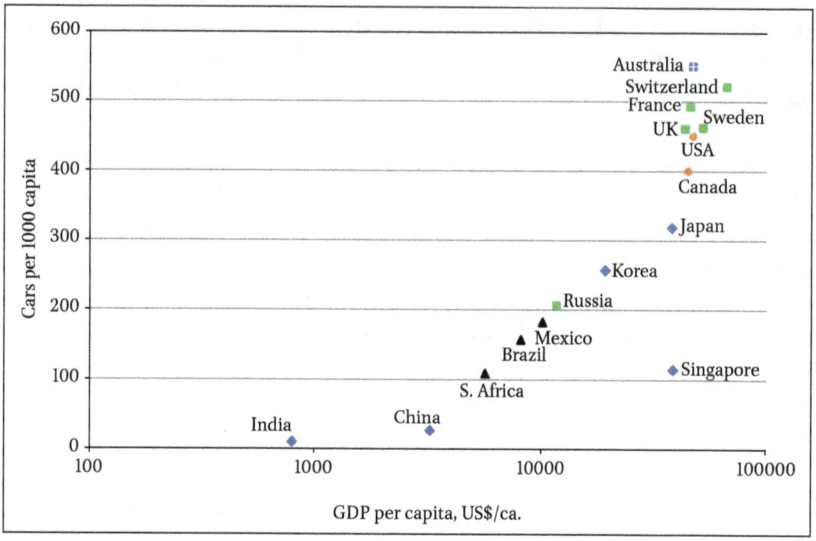

Note: In the figure, the rounds represent North American countries, the rhombuses represent Asian countries, the diamonds represent European countries, and the triangles represent Asian, African and Latin American countries.

Sources: World Bank. World Development Indicators database (July 1, 2009); National Bureau of Statistics, 2011, Table 10.7; Cai Yifeng, Ruan Zheming 蔡逸峰, 阮哲明, "Siren xiaoqiche fazhan dui chengshi jiaotong guihua de tiaozhan 私人小汽车发展对城市交通规划的挑战 [The Challenges of Development of Private Cars on City Transportation Planning]," Guihuashi luntan 1 (2002): 11–13; Analysis by Climate Policy Research Center, Tsinghua University.

Figure 9.12. Comparison of GDP per capita and the private cars-owning number per 1,000 people of major countries in 2008.

9.3.2.3 *Challenge: It is too Heavy to be Withheld for China's Low-Carbon Development to Walk up Old Paths of Developed Countries*

Along with the development process displayed in Figure 9.12, car consumption will show excessive growth, consuming large amounts of non-renewable fossil energy and land resources. This is unviable for China.

On the one hand, petroleum consumed by transport, conveyance, and postal industries (130 Mt) accounted for 35.6% of the country's total petroleum consumption[32] and grew 20 Mt or nearly three percentage points over 2005 (about 110 Mt, accounting for 32.9%). If no positive and effective solutions are taken, China's automobile energy demands

[32] China National Bureau of Statustics, China Statistics Yearbook 2010, China Statistics Press: Table 7.4.

will reach 425 Mt by 2030 to account for more than 69% of total petroleum demands.[33] This will form a serious threat to the security of energy supply.

On the other hand, excessive use of cars will occupy a lot of land resources. Only in the 11th FYP period, newly-added roads for private cars reached up to 5 billion m^2. If the number of private cars stocked reaches United States level (451/1,000 people), 125 million *mu* of land area (accounting for 7% of China's cultivated land area and able to feed more than 100 million people) will be occupied only to meet the needs of parking and driving private cars.[34] Even if the number of private cars stocked reaches the lower Singapore level (114/1,000 people), an additional land area equivalent to the size of a Beijing needs to be occupied.

Land occupied by urban transport accounted for only 1.2% of the urban area in China,[35] compared with 30% to 50% in developed countries.[36] Especially, in a car society like the United States, "cities' borders almost disappeared and land resources were extremely wasteful."[37] "Based on China's current per capita land resources, it is difficult for the country to build roads following the model provided by developed countries."[38]

Cars also caused urban traffic jams and environmental pollution. The increase of stocked cars can greatly seize road resources[39] of public transport and non-motorized transport to result in: (1) Aggregated urban

---

[33] Zhang Jihong 张继宏, "Dui Zhongguo weilai cheyong nengyuan fazhan de jidian sikao 对中国未来车用能源发展的几点思考 [China's vehicle-energy future]," *Guoji shiyou jingji* 7 (2010): 16–19.

[34] This does not include inter-city long-distance roads, such as highways, as well as other types of motorized vehicles, such as public service vehicles, buses and trucks.

[35] Zhang Yang 张洋, "Jiyu kechixu fazhan de chengshi keyun jiaotong moshi youhua yanjiu 基于可持续发展的城市客运交通模式优化研究 [Research on Optimization of Urban Passenger Transportation Mode Based on Sustainable Development. Master thesis]" (Chongqing: Jiaotong Daxue, April 2009): 3.

[36] Ding Weidong, Liu Ming, Du Shengpin 丁卫东, 刘明, 杜胜品, "Jiaotong fangshi yu chengshi lüse jiaotong 交通方式与城市绿色交通 [Transportation Means and Urban Green Traffic]," Wuhan keji daxue xuebao 26, no. 3 (2003): 50–53.

[37] Yang Tao, He Ning, Sun Jun 杨涛, 何宁, 孙俊, "Chengshi jiaotong zhengce yanjiu 城市交通政策研究 [A Study on the Policy for Urban Traffic]," *Guihuashi* 21 no. 8 (2005): 10–13.

[38] Du Lei 杜蕾, *Zhongguo jiaoche gongye fazhan yanjiu* 中国轿车工业发展研究 [Investigation on China's auto industry development] (Chengdu: Xinan caijing daxue chubanshe, 2006).

[39] The passenger and cargo transport effectiveness of cars is much lower than that of the public transport: in Beijing in 2007, the public transport only took 16% of road resources but undertook 77% traffic volume of passengers and cargos, but cars occupied more than 80% of road resources but only undertook less than 30% traffic volume of passengers and cargos (Guo Shuxia, etc., 2010).

congestion: Average vehicle speed is only about 20km/h in trunk roads of Beijing currently, about 50% lower than 10 years ago; in morning and evening peak periods, nearly 60% of 183 major intersections in the urban area were severely blocked.[40] (2) Deterioration of the urban environment: Cars are responsible for a seriously polluted urban air quality. Surveys show that 36% of total air pollutants in Beijing were emitted by cars.[41]

9.3.3 *Opportunity to Guide Low-Carbon Travel Patterns*

Given the dual constraints posed by energy and resources, China must meet the needs of economic development and upgrade residents' travel needs. Currently, China' economic structure is in a critical period of transformation, residential lifestyles are starting to change dramatically, and it is a key opportunity to guide residents to change their means of travel towards a low-carbon mode. The average share of car travel in the country is about 10% currently, still lower than that of developed countries (25% in Japan, and 40% in Western European countries).[42] The gap with developed countries is a space for continuing expansion of car travel as well as a space for guiding residents to low-carbon travel.

Car-based travel for residents includes two parts, purchase and use of cars. Although the number of privately-owned cars will continue to grow (Figure 9.12), the use of cars is different. A comparison of different countries and regions in Figure 9.13 shows that the share of car travel grew alongside the rise in per-capita GDP and basically followed two curves: one is starting from Indonesia and going through Beijing and Shenzhen to reach the level of the United States; the other is starting from India and going through China and Mexico to reach the level of Japan. The proportion of car travel of the former was 10% to 20% higher than that of the latter and the gap was widening along with the growth of per-capita GDP.

It is thus clear that in addition to the United States model, the increase of China's car travel share in the future also has a lower-proportion development mode. These two modes are usually defined as scattered and

---

[40] Chen Shanghe et al., 2008: 53–56.

[41] Ibid.

[42] Wang Fengbao 王逢宝, "Chengshi jiaotong jieneng jianpai celüe yanjiu. Cong chengshi gongjiao yu xiaoqiche jiaotong de bijiao kan 城市交通节能减排策略研究—从城市公交与小汽车交通的比较来看[Investigation on urban transport energy efficiency strategy]," *Chengshi cheliang* 8 (2008): 30–33.

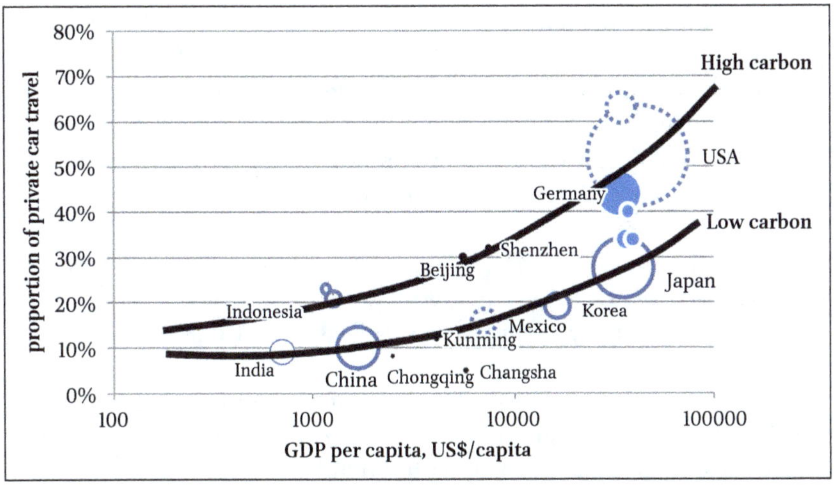

Note 1: The blue dotted-line hollow circles represent American countries, the blue solid circles represent European countries, the blue full-line hollow circles represent Asian countries, and the black points represent Chinese cities.
Note 2: The size of the circle area represents the GDP of each country.

Sources: Beijing Municipal Bureau of Statistics, *Beijingshi tongji nianjian 2006* 北京市统计年鉴 2006 [2006 Beijing Statistics Yearbook 2006] (Beijing, Zhongguo tongji chubanshe, 2006); Chongqing Municipal Bureau of Statistics, *Chongqingshi tongji nianjian 2006* 重庆市统计年鉴 2006 [Chongqing Statistics Yearbook 2006] (Beijing, Zhongguo tongji chubanshe, 2006); Shenzhen Municipal Bureau of Statistics, *Shenzhenshi tongji nianjian 2006* 深圳市统计年鉴 2006 [2006 Shenzhen Statistics Yearbook] (Beijing: Zhongguo tongji chubanshe, 2006); Changsha Municipal Bureau of Statistics, *Changsha tongji nianjian 2006* 长沙市统计年鉴 2006 [2006 Changsha Statistics Yearbook] (Beijing: Zhongguo tongji chubanshe, 2006); Mao Haixiao 毛海虓, *Zhongguo chengshi jumin chuxing tezheng yanjiu* 中国城市居民出行特征研究 [Research on Person Trip Characterstics on Chinese Citizens, Doctoral dissertation] (Beijing: Polytechnic Universtiy, March 2005); World Bank, 2006; Guan Juxiang, 2007; Zhang Yang, 2009; Liu Hongxia, Xia Jixing 刘红霞，夏继星, "Shenzhenshi jiaotong xuqiu fazhan bianhua qianxi 深圳市交通需求发展变化浅析. [Analysis of Development and Change of Traffic Demand in Shenzhen City]," *Youse yejin sheji yu yanjiu* 12, no. 4 (2004): 48–51; Meng Fei 孟菲, *Chengshi jiaotong chuxing moshi dui nengyuan he huanjing de yingxiang* 城市交通出行模式对能源和环境的影响. [The Impact of Urban Transportation Modes on Energy Consumption and the Environment. Engineering master thesis] (Beijing: Qinghua daxue, June 2010). Analysis by Climate Policy Research Center, Tsinghua University.

Figure 9.13. Comparison of per capita GDP and car travel in major countries in 2005.

intensive mitigation modes,[43] i.e., a car-oriented high-carbon mode and a public transport-oriented low-carbon mode.

In the 11th FYP period, profound changes took place in traffic demands to exacerbate the demand for car consumption. Meanwhile, the rapid growth of residents' income provided a strong guarantee for meeting this demand. Both economic development and demand drove urban residents' means of traveling to be car-oriented. Accordingly, residents' means of travel can be shifted towards a low-carbon model through certain economic regulations and demand-driven measures.

9.3.3.1 *A Reasonable Guide of Traffic Demand can Promote Residents to Reduce Selecting Car Consumption*

Traffic demand usually includes two aspects, travel purpose and travel radius.

Within leisure-based travel, the share of travel by means of private vehicle rose rapidly. Leisure-based travel accounted for more than 30% of car travel by residents in Beijing in 2008,[44] compared to only 18% in 2000,[45] but still lower than in developed countries. In the United States, nearly 60% of car travel was leisure-based.[46] This is because leisure activities have higher requirements in terms of traveling distance and degree of freedom, and cars can best accommodate these requirements.

Meanwhile, the average radius of residents' travel also continued to increase. In 1990, the average radius of residents' travel was less than 3.6 km,[47] increasing by 1.5 times in 20 years to more than 8 km in 2008,[48] average radius of leisure-based travel grew fastest and reached more than 13 km in 2008[49] from less than 7 km in 2001,[50] almost doubling; long-distance travel usually needs be met by buses and cars, but per-capita

---

43 Yang Tao et al., 2005: 10–13.

44 Zhang Shengyuan, 2010: 58.

45 Mao Haixiao, 2005: 109.

46 Liu Mingjun, Guo Jifu, Gao Liping, Zhang Dexin, Mao Baohua 刘明君, 郭继孚, 高利平, 张德欣, 毛保华, "Siren xiaoqiche chuxing xingwei tezheng fenxi jianmo 私人小汽车出行行为特征分析与建模, *Jilin daxue xuebao* 39, no. 9 (2009): 25–30; Analysis by Climate Policy Research Center, Tsinghua University.

47 Mao Haixiao, 2005: 116.

48 China's Urban Energy Consumption Status and Energy Conservation Policy Research Group, *Chengshi xiaofei lingyu de yongneng texing yu jieneng tujing* 城市消费领域的用能特性与节能途径 [Energy Characteristics and energy-saving approach of urban consumption field] (Beijing: Zhongguo jianzhu gongye chubanshe, July 2010: 133–136.

49 Zhang Shengyuan, 2010: 152.

50 Liu Hongxia, Xia Jixing, 2004: 48–51.

distance of bus travel (less than 5 km for 70%)[51] was significantly shorter than that of car travel (about 12.6 km).[52] Cars have an advantage to adapt to long-distance travel, especially leisure-based travel, resulting in the increase of the proportion of car travel.

It is thus obvious that cars can better accommodate the changes in residential transport needs so that proportion of travel by car continues to increase. Surveys show more than 62% of residents having cars still regard cars as their first choice for travel.[53] Therefore, to increase the advantages of other modes of transportation through appropriate demand-led measures can lead residents to reduce the usage of cars.

9.3.3.2 *To Improve Economic Costs of Car Consumption can Slow the Speed of Car-Oriented Process*

In the 11th FYP period, urban per-capita disposable income increased by more than 80% to RMB 19,109 per person in 2010 (up from RMB 10,493 per person in 2005), which increased the purchase and use of the number of cars.[54]

On the one hand, the purchasing power of urban residents increased. Both rising incomes and falling costs and prices of cars promoted by the automobile industry incentivized residents to buy cars. Usually, the higher the income of families, the more likely cars are bought. Surveys (Figure 9.14) show that the higher the income levels of residents, the bigger the number of privately-owned cars per 1,000 residents. The number of privately-owned cars per 1,000 residents in the lowest-income resident group is merely less than one-fifth of that of the highest-income resident group.

On the other hand, although gas prices increased year by year, the proportion between the gas price and disposable incomes of residents remained at about 42%,[55] and residents accordingly can afford the cost

51 Guo Shuxia, Yu Lei, Chen Xumei, Zhang Yu 郭淑霞, 于雷, 陈旭梅, 章玉, Beijing chengshi gongjiao qiche chuxing tezheng fenxi ji duice 北京城市公交汽车出行特征分析及对策 [Analysis on Beijing Urban Bus Transport]," *Yunshu guanli* 4 (2010): 58–61.

52 Meng Fei, 2010.

53 Zhou Min, Zhen Feng 周敏, 甄峰, "Kuaisu zengzhang de siren xiaoqiche jiaotong dui Nanjingshi juzhu kongjian fenyi de yingxiang yanjiu 快速增长的私人小汽车交通对南京市居住空间分异的影响研究 [The Impacts of Rapidly Increasing Private Car Traffic on the Differentiation of Urban Residential Space in Nanjing]," *Xiandai chengshi yanjiu* 10 (2009): 67–71.

54 China National Bureau of Statistics, 2011, Table 10.2.

55 From 2003 to 2010, China's per capita disposable income grew to RMB 19,109 from 8,472 (China National Bureau of Statistics, 2011) and meanwhile, the gas price grew to

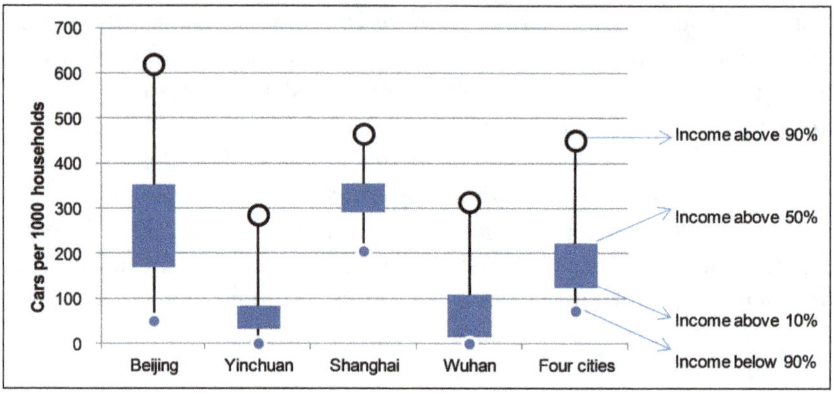

Source: Zhang Shengyuan, 2010.

Figure 9.14. The private cars-owning number per 1,000 persons for groups with different income levels in surveyed cities in 2008.

of gas consumption to a certain extent and choose cars for means of transport or travel by taxi, rather than take other modes of transportation.[56] According to a survey of 11,539 samples in 2008, the proportion of car travel of samples covering per capita income of less than RMB 2000 monthly was 11.1%, while the proportion of car travel of samples covering per capita income of more than RMB 10,000 monthly surged nearly 100% to 20.3%.[57] Generally, the higher the income, the stronger the capacity to afford cars. To increase residents' car consumption costs through appropriate economic means can slow down the growth of car consumption.

RMB 8,720 / ton from RMB 3,574 / ton (re-adjusted to RMB 9,070 / ton in February 2011) (www.cntv.cn, June 1, 2010; carschina.com, February 21, 2011). In the period, the proportion between the gas price and residents' per capita disposable income maintained about 42%.

[56] Ye Zupan, Wang Liping 叶祖盼, 王莉萍 "Cong jingjixue de jiaodu qiantan chengshi jiaotong jiegou de youhua 从经济学的角度浅谈城市交通结构的优化 [On optimization of urban traffic structure from economics angle]," *Shanxi jianzhu* 35, no. 5 (2009): 43–45.

[57] Liu Ailing, Hu Xiaoqi, Li Yanping et al. 刘爱玲, 胡小琪, 李艳平等, "Woguo feinongye zhiye renqun shangxia ban jiaotong fangshi de xianzhuang fenxi 我国非农业职业人群上下班交通方式的现状分析 [Modes of Transportation by Chinese Non-agriculture Employed Population]," *Zhongguo manxingbing yufang yu kongzhi* 17, no. 4 (2009): 114–116.

9.3.3.3 *Impact on Reducing Car Travel in Suburbs by Optimizing the Pattern of Urban Planning*

Mono-centric city expansion played an important role in the increase of travel radius and proportion of car travel. Mono-centric city expansion and changes in urban passenger transport modes mutually influenced and promoted each other. In developed countries, city expansion not only caused urban populations to migrate to suburbs,[58] but also greatly increased the proportion of residents' car travel. In Paris, the proportion of residents' car travel in the surrounding area was more than 50% higher than that in the central area.[59]

In China, the phenomenon of suburbanization is relatively common. The experience shows that beginning in the 1980s, the population in the central area of cities, such as Beijing, Shanghai, Shenyang, Hangzhou, Xi'an, Suzhou, Guangzhou, and Dalian continued to decline, while the population in the suburbs grew rapidly.[60] This was accompanied by the phenomenon of increased travel by car and significantly increasing proportion of car travel in suburbs. The proportion of car travel in the suburb of Shanghai was 11.8%, slightly higher than the central area (9.4%).[61] In Beijing in the morning peak hours, "one of every 1.68 persons enters the direction of the Fourth Ring Road and about 60% use private cars as travel mode."[62] This phenomenon directly resulted in an increase of the proportion of car travel for residents in suburbs.

Mono-centric city expansion caused suburbanization and promoted the proportion of car travel to grow. Therefore, a multi-center urban planning

58 Yang Tao et al., 2005: 10–13.

59 China's urban energy consumption status and energy conservation policy research group, 2010: 201.

60 Chen He 陈和, *Jiaotong fangshi dui chengshi jiaoquhua moshi de zuoyong jizhi yanjiu* 交通方式对城市郊区化模式的作用机制研究 [Research on the Mechanism of the Mode of Transportation to the Pattern of Suburbanization, Doctoral dissertation] (Beijing: Jiaotong University, July 2008); Li Junli, Wang Hui, Cao Mingming 李俊莉, 王慧, 曹明明, "Xi'anshi renkou kongjian bianhua yu jiaoquhua dongxiang [Demographic Spatial Distribution and Direction of Suburbanization in Xian in the 1990s]," *Zhongguo renkou ziyuan yu huanjing* 16, no. 3 (2006): 66–70; Zhang Wenxin 张文新, "Zhongguo dachengshi renkou juzhu jiaoquhua xianzhuang, wenti yu duice 中国大城市人口居住郊区化现状、问题与对策 [The Current Situation, Problems and the Countermeasures of the Suburb Inhabitance of the Urban Population in Big Cities of China]," *Renkou xuekan* 3 (2003): 15–20.

61 Mao Haixiao, 2005: 49.

62 Beijing Traffic Development Research Center, "Beijing disanci jiaotong zonghe diaocha jieguo fenxi 北京第三次交通综合调查结果分析 [The third survey on Beijing traffic situation]," February 2007, http://2006.moc.gov.cn/06beijing/fazhangh/200702/t20070207_177921.html [2009-08-08].

pattern will be helpful to reduce this phenomenon. To build convenient and available suburban public transport systems also can slow down the rapid rise of the proportion of car travel.

9.3.4 *Transport Policies Seek Balance between Industrial Development and Low-Carbon Travel*

The excessive trend towards private cars has had an adverse effect on China's low-carbon development, but the economic contribution of the automobile industry cannot be ignored. The trend towards the car greatly boosted the development of the automobile industry, pushed forward the development of upstream and downstream industries, and made an important contribution to GDP growth. In 2008, the automobile industry and related industries contributed about 5% to GDP, including 1.3%[63] by the automobile industry;[64] and 3.7% by related industries.[65] In addition, compared with the automobile industry, automobile consumption itself made a more significant contribution to GDP. It is estimated[66] that an increase in private cars by 1% can drive GDP to grow by 0.97 percentage[67] points, higher than that by the automobile industry (1% increase in investment can bring along 0.20 percentage points).

Therefore, how to balance low-carbon and development has become a central issue. In the 11th FYP period, China promulgated a series of policies concerning purchase and use of private cars (Table 9.3). The national policies' main purpose was to revitalize the automobile industry and coordinate low-carbon development and local policies based on local conditions. These policies mainly address issues such as traffic congestion, pollution emission, etc., and play a promoting role in low-carbon

[63] The proportion is calculated based on industrial added value of the automobile industry or it is calculated based on total output value of the automobile industry to be 6.0%.

[64] China Nationa Bureau of Statustics, China Statistics Yearbook 2010, China Statistics Press: Table 2.11.

[65] Usually, the automobile industry can drive the GDP of related industries (such as iron and steel, rubber and auto parts) to surge by more than 2.7 times (Ren Dongmei, 2007).

[66] Zheng Shuai 郑率, *Woguo qiche chanye yu jingji zengzhang lianxing de jihua fenxi* 我国汽车产业与经济增长关联性的计量分析 [The Econometric Analysis of the Relevance between Automotive Industry and Economic Growth in China, Master thesis] (Changchun: Jilin University, May 2010).

[67] The numerical value considers the consumption directly related to cars and the improvement of other consumption levels boosted by rising incomes (so, residents can afford to buy cars).

Table 9.3. Summary of car consumption-related policies and evaluation of emission reduction contribution in China in the "eleventh five-year plan" period.

Policy Purpose		Policy Content	Time	Place	Corresponding Means	Review of Implementation Results
Car stock	Regulate and control growth	Purchase restriction order	2011	Beijing	Economy	The growth of private cars-owning was under strict control.
		License plate auction	1986	Shanghai	Economy	The number of plated cars in Shanghai was under control. The number of cars running on roads kept growing.
	Adjust structure	Reduce the purchase tax of small-emission cars	2009	Nation-wide	Economy + demand	Small-emission car consumption grew rapidly. The proportion of small-emission cars increased. Residents' consumption wishes are related to policies.
		Subsidize small-emission cars	2010	Nation-wide	Economy + demand	
		Subsidize new energy cars	2010	Nation-wide	Economy + demand	Difficult to issue subsidies The market needs to be cultivated.
		Reduce the consumption tax of small-emission cars Increase the consumption tax of large-emission cars	2008	Nation-wide	Economy + demand	Temporarily, no exact data indicates the effect to encourage residents to buy small cars rather than large cars.

Table 9.3 (*cont.*)

Policy Purpose		Policy Content	Time	Place	Corresponding Means	Review of Implementation Results
Cars used	Control growth	Fuel tax	2009	Nation-wide	Economy	Residents paid most attention to fuel performance. Temporarily, no exact data indicates the car travel reduced.
		Increase parking fees	2011	Beijing	Economy	Car traffic flow fell 12%. Residents' car travel wills reduced.
	Encourage substitute	A series of policies to give priority to public transport	2005	Nation-wide	Economy + demand	The proportion of bus travel increased. Residents paid more attention to congestion than ticket prices.
		Low-cost bus travel	2007	Beijing	Economy + demand	
		Travel limit order	2008	Nation-wide	Demand	Cut down official vehicle travel by 20% and car travel by 20%, but the emission reduction was offset by the growth of cars.
			2008	Beijing	Demand	
		Guarantee non-motorized vehicle travel	2010	Beijing	Demand	The projects in demonstration transformation regions were completed. Bicycles' road right was protected.

development. Generally, all these policies regulate either the production or the demand side:

(1) The purchase side: a) regulation of growth: affects residents' car purchase costs in order to reasonably regulate the growth of the number of privately-owned cars; b) structural adjustment: creates cost differences among different types of cars in order to reasonably regulate the share of different types of private cars, so as to encourage the consumption of small size cars and clean energy vehicles;
(2) The usage side: a) regulation of growth: affects residents' car purchase costs in order to reasonably regulate of the use cars to avoid excessive use; b) encouragement of substitutes: guides residents to use public transport or other modes of transportation through low-price competition to reduce the use of private vehicles.

In terms of policy results, as shown in Table 9.1, many policies had a certain effect in various aspects, including controlling the excessive growth of private car ownership, enhancing the share of low-emission cars, adjusting the use of low-emission cars, and encouraging public transport and non-motorized transport. This indicates the possibility to guide residents' means of traveling towards low-carbon models through governmental policies. However, the number of privately-owned cars in China has taken on an exponential growth trend, indicating residents' strong demand for private cars. Therefore, it is the focus of future work on how to incentivize residents to lean to low-carbon models in the process of selection of car use through strong economic and demand constraints provided by policies under the premise of guaranteeing economic development and improvement of people's living standards.

9.4 The Public Has the Will, Ability, and Means to Fulfill Low-Carbon Policies

In the 11th FYP period, the public's participation in low-carbon business received more attention and fulfillment. The characteristics of public consumption behavior decide that the public has the responsibility for low-carbon development in the field. Based on the practices in the 11th FYP period, the public also had the will, ability, and means to respond and contribute to low-carbon development, which was mainly reflected in two aspects: one is policy participation, the other is low-carbon actions.

The 11th FYP period was a very important stage in which the public participated in the development of the public policy field. In the implementation of low-carbon policies, government departments may have behavioral biases, such as detachment from the situation on the ground or divided motives and loyalties, while enterprises generally lack the power, capacity, and leverage to influence public policy, hence public participation can be a viable tool to articulate and correct these problems in the stage of policy implementation to a certain extent. Since the beginning of this century, China began to develop policies and actively promoted and guaranteed public policies, especially public participation in low-carbon development. These policy documents provided an institutional foundation for the public in aspects such as laws and regulations related to low-carbon, low-carbon development planning, major low-carbon projects, and implementation and monitoring of low-carbon policies. The enthusiasm of the public to participate in the development of low-carbon policies and supervise the implementation of low-carbon policies continued to improve, playing a significant role in promoting public affairs. For example, on December 5, 2008, the National Development and Reform Commission, the Ministry of Finance, the Ministry of Transport, the State Administration of Taxation, and other departments solicited comments from the public for the "Refined Oil Tax Reform Program" (draft) for a week and received all kinds of comments totaling 48,643 individual inputs via Internet, fax, mail, and letter. The public demonstrated a high degree of attention and active participation in a short time, showing that the public's enthusiasm to participate in public affairs concerning their own interests and rights. Taking another example, from September 2010, "power rationing" took place in many places of the country, including Anping, Hebei, and Jiaozuo (Henan) which caused a large number of complaints from the public. Exactly due to complaints and supervision by the public and the media, the National Development and Reform Commission quickly issued the "Emergency Notice on Further Doing Well in Current Energy Saving and Emission Reduction Work" and made great efforts to correct the wrong approach to stop and limit electricity used by residents and public service units.

The public mainly engaged in low-carbon related activities through NGOs and individual efforts to make an important contribution to advance low-carbon issues. At the NGO level, taking the WWF as an example, the WWF officially launched the "China's low-carbon city development project" in Beijing in January 2008 and Shanghai and Baoding were selected to be among two of the first batch of pilot cities and signed memorandums of cooperation with the WWF. Under the assistance of the WWF, the two cities quickly became the focus of the entire nation and promoted the spread of the concept of low-carbon cities in China. At the individual level, more and more publicity helped to develop awareness for low-carbon issues amongst the populace, which was reflected in actions such as active participation in various low-carbon-related activities and leading low-carbon lifestyles. For example, organizations such as Global Village Beijing launched the "26 degree air conditioning energy saving action," the WWF advocated the "Earth Hour," and the promotion and degree of "green travel" launched by the China Association for NGO Cooperation and other organizations was increased, and more and more people began to consciously reduce the use of disposable items, started to use energy-saving products, and avoided extravagance and waste, etc. These changes not only reflected the diversity and creativity of the public, but also demonstrated important role the public plays in this regard, as well as the potential for future low-carbon issues.

In the 11th FYP period, the overall changes of public behavior in the low-carbon field alongside macro-level environmental policies took on a momentum, but still had deficiencies:

(1) Policy participation: In terms of participation, social organizations in the low-carbon field (such as NGOs) were very active, but actions by individual citizens did not yet have not sufficient impact. The main reason is that low-carbon development issues are often based on the positive orientation and the nature of public goods and are lacking incentives for ordinary people. In terms of specific issues involved, only issues that may hurt the interests of certain individuals are able to receive attention and create active participation from the public (NIMBY phenomenon), but constructive public participation is relatively insufficient. The issues mentioned above, including the refined oil tax reform and instances of power rationing, caused active public participation because they were closely related to public interests. Because of insufficient decision-making mechanisms and participational awareness of the general public, the process of constructive participation often was supplemented by social elites through organizations, such as NGOs, the people's congresses, and the CPPCC.

(2) The low-carbon awareness was further popularized on the individual level, but the form and diversity of the implementation of low-carbon actions was extremely wide, including above-mentioned collective

actions, Internet micro-campaigns in the field of new media, and indirect impact of practices and demonstrations. As far as low-carbon development is concerned, these approaches demonstrate the diversity and creativity of the public, but many avenues of participation have not yet been fully explored because of the inadequacy of related public knowledge.

CHAPTER TEN

OUTLOOK OF CHINA'S ENERGY CONSERVATION AND EMISSIONS REDUCTION IN THE 12TH FYP PERIOD

Abstract: The 12th FYP's low-carbon target is greener and more rationally decomposed and allocated with clearer orientation than that of the 11th FYP. In regards to policy execution, the 12th FYP takes into account total energy consumption control, low-carbon city and low-carbon province pilot projects, experimental carbon trading schemes, and other innovative regimes and programs such as the "pioneer" standard system.

China's low-carbon development will face more rigorous challenges in the 12th FYP period; that is to say, it will become more difficult for China to achieve the mission of energy conservation. This is not only because energy-saving indicators are more stringent in the 12th FYP, but also because energy consumption per unit of GDP has limited capacity to decline and the marginal cost of energy conservation and emissions reduction will continue to rise. Moreover, the fact that China is undergoing a rapid industrialization and urbanization will continue to drive the growth rate of total energy consumption; the increase of export and household energy consumption will further accelerate this process. Local governments' ambitious GDP target will cause their energy consumption levels to be considerably higher than the national total energy consumption control targets.

Furthermore, non-fossil energy target in the 12th FYP faces demanding challenges as well. Since the rapid development of non-fossil energy is not in accordance with its quality and cost-effectiveness, many problems and constraints have emerged during the development process. The immense amount of energy consumption demand will further impede the share of non-fossil energy in total energy consumption from scaling up.

The gap between China's energy consumption and carbon emissions and its US counterpart will continue to widen in the 12th FYP. This may raise an unprecedented amount of international pressure; therefore, China's low-carbon development calls for more strategic action.

The 12th FYP period is a crucial period for China to build a moderately prosperous society, deepen its reform, and accelerate the transformation of its economic growth. The international and domestic political environment will continue to undergo tremendous changes in the 12th FYP period. From an international perspective, the financial crisis had far-reaching impact and slowed down the global economic recovery. Significant structural changes in global demand, climate change, and energy security issues have become more prominent and given rise to a more complex external environment for China's development. Domestically, as China's industrialization and urbanization process continues, many issues remain unsolved, such as resource shortages and environmental constraints, the imbalance between investment and consumption, and an unreasonable sectoral structure. The 12th FYP period faces a number of predictable and unpredictable risks and challenges, as well as a historical opportunity. China needs to adapt to new changes in the domestic and international environment, accelerate the transformation of its economic development, and adhere to constructing a resource-saving and environment-friendly society.

10.1 Comparison of Targets and Policies in the 12th FYP and the 11th FYP

Looking back at previous FYPs, the adjustment of China's economic structure was first mentioned in the 9th FYP and continued to be elevated to an ever higher status ever since. The "National Economic and Social Development 12th FYP," hereafter referred to as the 12th FYP, proposes important guidelines such as to "regard the strategic economic structure adjustment as the main direction to accelerate the transformation of economic development" and "adhere to build a resource-saving and environment-friendly society as an important way of changing the pattern of economic development." The 12th FYP makes overall arrangements for energy saving and emissions reduction. Moreover, based on the basic state policy of resource conservation and environmental protection, the plan proposes to "improve incentives and restriction mechanisms for energy saving" as a policy direction and carry out energy conservation and greenhouse gas emissions intensity reduction throughout the production, circulation, consumption, and construction processes of various fields. This highlights the importance, urgency, and difficulty of energy saving and emissions reduction.

The "12th FYP Comprehensive Scheme of Energy Saving and Emission Reduction" specified the detailed deployment of energy conservation and emissions reduction, including strengthening the target accountability system, adjusting and optimizing sectoral structures, implementing key projects of energy saving and emission reduction, strengthening energy conservation management mechanisms, developing a circular economy, accelerating energy-saving technology development and applications, improving energy-saving economic policies, and strengthening supervision and inspection of energy conservation, as well as strengthening energy conservation capacity building and mobilizing the whole society to participate in energy-saving and emissions reduction work.

10.1.1 *Target Comparison*

Compared to the 11th FYP, the 12th FYP adds three new binding targets: carbon dioxide emissions per unit of GDP, the share of non-fossil energy in primary energy consumption, and forest reserves (Table 10.1). The 12th FYP's low-carbon target is greener and more rationally decomposed and allocated with clearer orientation than that of the 11th FYP.

10.1.1.1 *Clearer Orientation*

China's average annual GDP growth rate was 11.2% during the 11th FYP; the expectation of a 7% average annual GDP growth rate for the 12th FYP fully demonstrates China's determination to shift its economic development mode from rapid growth to sustainable growth. Controlling the economic growth, which means to control energy consumption without restricting economic growth, signals China's transition from emphasizing the speed of economic growth towards the quality of economic growth.

Table 10.1. Comparison of low-carbon targets between the 11th FYP and the 12th FYP.

Targets	11th FYP	12th FYP	Nature
Annual GDP Growth	7.5%	7%	Projected
Decrease in energy consumption per unit of GDP	20%	16%	Binding
Decrease in CO_2 emissions per unit of GDP	NA	17%	Binding
Share of Non-fossil fuel in primary energy consumption	NA	11.4%	Binding
Forest coverage (%)	20%	21.66%	Binding
Forest growing stock	NA	Increase from 600 million m^3 to 14.3 billion m^3	Binding

Sources: "National Economic and Social Development Twelfth Five-Year Plan" and "National Economic and Social Development Eleventh Five-Year Plan."

The slowing down of GDP growth helps to adjust China's sectoral structure and economic development mode. Many imbalances exist in China's economic structure right now: the development of the tertiary industry is lagging behind; economic growth excessively relies on the secondary industry; insufficient economic growth is driven by consumer consumption; the investment structure is irrational and consumption of material resources is excessive; energy security and environmental issues have become bottlenecks for sustainable economic development. Reasonable economic growth is conducive to adjusting the economic structure and alleviating aforementioned imbalances.

10.1.1.2 *Greener Targets*

The 11.4% non-fossil energy share in total primary energy consumption was included in the 12th FYP as a binding target for the first time in history, which shows China's commitment to transform its energy structure towards a more diversified and cleaner direction. In addition, the 12th FYP included a new target: reducing carbon dioxide emissions per unit of GDP by 17%. This target indicates that China has effectively integrated climate change issues into its economic and social development planning.

10.1.1.3 *More Rationally Decomposed and Allocated Targets*

Different from the "one size fits all" target decomposition method of the 11th FYP, the 12th FYP divided the country into five regions on the basis of the level of economic development, sectoral structure, energy-saving potential, environmental capacity, the national sectoral layout, and other factors. The national energy intensity reduction target was broken down

Table 10.2. Energy intensity target allocation of the 12th FYP.

Energy intensity reduction targets	Regions
18%	Tianjin, Shanghai, Jiangsu, Zhejiang and Guangdong
17%	Beijing, Hebei, Liaoning and Shandong
16%	Shanxi, Jilin, Heilongjiang, Anhui, Fujian, Jiangxi, Henan, Hubei, Hunan, Chongqing, Sichuan and Shaanxi
15%	Inner Mongolia, Guangxi, Guizhou, Yunnan, Gansu and Ningxia
10%	Hainan, Tibet, Qinghai and Xinjiang

Source: "12th FYP Comprehensive Scheme of Energy Conservation and Emission Reduction" (2011: 26), http://www.gov.cn/zwgk/2011-09/07/content_1941731.htm [accessed: 2011-09-14].

into various regions and sectors. Regions in different developmental stages were assigned different binding targets. More specifically, less developed areas were granted some leeway to allow for their economic development (Table 10.2).

10.1.2 *Innovation, Development, and Inheritance of Low-Carbon Policies and Institution*

The innovation with regards to low-carbon policies and institution in the 12th FYP is reflected in four aspects: the rational control of energy consumption, low-carbon cities and low-carbon provinces pilot projects, a carbon emission trading pilot scheme, and the "pioneer" standard system. Table 10.3 summarizes the specific measures of the 12th FYP's policy innovations.

Table 10.3. Innovation of policy and institution in the 12th FYP.

Policy innovation	Specific measures
Rational control of energy consumption	Control total energy consumption, allocate target to provincial governments Implement a target accountability system Conduct an energy conservation assessment review for fixed asset investment projects Establish a total energy consumption forecast and warning mechanism Develop total coal consumption control pilot projects in atmospheric-pollution joint prevention and control areas
Low-carbon cities and low-carbon provinces pilot project	Implement the "five provinces and eight cities" low-carbon pilot project Develop low-carbon industries Establish greenhouse gas emission data accounting and management system Promote a low-carbon lifestyle
Carbon emissions trading pilot project	Carry out a carbon emissions trading pilot project Establish mechanisms for emissions reduction Promote the construction of a carbon emissions trading market
"Pioneer" standard system	Determine the advanced level of products energy efficiency Combine the "pioneer" energy efficiency standards with the new project review assessment, and promotion and application of energy-saving products

Sources: Summarized from "12th FYP Comprehensive Scheme of Energy Conservation and Emission Reduction" and "National Economic and Social Development Twelfth Five-Year Plan."

10.1.2.1 *Rationally Control Total Energy Consumption*

The Fifth Plenum of the 17th Communist Party of China Central Committee and the Central Economic Conference proposed the rational control of total energy consumption for the first time in history. The National Energy Bureau has initiated a tentative plan to control fossil fuels consumption, especially coal consumption, and initially proposed the decomposition method of "verify base number, and decompose incremental consumption." The other two alternatives are direct control of energy consumption and limiting total fossil energy consumption as well as the share of fossil energy consumption in total energy consumption.[1] The control of total energy consumption is a necessary complement to the targets of energy intensity, which will affect the GDP growth rate, energy structure, energy prices, GDP sectoral structure, migration of energy-intensive industries, and other important economic and social aspects. A fixed target of total energy consumption will encourage local governments to focus on energy-intensive industries for GDP growth, and form a mechanism that prevents excessive growth of energy-intensive industries, which will contribute to the transformation of the economic development model.

10.1.2.2 *Low-Carbon Cities and Low-Carbon Provinces Pilot Project*

The NDRC issued a "Notice on Low-carbon Provinces and Cities Pilot" (NDRC [2010] No. 1587) in July of 2010 which explicitly specified the construction of low-carbon cities, in the "five provinces and eight cities," namely Guangdong, Liaoning, Hubei, Shaanxi, and Yunnan provinces and eight municipalities including Tianjin, Chongqing, Shenzhen, Xiamen, Hangzhou, Nanchang, Guiyang, and Baoding. These pilot areas will develop low-carbon industries, establish greenhouse gas emissions data management systems, and advocate low-carbon lifestyles. The implementation of this pilot project will help mobilize enthusiasm for low-carbon lifestyles and gather experience in different regions, which will be an important measure to control greenhouse gas emissions.

10.1.2.3 *Carbon Emissions Trading Pilot Project*

The implementation of energy conservation and emissions reduction during the 11th FYP, particularly the decomposition and allocation of energy intensity indicators among regions and the cross-level assessment

[1] British Petroleum Company, *Statistical Review of World Energy 2011*. http://www.bp.com/sectionbodycopy.do?categoryId=7500&contentId=7068481 [accessed 2011-08-05].

approach have achieved great results. However, the main policy instruments are still government-led, and market mechanisms did not play a significant role in the rational allocation of resources. Carbon emissions trading will be an important entry point for the introduction of more market-based approaches in the 12th FYP period.

At present, economically developed regions and less developed regions face different pressures in terms of emissions reduction. Economically developed regions have lower energy intensity but less potential for further energy savings, while less developed areas have a large potential to reduce carbon emissions but also are in need of economic development.

The essence of carbon emissions trading mechanisms is to allocate resources (capital, technology) to places with the highest potential for emissions reduction and the lowest opportunity costs, so as to make the best use of the emissions reduction potential of less developed regions, share some emissions reduction responsibility for developed regions, and promote the optimization of a variety of resources as well as improve economic efficiency.

An effective operation of carbon emissions trading requires a range of conditions, such as the establishment of regional and business carbon dioxide emissions accounts, and the monitoring, verification, and reporting of carbon dioxide emissions. With the help of key enterprises and government sectors, energy conservation of the 11th FYP has laid some foundations in terms of technology, equipment, and human resources for energy monitoring, reporting, and auditing. Relevant ministries and local governments also accumulated some experience in energy accounting, examination, verification, and monitoring, established energy consumption accounting, examination and verification systems, and cultivated the relevant organizations, institutions and personnel. Given the existing organizational, technical, and personnel reserve structures, the carbon emissions trading system of the 12th FYP will not be a project built from scratch.

10.1.2.4 *"Pioneer" Standard System*

The "Pioneer" standard system sets the reference level of energy efficiency leaders in energy-intensive industries as well as refrigerators, washing machines, tablet PCs, and other end-use products with a clear time frame for implementation. The "pioneer" system will also be an important basis for the energy evaluation and promotion of energy-saving products for a new project. The establishment of the system means that China will enhance its energy-saving standards system and strictly enforce the labeling and authentication of low-carbon products.

In addition to these policy innovations, the 12th FYP will continue to deepen the price reform of resource products, rationalize pricing mechanisms for coal, electricity, oil, gas, water, mineral and other resources, accelerate the implementation of a graduated electricity tariff for residential housing, improve peak-trough TOU power pricing, enhance differential power pricing and the intensity of punitive tariffs, so as to form a price mechanism that will objectively reflect the scarcity of resources and the relationship between supply and demand. Moreover, the 12th FYP will actively promote the reform of resource taxes to shift taxes on crude oil, natural gas, and coal from being levied according to quantity towards ad valorem taxation, while the tax rate will be increased to an appropriate level. It also proposes to study and formulate favorable tax incentives for necessary key components and raw materials of large energy-saving equipment manufacturing. The 12th FYP will maintain the 11th FYP's operating standards and level of strictness regarding the target responsibility system. The energy-saving measures of the "1,000 Enterprises Campaign" will be extended to encompass millions of enterprises. The four key projects of energy conservation of the 12th FYP are energy-saving technology industrialization demonstration projects, energy-saving technological improvement project, subsidizing consumers for energy-saving products projects, and energy management contract projects.

10.2 A More Difficult Mission of Energy Conservation

Compared to the 11th FYP, the target of the 12th FYP is harder to achieve: First, targets are more stringent; secondly, energy consumption per unit of GDP has a limited capacity for further reduction, and the marginal cost of energy conservation and emissions reduction will continue to rise; third, the fact that China is undergoing rapid industrialization and urbanization will continue to drive the growth rate of total energy consumption in the 12th FYP, and the increase of export and households' energy consumption from construction and transportation will accelerate this process. These factors will form a substantial challenge to the control of total energy consumption. In addition, local governments' ambitious GDP targets will cause their energy consumption levels to be considerably higher than the national total energy consumption control target, which will lead to conflicts with the central government which wants to optimize China's economic structure.

10.2.1 *Targets are More Stringent*

From a purely numerical point of view, the 12th FYP's target to reduce the energy consumption per unit of GDP by 16% is not as ambitious as the 20% target of the 11th FYP, but in fact achieving the target in the 12th FYP will be much more difficult than that of the 11th FYP.

If the average annual GDP growth rate of the 12th FYP is higher than 8.9%, a completion of the 16% energy intensity reduction targets would require energy savings of 670 Mtce; while the 11th FYP achieved 630 Mtce in energy savings to fulfill its energy intensity reduction target of 20%. During the 11th FYP, the average annual energy consumption growth rate of 6.6% supported the national average economic growth rate of 11.2%, while in the 12th FYP, an average of 3.3% growth rate of energy consumption needs to support 7% of the national economic growth, which means that the energy consumption elasticity coefficient will decrease from 0.59 down to 0.47 (Table 10.4), making the task more difficult.

As far as emissions reductions are concerned, the 12th FYP will reduce 1.6 billion tons of CO_2, more than the 1.49 billion tons of CO_2 reduced during the 11th FYP. According to the centrally planned targets, the expected energy savings and emissions reduction of each year during the 12th FYP is shown in Figure 10.1.

Table 10.4. Comparison of energy elasticity indicator between the 11th FYP and the 12th FYP.

Facts and Calculation	11th FYP (actual)	12th FYP (expected)
Energy Intensity Reduction	19.1%	16%
Annual GDP Growth	11.2%	7%
Annual energy consumption growth	6.6%	3.3%*
Energy consumption elasticity	0.59	0.47*
Cumulative energy savings (10^8 tce)	6.3	6.4* (6.7 from "Scheme Plan")
Cumulative CO_2 reduction (10^8 ton)	14.9	16.0*

Note: * dictates calculated figure from 7% annual GDP growth and 16 percent of the unit GDP energy consumption reduction.

Source: National Bureau of Statistics, *Summary of 2011 China Statistical Yearbook* (Beijing: China Statistics Press, 2011).

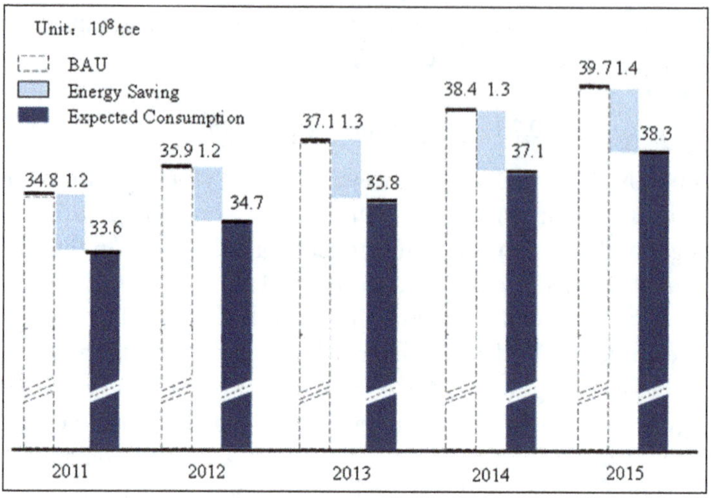

Figure 10.1a. Year-on-year energy savings 2011~2015 calculated according to the 12th FYP targets.

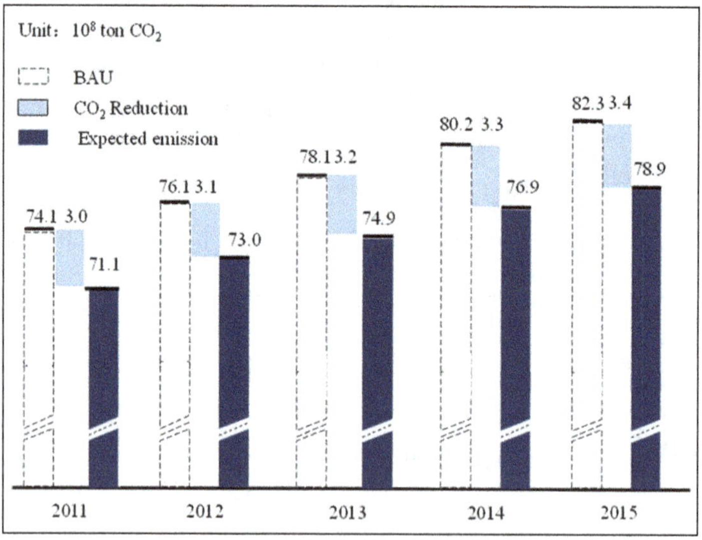

Note: The expected energy consumption and CO_2 emissions are calculated according to the 12th FYP: 7% average annual GDP growth and 16% energy consumption per unit of GDP reduction. The BAU is calculated on the basis of previous year's unit energy consumption and previous year's unit CO_2 emissions respectively; energy saving is the difference between the BAU and the expected energy consumption; and the CO_2 reduction is the difference between the BAU and the expected CO_2 emissions.

Figure 10.1b. Year-on-year CO_2 reduction 2011~2015 calculated according to 12th FYP targets.

10.2.2 *Marginal Cost Rise*

As low-carbon technologies underwent rapid development, the energy consumption of main energy-consuming products decreased significantly. The gap between China and the advanced international level in terms of energy consumption per unit of industrial products narrowed; clean coal combustion technologies, renewable energy technologies, modern coal chemical processing technologies, large complete sets of equipment of cement production technology with daily production up to 5,000 tons, the new electrolytic aluminum cathode cell manufacturing process and some other energy-saving technologies have reached advanced international standard.[2]

A number of low-carbon technologies with low or even negative costs have already been applied in the 11th FYP (see Chapter 2 of this report). To further reduce energy consumption on the basis of existing achievements, technologies with relatively high costs and low utilization rates will need to be adopted, which will involve higher costs. For the 12th FYP, as the "1,000 Enterprises Campaign" is extended to millions of enterprises, the cost of emissions reduction through technological improvements will rise.

During the 11th FYP, the "low-hanging fruits" have been picked for projects such as elimination of obsolete production capacities and replacement of small-scale thermal power units with large-scale ones. China has eliminated a large number of obsolete production capacities, including outdated iron production capacity of 110 Mt, steel around 68 Mt, cement 330 Mt, coke 100 Mt, paper manufacturing 10.3 Mt, and glass 38 million weight cases, which accounts for about 50% of the total outdated production capacity.[3] For the 12th FYP, the room for improvement in energy efficiency through elimination of outdated production capacity has shrunk and the costs of elimination have become greater.

Compared to the scale of small thermal power capacity that has been shut down in between 2006 and 2010 (Figure 10.2), it is easy to find that from 2006 to 2008, 61% total eliminated capacities are less than 50,000 kW, while in 2010, only 32% of total eliminated capacities were

---

[2] National Development and Reform Commission, "'Shiyiwu' shiqi woguo jingji jiegou tiaozheng qude zhongyao jinzhan '十一五'时期我国经济结构调整取得重要进展 [China's Economic Restructuring Made Important Progress during the 11th FYP]" (2011a), http://www.gov.cn/gzdt/2011-03/11/content_1822557.htm [accessed: 2011a-07-20].

[3] Ibid.

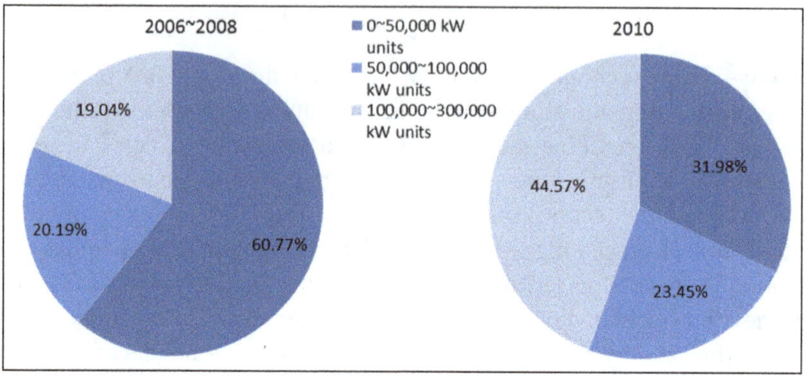

Sources: CPI, 2011.

Figure 10.2. Share of different power units in replaced obsolete coal power plants for the 11th FYP period.

less than 50,000 kW. Meanwhile, from 2006 to 2008, 20% of total eliminated capacities were between 100,000 to 300,000 kW, while in 2010, 45% of total eliminated capacities were between 100,000 to 300,000 kW. This shows that the potential to save energy and reduce carbon emissions by shutting down small thermal power capacities is declining.

The deputy director of the National Panel on Climate Change, He Jiankun, has pointed out that compared to the 11th FYP, the task of energy conservation and emissions reduction becomes more and more difficult and more financial and technical inputs are required.[4] The deputy director of the NDRC, Xie Zhenhua, also said that China's energy consumption per unit of GDP fell by 47% between 1990 and 2005, and now the target is to reduce the carbon intensity by 40–45% between 2005 and 2020, which means that energy conservation and the improvement of energy efficiency will be more and more difficult in the future.[5]

[4] Yi Rongrong, Li Xiaoming 易蓉蓉, 李晓明, "He Jiankun: Nengyuan xiaofei zongliang buyao tupo 40 yi dun biao mei de tianhuaban 何建坤: 能源消费总量不要突破 40 亿吨标煤的天花板 [He Jiankun: The Total Energy Consumption Must Not Exceed the Ceiling of 40 million Tons of Coal]," *Kexue Shibao* (2011-04-11), http://www.cas.cn/xw/zjsd/201104/t20110411_3109957.shtml [2011-05-25].

[5] Xinhua, "Danwei GDP tan qiangdu jiangdi 40–45% shi woguo ziyuan、 zizhu de xingdong 单位 GDP 碳强度降低 40–45% 是我国自愿、自主的行动 [A 40–45% Reduction in Carbon Intensity Per Unit GDP Is A Voluntary, Independent Step by China]" (2009-11-26), http://news.xinhuanet.com/video/2009-11/26/content_12546090.htm [accessed: 2011-5-25].

10.2.3 *Total Energy Consumption Continues to Grow*

In recent years, China has emerged as one of the world's biggest energy consumers, ranking first in terms of both total primary energy production and consumption. Coal production has surpassed three billion tons. China's dependence on foreign oil resources exceeded the 50% warning level in 2009, reached 52% for the first time and further rose to 53.8% in.[6] China's dependence on foreign oil was 54.8% in the first half of 2011, 1 percent higher than the 2010 level.[7] In the 12th FYP, electricity consumption will continue to grow with an average annual growth rate of around 8–9%; construction and transportation energy use will continue to grow as well.[8] Overall, the widely-accepted prediction that China will continue to experience rapid economic development for a long period of time implied the arising energy demand for the country in the next decades.

10.2.3.1 *Industrialization and Urbanization*

China is in a stage of rapid industrialization and urbanization. Rapid industrialization will lead to a dramatic change of the economy in the central and western regions and shape China's regional economic structure during the 12th FYP. The eastern region has been developed to a certain stage of industrialization during the 11th FYP, and the heavy industry development has started to slow down, while the central and western regions continue to attract heavy industries due to low labor costs and abundant resources. China's regional energy intensity decreases when moving from the west, to the northeast, to central and finally eastern parts of China, which means that China's energy demand is bound to rise rapidly since the heavy industry will be the major driving force behind the economic development of the western region.

---

6 Zhongguo Nengyuanbao, "2010 nian woguo jinkou yuanyou 2.39 yi dun, duiwai yicundu bijin 55% 2010 年我国进口原油 2.39 亿吨, 对外依存度逼近 55% [2010 China Imported 239 Million Tons of Crude Oil, Dependence On Foreign Oil Approaching 55%]" (2011-01-17), http://paper.people.com.cn/zgnyb/html/2011-01/17/content_725182.htm?div=-1 [accessed: 2011-01-17].

7 National Development and Reform Commission, "7 yuefen chengpin you xiaofei jixu baochi jiaogao shuiping 7 月份成品油消费继续保持较高水平 [Refined Oil Consumption Continues to Maintain a High Level During July]" (2011b), http://www.ndrc.gov.cn/jjxsfx/t20110815_428379.htm [accessed: 2011-08-16].

8 Industrial Economics Research Department of the Development Research Center of the State Council, *'Shierwu' tan qiangdu jianpai mubiao、 tujing he zhengce yanjiu* '十二五' 碳强度减排目标、途径和政策研究 [Research on Carbon Intensity Reduction Targets, Methods and Policies During the 12th FYP] (Beijing: Nengyuan jinjihui, 2011).

The 11th FYP underwent a rapid development in urbanization. According to the sixth census announced in April 2011, China's urban population is now close to 666 million, with an urbanization rate reaching 49.68%, 6.68 percent higher than that of 2005. From the scale of China's urbanization, both of the amounts of annual net increase and the total urban population will be the largest in the world for a long term.[9] The 12th FYP requires the urbanization rate to increase by 4 percent to achieve 53.68% urbanization rate by 2015. Large-scale urbanization means more demand for new housing and urban infrastructure which will ultimately increase energy demand.

10.2.3.2 *Export*

Although China is the world's largest exporter, China's exports of products are mostly low-end manufacturing goods with high energy consumption and low added value. The unit GDP energy consumption of export products is higher than that of the national level. The heavy dependence on an export-dominated economic development mode and the irrational structure of export products seriously prevented China from being a strong economic power. However, the adjustment of export structure requires a long time period. Although China's export growth will slow down due to the economic crisis in the U.S. and the Euro zone, the absolute amount of export will maintain its rising trend (Figure 10.3) which means that the export-led pressure on energy demand will remain high during the 12th FYP.

The 12th FYP proposed to increase the proportion of low energy-consuming services in foreign trade. Affected by global economic weakness and the impact of trade protectionism, export growth slowed down;[10] therefore, China should seize the opportunity to accelerate the process of improving and balancing its economic development structure, and reduce the energy consumption per unit of export.

---

[9] CASS Urban Development and Environment Institute, *2010 nian chengshi lanpishu* 2010 年城市蓝皮书 [The City Blue Book 2010] (Beijing: Shehui kexue wenxian chubanshe, 2010).

[10] Zhang Ping, Liu Xiahui, Yuan Fuhua, Zhang Ziran 张平, 刘霞辉, 袁富华, 张自然, *Zhongguo jingji zengzhang baogao 2010–2011: 1–2* 中国经济增长报告 2010–2011 [Report on China's Economic Growth, 2010–2011] (Beijing: Shehui kexue wenxian chubanshe 2011): 1–2.

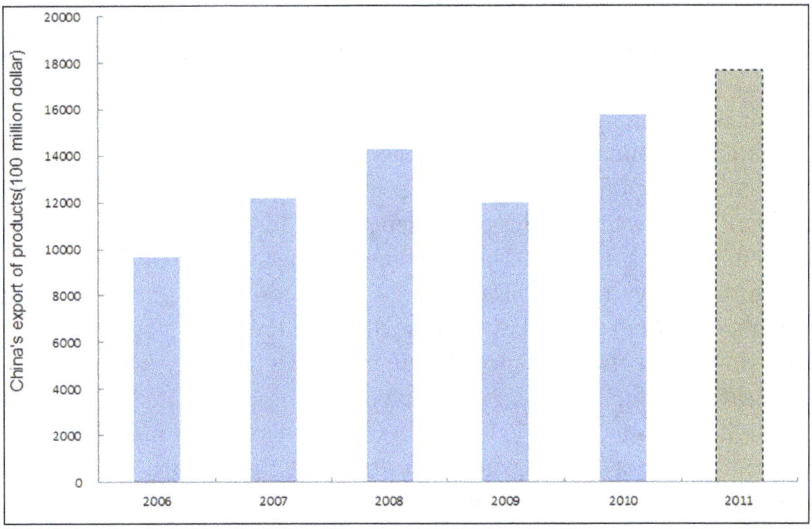

Note: The 2011 export volume is forecasted according to the actual exports during the first half of 2011. Source: National Bureau of Statistics, *China Statistical Yearbook Summary 2011* (Beijing: China Statistics Press, 2011); General Affairs Ministry of Commerce of the People's Republic. Import and Export Situation in June 2011, http://zhs.mofcom.gov.cn/aarticle/aa/201107/20110707640397.html [accessed: 2011-07-20].

Figure 10.3. China's export volume.

10.2.3.3 *Household Energy Consumption*

People's daily behavior is accompanied by energy use and carbon emissions. Since residents have very small incentives to reduce carbon emissions, their behavioral patterns and carbon-intensive lifestyles will cause a sharp rise in energy demand. A recent report noted that "the life-cycle-consumption embedded energy of residents" will have more than doubled by 2030, reaching 2 billion tce.[11]

Currently, China is moving from a low-income country to a middle-income country. Their rising spending power allows residents to meet basic living demands and improve their quality of life. The requirement of living space has increased, and car travel has constantly been replacing public transport, both of which will seriously affect China's low-carbon development. China's energy consumption per capita was 2.4 and 2.3 tce

[11] Xiong Huawen 熊华文, "Jiyu touru chanchu fangfa de jumin xiaofei quan liucheng nenghao fenxi 基于投入产出方法的居民消费全流程能耗分析 [Analysis of Share of Household Energy Consumption Based On Input-Output Method]," *Zhongguo nengyuan* 7 (2009): 17–20.

in 2010 and 2009, respectively, only one-fifth of that of the United States.[12] Therefore, regardless of the rapidly growing energy demand or the potential increase in per capita energy consumption, China will have a high growth rate of energy consumption in the next five years.

To sum up, China's total energy consumption will continue to grow fast for a long period of time. The rapid growth in energy consumption of the 12th FYP will become a direct challenge for China to meet its target of total energy consumption. The total energy consumption of the 12th FYP is determined by the decrease of energy consumption per unit of GDP and annual growth rate of GDP, whereas when the total energy consumption is fixed, if either the GDP growth rate or energy intensity decline rate is fixed, the third factor will be inevitably fixed.

The shadow area of Table 10.5 shows the scenarios that meet 4.1 billion tce primary energy consumption controls.[13] If the average annual GDP growth is kept at 10%, then to achieve total energy control goal of 4.1 billion tce will require a decline in energy intensity by more than 21%; if energy intensity falls by 16%, the average annual GDP growth rate must be controlled to below 8.5%. Reasonable total energy consumption control is a rigid and difficult task, but it may be achieved through the following approaches: (1) restrained GDP growth, which will be helpful for the structural adjustment and quality improvement of its economy; Secondly (2) further decline of energy consumption per unit of GDP. If the 4.1 billion tce energy consumption control objective can be achieved, then the incremental amount of energy consumption in comparison to 2010 (3.25 billion tce) must be within 8.5 Mtce by the end of 2015.

2011 is the first year of China's 12th FYP. According to the National Bureau of Statistics, China's GDP still amounted to 20.4459 trillion Yuan in the first half of 2011, 9.6% higher than the previous period if calculated at comparable prices. The GDP in the first and second quarter increased by 9.7% and 9.5%, respectively. Assuming that the 12th FYP period maintains the momentum of the economic development as in the first half of 2011 (9.6% GDP growth), then in order to control the energy consumption to below 41 Mtce by 2015, the decline in energy consumption per unit of GDP must reach or exceed 20%. To achieve the 16% decline of energy

[12] United States Energy Information Administration, *U.S. Primary Energy Flow by Source and Sector, 2009* (2010), http://www.eia.gov/totalenergy/data/annual/pecss_diagram.cfm [accessed: 2011-08-10].

[13] This total energy consumption control assumption corresponds with the 16% decrease of energy consumption per unit GDP in 2015 compared with 2010, and an average annual GDP growth of around 8.5%.

Table 10.5. Sensitivity analysis: 2015 Primary Energy Consumption under different Scenarios (10^8 tce).

Annual GDP Growth \ Energy Intensity Decline	16%	17%	18%	19%	20%	21%	22%
7.0%	38.3	37.8	37.4	36.9	36.5	36.0	35.6
7.5%	39.2	38.7	38.3	37.8	37.3	36.9	36.4
8.0%	40.1	39.6	39.2	38.7	38.2	37.7	37.2
8.5%	41.0	40.6	40.1	39.6	39.1	38.6	38.1
9.0%	42.0	41.5	41.0	40.5	40.0	39.5	39.0
9.5%	43.0	42.5	42.0	41.4	40.9	40.4	40.0
9.6%	43.2	42.7	42.1	41.6	41.1	40.6	40.1
10%	44.0	43.4	42.9	42.4	41.9	41.3	40.8

intensity as proposed in the 12th FYP is not easy, not to mention the more stringent goal of a 20% decline.

10.2.4 *Local Governments' High GDP Growth Expectation Causes Gap*

Local governments have strong incentives to develop their economies; therefore, national and local governments have very different goals of economic development, where the weighted local average annual GDP growth rate is far higher than the national GDP growth rate. As the total energy consumption is not only impacted by the per unit GDP energy consumption, but also by the impact of total GDP, if local governments all follow their own local goal of economic growth, the sum of all provincial GDP will be much larger than the national goal of GDP set by the central government. That is to say, there will be a great difference in energy consumption, CO_2 emissions, and the corresponding non-fossil energy demand under different circumstances of national and local planning.

China's national average annual GDP growth of the 12th FYP is expected to be 7%, which served as a strong guiding indicator, reflecting China's determination to optimize the economic development mode and sustainable outlook on development to plan economic development and emissions reduction as a whole. But all of China's 31 regions' annual economic growth targets greatly exceed the national target. Among them, the lowest still reaches 8%, and the average is around 10.6% (Table 10.6).

Table 10.6. GDP growth rate expectation for different provinces (%).

Nationwide	7	Jiangxi	11
Beijing	8	Liaoning	11
Guangdong	8	Gansu	12
Shanghai	8	Guizhou	12
Zhejiang	8	Heilongjiang	12
Hebei	8.5	Jilin	12
Henan	9	Inner Mongolia	12
Shandong	9	Ningxia	12
Anhui	10	Qinghai	12
Fujian	10	Shaanxi	12
Guangxi	10	Sichuan	12
Hainan	10	Tianjin	12
Hubei	10	Tibet	12
Hunan	10	Yunnan	12
Jiangsu	10	Chongqing	12.5
Xinjiang	10	Shanxi	13

Source: Summary of the provincial 12th FYP of the 31 regions.

The weighted averages of regional expected GDP growth target are calculated by the weight of local GDPs. And the result shows that the weighted averages of the regional GDP growth targets are between 9.86–9.95% (Table 10.7) between the years 2011 and 2015, much higher than the 7% of the GDP growth target of the central government. Whereas during the 12th FYP the targets for reduction in weighted average of regional energy consumption per unit GDP by 16.8–16.9%, equaling to 3.63–3.62% per year, are slightly higher than the 3.43% target by the central government (16% decline for five years).

As shown in Figure 10.4, there exists a big gap between the weighted averages of local targets and the targets of central planning. If all the provincial governments achieved their own targets, the total energy consumption, CO_2 emissions and corresponding non-fossil energy demand will increase by 500 Mtce, 1.04 billion tons and 57.59 Mtce by the year 2015 than those under central planning. In the whole 12th FYP, the energy gap will be up to 1.4 billion tce, accounting for 40% of the total energy consumption in 2010. And the gap in CO_2 emissions during the 12th FYP will be up to 3 billion tons under different circumstances of national and local planning China's GDP grew by 9.6% in the first half of 2011 over the same period of the previous year, close to the weighted average of regional economic growth target. Since the regional targets of economic growth are all far above the expected economic growth rate of central planning,

Table 10.7. National and provincial development indicators.

	2011	2012	2013	2014	2015
GDP growth target of the central planning	7%	7%	7%	7%	7%
Weighted average of the regional GDP growth targets	9.86%	9.89%	9.91%	9.93%	9.95%
Decline target of the energy consumption per unit GDP of the central planning	3.427%	3.427%	3.427%	3.427%	3.427%
Decline targets of the weighted average of regional energy consumption per unit GDP	3.628%	3.625%	3.623%	3.620%	3.617%

Note: The annual decline target of the energy consumption per GDP of central planning is calculated based on the 16% decline rate of the 12th FYP; the regional weighted average annual GDP growth rate and the regional annual decline target of the energy consumption per GDP are calculated by the weight of regional GDP of that year.

regional pursuit of rapid economic development has not been changed, it is therefore inevitable that all regions will follow their high economic growth rate to guide all kinds of economic activities. The report "Targets, Pathways, and Policy Recommendation for Carbon Intensity Reduction during the '12th FYP' Period in China" expects GDP growth rate to be between 9 to 10%;[14] the report "Energy and Carbon Emissions Projections and Prospects of the 12th FYP" expects GDP growth rate to remain high as around 9.7%;[15] "Target Allocation Methodology for China's Provinces: Energy Intensity in the 12th Five-Year Plan" assumes a GDP growth around 8.5%.[16] If economic growth of the 12th FYP remains at 9.6% like what it is in the first half of 2011, the national energy consumption will

---

[14] Industrial Economics Research Department of the Development Research Center of the State Council, *'Shierwu' tan qiangdu jianpai mubiao、 tujing he zhengce yanjiu* '十二五 '碳强度减排目标、途径和政策研究 [Research on Carbon Intensity Reduction Targets, Methods and Policies During the 12th FYP] (Beijing: Nengyuan jinjihui, 2011).

[15] Liao Hua, Wei Yiming 廖华, 魏一鸣, " 'Shierwu' nengyuan he tan paifang yuce yu zhanwang '十二五'能源和碳排放预测与展望 [Energy and Carbon Emissions Projections and Prospects for the 12th FYP]," *Zhongguo kexueyuan yuankan* 26, no. 2 (2011): 150–152.

[16] Stephanie Ohshita, Lynn Price, Zhiyu Tian. 2011. Target allocation methodology for China's provinces: Energy intensity in the 12th Five-Year Plan: 28 中国省区可用的目标分配方法: 第十二个五年规划的能源强度. 伯克利: 劳伦斯伯克利国家试验室: 28.

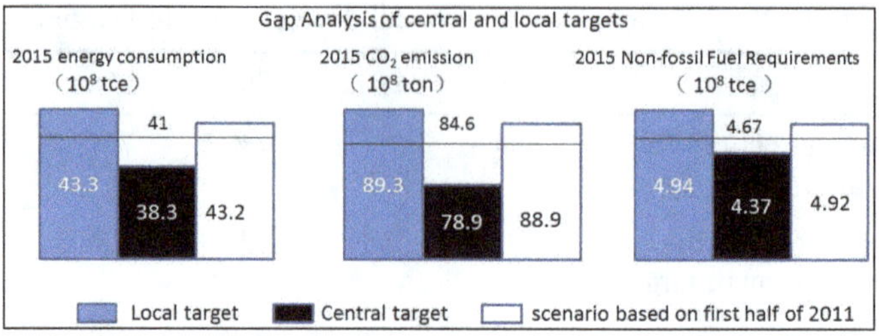

Figure 10.4. 2015 energy saving, CO_2 emission and non-fossil energy demand under different scenarios.

reach 4.32 billion tce by the year 2015; CO_2 emissions will reach 8.89 billion tons and the corresponding non-fossil energy supply needs to reach 492 Mtce so as to meet the non-fossil energy target of 11.4%.

Meanwhile, if the target of total energy consumption control is set at 4.1 billion tce, which corresponds to 84.6 billion tCO_2 emissions and 467 Mtce of non-fossil energy demand, as shown in Figure 10.4, the energy consumption will be far beyond the total energy control target under both scenarios of regional targets and the continuation of economic growth of the first half of 2011.

10.3 Non-Fossil Energy Challenges

The binding target of non-fossil energy has been clearly stated in the 12th FYP; the non-fossil energy should account for 11.4% of primary energy consumption by 2015. To achieve this target, China needs to vigorously adjust its coal-dominated energy structure, support non-fossil energy, and promote diversification of clean energy development. However, there are many problems and constraints on the way towards non-fossil energy development. In addition, the unpredictable nature of the rapid growth in energy use and the gap between non-fossil energy demand and its actual development scale make the challenge greater.

10.3.1 *Problems and Constraints of Non-Fossil Energy Development*

Non-fossil energy has undergone rapid development during the 11th FYP; however some problems were exposed during the process. Hydropower will play a very important role to achieve the non-fossil fuels target, but the environmental problems caused by the construction of hydropower stations become more and more acute. The nuclear accident in Fukushima,

Japan, highlighted the issue of safety in nuclear power development and slowed down the global nuclear power development process. The Chinese government also took relevant measures to eliminate hidden dangers of China's nuclear power industry, and put in place strict supervision on existing nuclear power projects, and suspend approval of new nuclear power projects before the promulgation of nuclear safety regulation. The problem of grid connection of the intermittent wind power bottlenecks further development of the wind power. As a result of the lack of coordination between the smart grid construction and wind farms development, a large number of wind power projects have been disconnected from the grid. Solar photovoltaic (PV) not only faces the same grid connection problem as the wind power does, but also the problem of high cost which prevents large-scale PV from dynamic development. Since both the upstream and downstream industries of China's PV industry are outside China, the national policy is insufficient to boost its domestic PV market. The NDRC's recent issuance of the solar photovoltaic electricity feed-in-tariff policy will facilitate the market development, however, the time frame of this policy is not clear, and many details of this policy need to be improved. Non-fossil energy should be developed on condition of harmonious coordination with China's developmental speed, quality, and efficiency without blind expansion.

The target of non-fossil energy accounting for 11.4% of the primary energy consumption has put forward higher requirements for non-fossil energy development in the 12th FYP. Various restrict factors make the meeting of the target a huge challenge. From the perspective of natural resources, hydropower and biomass are limited by their storage, while wind, solar, and nuclear energy consume large areas of land resources. With regards to human resources, shortages of experienced personnel, outdated management capability, and lacking R&D in the new and renewable energy sector have led to the lack of technological innovation. In addition, the development of non-fossil energy also faces a lack of capital, inadequate facilities, and many other related issues which render the development of high quality non-fossil energy sources an arduous task.

10.3.2 *The Unpredictability of Total Energy Consumption Makes the Non-Fossil Energy Target Hard to Achieve*

The scaling-up of non-fossil energy share in primary energy consumption and appropriate control of the excessive growth of fossil energy consumption are realistic measures to adjust energy structure and reduce China's

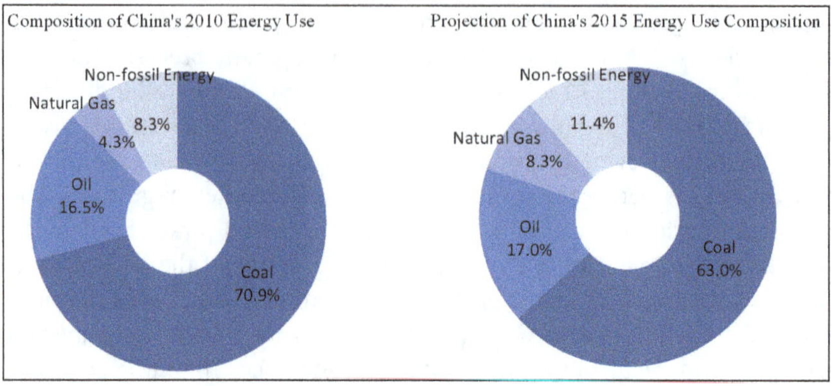

Source: Energy structure in 2010: 2011 National Bureau of Statistics, *China Statistical Yearbook 2011*, Beijing: China Statistics Press, 2015 forecast: Eurasia Group, 2011. China's great rebalancing act. Washington D.C.

Figure 10.5. Projection of China's primary energy structure.

carbon emissions. If the energy structure of the 12th FYP is successful adjusted according to plan, it will be possible for China's energy structure to achieve positive changes shown in Figure 10.5, where the proportion of non-fossil energy consumption will rise from the current 8.6% to 11.4% in the end of 2015.

The biggest challenge in achieving non-fossil energy target of the 12th FYP is the uncertainty of total energy consumption. The energy consumption will be only 3.8 billion tce if the national GDP growth is 7% and energy intensity drop by 16% during the 12th FYP as the central plan has set. In this case, 430 Mtce of non-fossil energy will be needed to meet the 11.4% non-fossil energy target. However, if GDP growth turns out to be between 8.5% and 10%, the energy consumption then will reach 4.1 to 4.4 billion tce in 2015. To meet the 11.4% non-fossil energy consumption goal, the corresponding non-fossil energy needs to increase to about 467–502 Mtce. The uncertainty and unpredictability of total energy consumption implies that even if non-fossil energy sources all achieve their intended target of absolute volume, it is still insufficient to ensure the realization of the 11.4% non-fossil energy target.

10.3.3 *Shortfall of Non-Fossil Energy*

The NDRC promulgated the "Mid- and Long-Term Development Plan for Renewable Energy" in 2007 which set up the development goals for hydro, biomass, wind, solar and other renewable energy sources for 2010 to 2020, however the 11th FYP has seen great development of non-fossil energy sources. The cumulative power generation of hydropower, nuclear power, and wind power reached more than 3 trillion kWh in the past five years.[17] The goal of the "Mid- and Long-Term Development Plan for Renewable Energy" fell behind the rapid development of some renewables, therefore the "emerging energy industry development plan" and the "12th FYP renewable energy development plan" were drafted accordingly, and are currently pending approval by the State Department. The "Emerging energy industry development plan" will clarify the implementation path, scale of development, and major policy initiatives for new energy development and utilization, and traditional energy upgrades. The plan is expected to put into place at the end of 2011, and will give rise to a boom in new energy investment. It is estimated that China's new energy investment will reach 5 trillion RMB in the next decade.

According to the literature review and summary of relevant plan information, by the year 2015, the installed capacity of hydropower will be up to 250–284 GWs (Chinese Academy of Engineering, 2010, and the tentative power industry 12th FYP); nuclear power will reach 40 GWs;[18] wind power

Table 10.8. Electricity generation by major non-fossil energy type by the end of the 12th FYP.

Non-fossil energy type	Effective full power hours	Installed capacity (GW)	Generating capacity (TWH)
Hydro-power	3256[19]	250–284	814–925
Nuclear power	7100	40	284
Wind power	1612	100–150	161–242
Solar PV	1198	5–10	6–12

Note: the effective full power hours of various non-fossil energy sources are calculated according to the new installed capacity and new generating capacity of 2010, so the effective full power hours already include the efficiency factor. Also, the generating capacity = installed capacity × effective full power hours.

[17] Newenergy.com, "Zhongguo lizheng dao 2015 nianmo feihuashi nengyuan zhan bi da 11.4% 中国力争到 2015 年末非化石能源占比达 11.4% [China Strives to Let Share of Non-Fossil Energy Sources to Reach 11.4% by End of 2015]" (2011-01-07), http://www.newenergy.org.cn/html/0111/171138162.html [accessed: 2011-6-23].

[18] State Power Information Network, "Zhongguo hedian zhuangji rongliang 2015 nian jiang guodu 4000 wan qianwa 中国的核电装机容量 2015 年将超过 4000 万千瓦 [Installed Capacity of China's Nuclear Power Will Exceed 40 Million Kilowatts by 2015],' (2011-05-17), http://www.sp.com.cn/dlzj/zc/201105/t20110517_178464.htm [2011-07-20].

[19] Effective full power hour of hydropower is calculated on the basis of 19:31 ratio of small scale hydropower and medium-to-large hydropower. The effective full power hours of small hydropower and mid-large hydropower are 3464 hours and 3129 hours respectively.

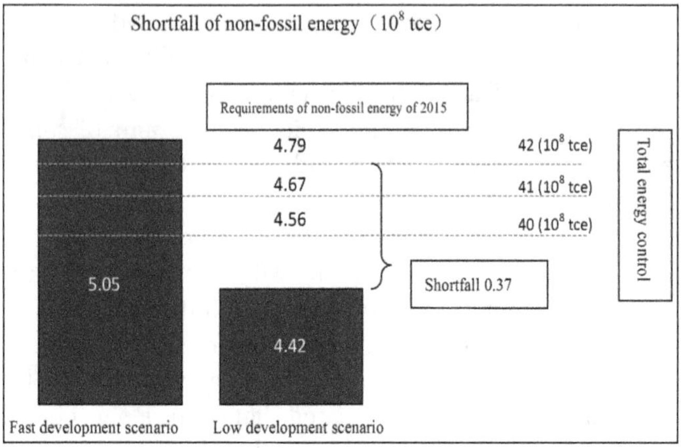

Figure 10.6. Non-fossil energy gap under different scenarios of total energy control targets.

will be up to 100–150 GWs and solar PV power generation will range from 5–10 GWs.[20] The power generated by the above mentioned four types of non-fossil energy ranges from 1265 to 1463 TWh, which equals about 402–465 Mtce (Table 10.8). If we take the 40 Mtce of various types of biomass utilization into account,[21] then non-fossil energy capacity will total up to 442–505 Mtce by the end of 2015. The development of non-fossil energy is urgent, in that there will be a 37 Mtce shortfall of non-fossil energy if the scenario of slow development should come into effect when the total energy consumption reached 4.2 billion tce (Figure 10.6).

10.4 International Pressure on the 12th FYP

The 12th FYP is very important for China's low-carbon development. The central government attaches great importance to China's emissions reduction. A retrospect of previous FYPs shows that the Chinese government's awareness of emissions reduction increased year by year and the task of

[20] Li Junfeng, Cai Fengbo, Tang Wenqing, et al. 李俊峰, 蔡丰波, 唐文倩等, *Fengguang wuxian—Zhongguo fengdian fazhan bobao: 80*, 2011 风光无限—中国风电发展报告 2011 [Infinite Possibilities—2011 Report on China's Wind Energy Development] (Beijing: Zhongguo Huanjing kexue chubanshe 2011): 80.

[21] Li Yongqiang 李永强, "'Shierwu' mubiao jiben queding. Shengwu zhineng youwang yinglai fazhan liangji '十二五'目标基本确定 生物质能有望迎来发展良机 [12th FYP's Target Basics Are Determined. Great Development Opportunities For Biomass Expected]," *Zhongguo nengyuan bao* (2011-07-20).

energy conservation and emissions reduction was also placed in an increasingly important position. The time period of the 12th FYP overlaps with the election of China's next-generation leaders in 2012. Since the election is part of collective decision-making process, and China has always strived for policy continuity, the smooth transition of government will not affect China's efforts on energy conservation and emissions reduction. Internationally, China is facing great international pressure for its continued rise in carbon emissions. Under the framework of the "United Nations Framework Convention on Climate Change" and the Kyoto Protocol, following the "common but differentiated responsibilities" principle, China is taking proactive steps to promote China's low-carbon development.

10.4.1 *The Gap between China and the U.S. Carbon Dioxide Emissions Widens*

As a developing country, the fact of China being at the stage of rapid development determines that China's CO_2 emissions will continue to rise in the near future.

The global economy now is still in the recovery phase of the 2008 global financial crisis, and faces all kinds of challenges from global inflation to the debt crisis of the United States, and the Euro zone. The report "World Economic Situation and Prospects 2011" of the United Nations said that the growth of world economy grew by 3.6% in 2010 and projected that the growth rate will decline to 3.2% in 2011 and 3.5% in 2012. Developing countries are the main driving forces of future world economic growth.[22] The "Asia Economic Monitor," conducted by Asian Development Bank, estimated that China's economic growth will slow down gradually, but will remain at 9.6% and 9.2% in 2011 and 2012.[23] This implies that although the CO_2 emission of the United States is increasing, the growth of China's CO_2 emission is even faster. Since China surpassed the United States to become the world's largest CO_2 emitter in 2007, the carbon emissions gap between China and United States will continue to widen during the 12th FYP.

---

[22] Grigor Agabekian, Tarik Alami, Clive Altshuler, et al., *World Economic Situation and Prospects 2011* (New York: United Nations Publication, 2011): 147.

[23] Asia Development Bank, *Asia Economic Monitor-July 2010* (Mandaluyong City, Philippines: Asian Development Bank, 2010): 34. http://beta.adb.org/sites/default/files/pub/2010/Jul_AEM_complete.pdf [accessed: 2011-08-01].

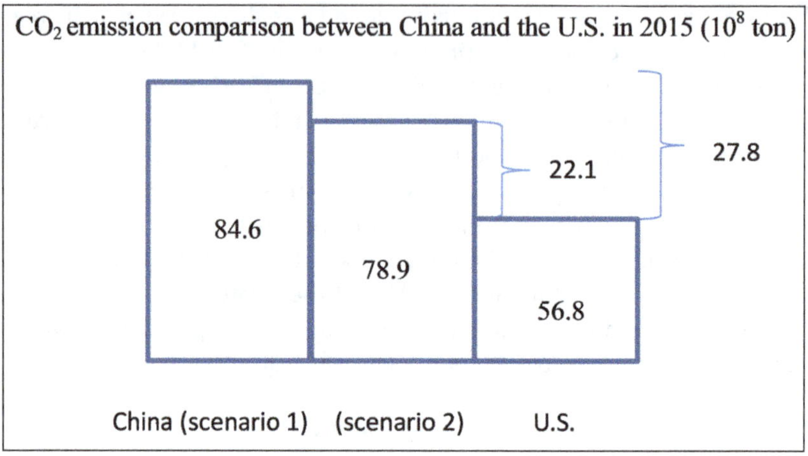

Figure 10.7. CO_2 Gap between China and the U.S. in 2015.

Shown in Figure 10.7, the U.S. Energy Information Agency projected that the CO_2 emissions of the United States will be about 5.68 billion tons in 2015,[24] while for China, if the total control target is set at 4.1 billion tce, the corresponding CO_2 emissions will reach 8.46 billion tons in 2015 (Scenario 1); and if calculated by relevant targets of China's 12th FYP, the CO_2 emissions will be 7.89 billion tons in 2015 (Scenario 2) (Annex 1—Table 1, relevant emissions reduction targets of the central government 12th FYP). Under Scenario 1 and Scenario 2, China's CO_2 emissions will exceed the U.S.' by 2.21 and 2.78 billion tons respectively in 2015. If the total energy consumption exceeded 4.1 billion tce, in the year 2015, the gap of CO_2 emissions between China and the United States would be even greater.

10.4.2 *China Emissions Reduction Effort Increases*

The commitment of energy conservation and emissions reduction, as well as low-carbon development, not only accords with China's own developmental needs, but also fulfills international obligations to address the issue of global climate change. Under the "United Nations Framework Convention on Climate Change" and the Kyoto Protocol, China is willing to take active actions following the "common but differentiated

[24] United States Energy Information Administration, 2011-04-26. Data tables: Reference case http://www.eia.gov/forecasts/aeo/data.cfm [2011-09-23].

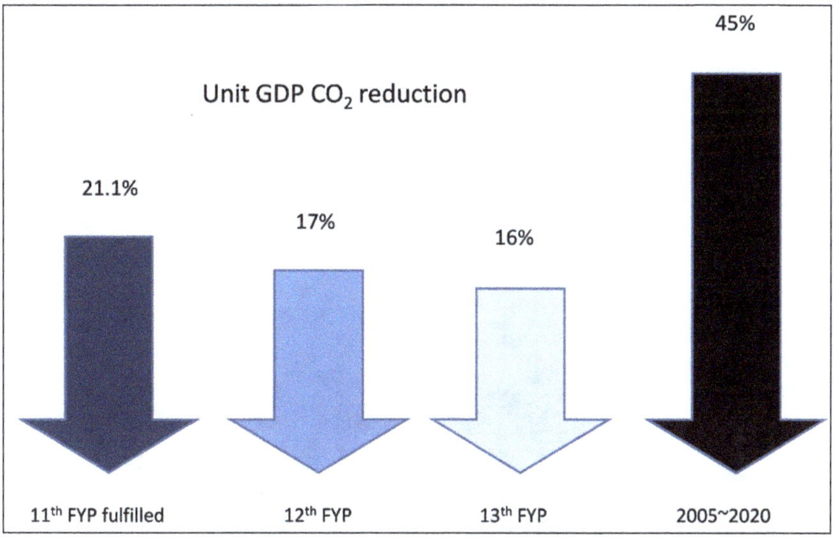

Figure 10.8. Decline rate of CO_2 reduction per unit of GDP.

responsibilities" principal. China faced much pressure during the 11th FYP when the Copenhagen and Cancun and other global climate negotiations were held. The Chinese government announced its voluntary emissions reduction target on international climate negotiations in Copenhagen and reiterated in Cancun that it will lower its carbon dioxide emissions per unit GDP by 40 to 45% by 2020, compared with the 2005 level. The binding targets of energy and carbon intensity declines of the 12th FYP is the concrete manifestation of the Copenhagen commitments, which shows China's pro-active attitude to address climate change issues. Prior to the arrival of climate negotiations in Durban of South Africa, China is acting to contribute to global emissions reduction.

Shown in Figure 10.8, the 11th FYP has accomplished 21.1% decrease in carbon dioxide emissions per unit of GDP, if the 12th FYP accomplishes the 17% decline rate as planned, the 13th FYP only needs to achieve another 16% decrease to allow China to successfully fulfill the upper bound of the voluntary emission reduction targets of 40% to 45% proposed at the global climate conference in Copenhagen. That is to say, by the year 2020, China's carbon dioxide intensity will decline by 45% on the basis of its 2005 level. China will use three FYPs to achieve an effective reduction of the core target of carbon dioxide intensity, and thus proceed vigorously towards a low-carbon society.

INDEX

www.ingramcontent.com/pod-product-compliance
Lightning Source LLC
LaVergne TN
LVHW021724021026
849161LV00004BA/679